Surface and Interfacial Aspects of Biomedical Polymers

Volume 2
Protein Adsorption

SURFACE AND INTERFACIAL ASPECTS OF BIOMEDICAL POLYMERS
Edited by Joseph D. Andrade

Volume 1 Surface Chemistry and Physics
Volume 2 Protein Adsorption

Surface and Interfacial Aspects of Biomedical Polymers

Volume 2
Protein Adsorption

Edited by

Joseph D. Andrade

University of Utah
Salt Lake City, Utah

PLENUM PRESS • NEW YORK AND LONDON

Library of Congress Cataloging in Publication Data

Main entry under title:

Surface and interfacial aspects of biomedical polymers.

Bibliography: v. 1. p.
Includes index.
Contents v. 1. Surface chemistry and physics.
1. Biomolecules. 2. Polymers and polymerization. 3. Surface chemistry. I. Andrade, Joseph D., 1941–
QD419.S94 1985 574.19′24 84-26601
ISBN 0-306-41742-1

A Division of Plenum Publishing Corporation
233 Spring Street, New York, N.Y. 10013

Printed in the United States of America

Contributors

Joseph D. Andrade ● College of Engineering, University of Utah, Salt Lake City, Utah 84112

Yu-Ling Cheng ● Department of Chemical Engineering, Stanford University, Stanford, California 94305. Present address: ALZA Corp., 950 Page Mill Road, Palo Alto, California 94303

Therese M. Cotton ● Department of Chemistry, Illinois Institute of Technology, Chicago, Illinois 60616. Present address: Department of Chemistry, University of Nebraska, Lincoln, Nebraska 68588

V. Hlady ● Ruder Bošković Institute, Zagreb, Yugoslavia

Herbert P. Jennissen ● Ludwig-Maximilians-Universität München, Institut für Physiologie, Physiologische Chemie und Ernährungsphysiologie, Veterinärstr. 13, D-8000 München 22, West Germany

Bernard Koltisko ● Department of Macromolecular Science, Case Western Reserve University, Cleveland, Ohio 44106

Brent K. Lok ● Department of Chemical Engineering, Stanford University, Stanford, California 94305. Present address: Chevron Research Company, Richmond, California 94802

J. Lyklema ● Laboratory for Physical and Colloid Chemistry, Agricultural University, De Dreijen 6, 6703 BC Wageningen, The Netherlands

W. Norde ● Laboratory for Physical and Colloid Chemistry, Agricultural University, De Dreijen 6, 6703 BC Wageningen, The Netherlands

Royston W. Paynter ● Laboratoire de Chirurgie Expérimentale, Pavillon Agriculture Service, Université Laval, Québec, P.Q. G1K 7P4, Canada

Buddy D. Ratner ● Department of Chemical Engineering, University of Washington, Seattle, Washington 98195

Channing R. Robertson ● Department of Chemical Engineering, Stanford University, Stanford, California 94305

A. Silberberg ● Weizmann Institute of Science, Polymer Department, Rehovot 76100, Israel

R. A. Van Wagenen ● Department of Bioengineering, College of Engineering, University of Utah, Salt Lake City, Utah 84112

Alan Walton ● University Genetics Co., 537 Newton Ave., Norwalk, Connecticut 06852

Preface

Although protein adsorption at liquid/air and solid/liquid interfaces has been studied for over a century and many excellent reviews are available, the subject is only superficially presented in practically all textbooks or monographs dealing with adsorption and interfacial chemistry. Proteins at interfaces are so important to a wide variety of industrial, biomedical, and scientific topics that the subject merits a detailed presentation. Recently a number of new methods and techniques have become available, as well as new conclusions, models, and theoretical treatments.

Volume 2 of *Surface and Interfacial Aspects of Biomedical Polymers* is intended to provide a firm basis for the study of proteins at interfaces. It emphasizes some rather new approaches, techniques, and models. In order to simplify the problem and obtain as full an understanding as possible of the field at this early stage in its development, adsorption from single protein solutions rather than mixtures is emphasized.

I am indebted to all the contributors for their cooperation and timely submissions. We are all indebted to those not represented here who have done so much for the field of protein adsorption: R. Bagnall, J. L. Brash, H. B. Bull, F. Eirich, T. A. Horbett, F. Macritchie, B. Morrissey, M. C. Phillips, A. Rothen, L. Vroman, and many others.

The contributors and I hope that this volume may aid the development of the field and stimulate others to help elucidate the complex process of protein adsorption.

J. D. Andrade, Salt Lake City

Contents

2. Total Internal Reflection Intrinsic Fluorescence (TIRIF) Spectroscopy Applied to Protein Adsorption
V. Hlady, R. A. Van Wagenen, and J. D. Andrade

3. Interactions of Macromolecules with Surfaces in Shear Fields Using Visible Wavelength Total Internal Reflection Fluorescence
Yu-Ling Cheng, Brent K. Lok, and Channing R. Robertson

4. Surface Enhanced Raman Spectroscopy of Biological Macromolecules

Therese M. Cotton

5. The Study of Interfacial Proteins and Biomolecules by X-Ray Photoelectron Spectroscopy
Royston W. Paynter and Buddy D. Ratner

6. Chromatographic Analysis of Protein Adsorption
Bernard Koltisko and Alan Walton

7. Proton Titration and Electrokinetic Studies of Adsorbed Protein Layers
W. Norde and J. Lyklema

8. Probing Protein Adsorption via Microcalorimetry
W. Norde

9. Protein Adsorption Hysteresis
Herbert P. Jennissen

10. Modeling of Protein Adsorption

A. Silberberg

Contents of Volume 1

Contributors to Volume 1

J. D. Andrade College of Engineering, University of Utah, Salt Lake City, Utah 84112

F. M. Fowkes Department of Chemistry, Lehigh University, Bethlehem, Pennsylvania 18015

D. E. Gregonis Department of Materials Science and Engineering, College of Engineering, University of Utah, Salt Lake City, Utah 84112

B. Hupfer Institut für Organische Chemie, Johannes Gutenberg-Universität, Johann-Joachim-Bercher-Weg 18–20, D-6500 Mainz, West Germany

M. S. Jhon Korean Advanced Institute of Science, P.O. Box 150, Chongyangni, Seoul, Korea

K. Knutson Department of Pharmaceutics, College of Pharmacy, University of Utah, Salt Lake City, Utah 84112

J. Lyklema Laboratory for Physical and Colloid Chemistry, Agricultural University, De Dreijen 6, 6703 BC Wageningen, The Netherlands

D. J. Lyman Department of Materials Science and Engineering, University of Utah, Salt Lake City, Utah 84112

Y. G. Oh Korean Advanced Institute of Science, P.O. Box 150, Chongyangni, Seoul, Korea

B. D. Ratner Department of Chemical Engineering, University of Washington, Seattle, Washington 98195

W. M. Reichert Department of Bioengineering, College of Engineering, University of Utah, Salt Lake City, Utah 84112

H. Ringsdorf Institut für Organische Chemie, Johannes Gutenberg Universität, Johann-Joachim-Bercher-Weg 18–20, D-6500 Mainz, West Germany

L. M. Smith Department of Bioengineering, College of Engineering, University of Utah, Salt Lake City, Utah 84112

1

Principles of Protein Adsorption

Joseph D. Andrade

1. INTRODUCTION

The understanding and control of the interaction of proteins with solid surfaces is important in a number of fields (Table 1). The blood compatibility or incompatibility of materials is generally considered to be related to protein adsorption processes, involving Factor 12 (Hageman Factor), Factor 11, high molecular weight kininogen, possibly complement C3, and others. There has been considerable work attempting to correlate the blood compatibility of materials with the adsorption of gamma-globulins, fibrinogen, fibronectin, albumin, and other proteins.[1-3] More recently, there has been activity in the adsorption of other plasma proteins and their possible relationship to blood compatibility, including hemoglobin, haptoglobin, lipoproteins, transferrin, prothrombin, and others.

A large number of review papers and several monographs on protein adsorption are now available (Table 2).[1-20] A large source of information is the literature on fundamental surface science at the air/water interface.[13] These studies have been going on since the early part of the century and have provided much of the information and most of the general "rules of thumb" on protein interactions at interfaces. The second major source of protein adsorption information is the solid-liquid chromatography literature.[21-26] The interaction of proteins with the solid supports used in chromatography and their retention times under different conditions are related to the surface characteristics of the proteins and of the solids, as well as the solution environment. Chromatographic studies have established the importance of electrostatic, hydrophobic, and charge-transfer interactions of proteins with solid supports. These interactions form the basis of ion exchange, hydrophobic, and charge-transfer chromatographies, respec-

---

Joseph D. Andrade • College of Engineering, University of Utah, Salt Lake City, Utah 84112.

TABLE 1
Processes and Fields of Study Where Protein Adsorption Is Important

Field	Role of Protein Adsorption
Medical devices—especially cardiovascular	Interaction of the surfaces of medical devices with surrounding fluids leads to protein adsorption and other interfacial processes. Protein adsorption is important in blood coagulation, complement activation, and probably in the reaction of tissues to artificial implants.
Pharmacy	Interaction of peptide and protein agents with container surfaces leads to depletion of the active agent from solution. The controlled delivery of peptides and proteins by drug delivery devices is also dependent on protein–solid interactions.
Protein processing	The separation and purification of proteins (from plasma or other complex media) is usually accomplished by solid–liquid chromatography. Protein chromatography depends on interaction of protein with the chromatographic support—generally by hydrophobic, ion exchange, or charge-transfer mechanisms.
Biotechnology and cell culture	Cell culture *in vitro* usually requires cell–surface adhesion which is mediated by an interlayer of adsorbed protein. The modern biotech industries depend in part on mass cell culture and on protein separation and purification by chromatographic means.
Soils and agriculture	Much of the pioneering work on protein adsorption considered enzyme interactions with minerals. Protein adsorption on mineral surfaces may be important in evolution, soil microorganisms, plants, etc.
Filtration	Water treatment: Clarification is often done by filtration processes. The long-term function and stability of the filters depends on the adsorption properties of proteins and other macromolecules in the solutions treated.
Fouling	The attachment of marine and microorganisms to materials is often mediated by protein interactions. It is often necessary to minimize fouling, such as on naval ships, underwater structures, etc.
Biochips	There is now much interest in the potential electronic and semiconductive properties of oriented arrays of proteins on solid supports.
Sensors and diagnostic products	Solid-phase immunoassays often depend on the adsorptive immobilization of antigen or antibody.
Directed drug delivery	The use of liposomes or other particles as "targeted" drug delivery devices depends on the adsorptive and/or covalent immobilization of tissue and organ-specific immunoglobulins.

tively. Another very major source of protein adsorption information has been in the literature dealing with soils and minerals.

In the last twenty years there has been considerable interest in protein interactions with plastics and other materials used in medical devices, particularly cardiovascular and blood-contacting devices. This biomaterials

TABLE 2
Reviews and Monographs on Protein Adsorption and Interfacial Behavior

Author	Year	Ref.	Focus	Comments
Brash	1983	4	S/L—competitive aspects	Driving forces, transport, blood.
Brash and Lyman	1971	5	A/W, O/W, S/W	Classic review on S/L, especially polymers.
Chessman and Davies	1954	6	A/W, O/W	Comprehensive review—proteins and poly(amino acids).
Cumper and Alexander	1951	7	A/W, O/W, pH, ionic strength	Considers adsorption process as diffusion, adsorption, denaturation, coagulation, removal.
Eirich	1977	8	S/L, theory, mechanism	Classic review on macromolecule adsorption.
Fraser	1957	9	A/W & O/W interfaces, enzymes at S/W interfaces	Biological activity, denaturation, enzyme-lipid interactions, cell membrane effects.
Gorter	1941	10	A/W	Early review on protein monolayers.
James and Augenstein	1964	11	A/W, O/W, S/W	Biological activity effects, allosteric effects
Horbett	1982	12	S/W—blood interactions	Good review on biomaterials aspects.
MacRitchie	1978	13	A/W kinetics, mechanisms, conformation, reactions	Classic, rigorous review.
Morrissey	1977	14	Ellipsometry and IR	Conformation and adsorbed film thickness studied.
Norde	1980	15	Thermodynamics and mechanisms	Reviews classic set of papers; concludes entropic effects dominate.
Phillips	1977	16	A/W, O/W, conformation	Albumin, lysozyme, casein.
Rothen	1947	17	Enzymes at A/W interface	Protein film transfer, immunologic reactions of protein monolayers.
Ter-Minassian-Saraga	1956	18	A/W and film characterization	Classic paper on preparation, characterization, and analysis of monolayers.
Walton and Koltisko	1982	19	Chromatography and CD methods	Kinetics and conformational change.
Zittle	1953	20	Enzyme chromatography	Effects of protein, solid, pH, solvent, temperature, etc. Irreversibility of adsorption on inorganic supports.

literature is another rich source of information on protein adsorption processes.[1-3] This literature and others have been reviewed by the authors listed in Table 2.

The objective of this chapter is to present the general principles and "rules of thumb" which are now available. We will focus on the adsorption of single proteins, i.e., we will not discuss competitive adsorption of proteins in any detail, not because it isn't interesting or important, but because the much simpler case of adsorption from mono-component protein solutions is not very well understood. Clearly it is the competitive processes which will eventually correlate with such phenomena as blood coagulation, complement activation, cell adhesion, etc. Unfortunately, our basic science data bank is too meager to address these complex interactions for now. Hopefully this chapter and this book will help provide a base for those more complex studies in the years ahead.

2. INFORMATION DESIRED AND METHODS AVAILABLE

Figure 1 summarizes the basic components of the protein adsorption process. We are concerned with the nature of the protein itself, the medium in which it is located, and the nature of the solid surface. The solution

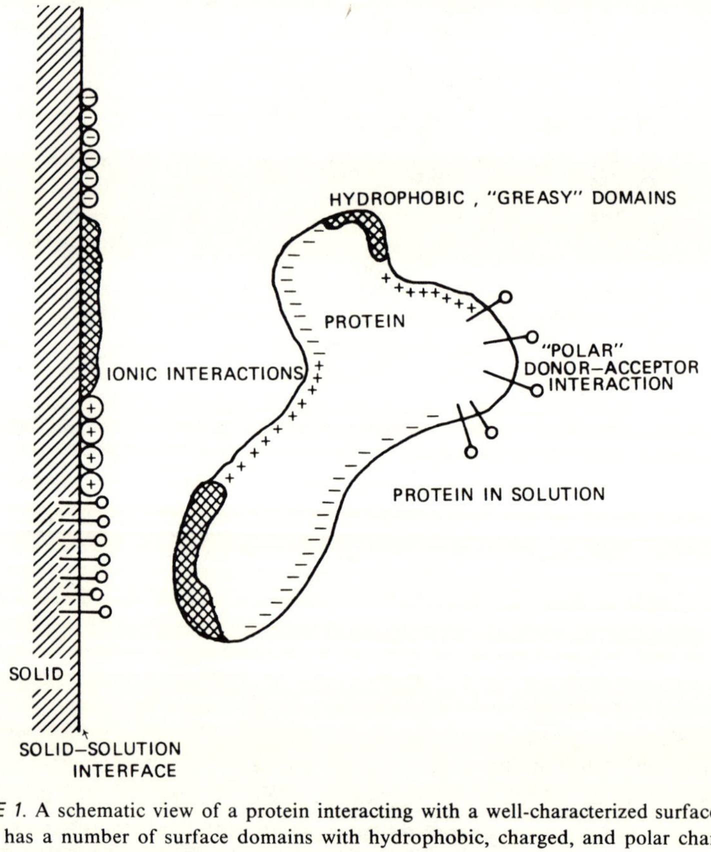

FIGURE 1. A schematic view of a protein interacting with a well-characterized surface. The protein has a number of surface domains with hydrophobic, charged, and polar character. The solid surface has a similar domain-like character.

TABLE 3
Protein and Solution Variables Important in the Study of Protein Adsorption

Protein	Solution
Isoelectric point	pH
Net charge and charge distribution	Ionic strength, ionic components, and buffer type
3-D structure in solution	Low MW species: amino acids, fatty acids, etc.
Placement and nature of hydrophobic patches	Protein type and concentration
Charge-transfer groups or regions	Temperature /pressure
Bound water	Hydrodynamics (flow)
Dimer content—other "impurities"	
Binding of low MW species	
Rigidity/shape—conformational variability and dynamics	
Effect of labeling on all of the above	

conditions which affect protein adsorption, such as pH, ionic strength, and other variables, are given in Table 3. The protein's own particular properties and characteristics *in* the solution medium selected are all of vital importance and must be well characterized and understood (Table 3). The nature of the solid surface is of course important and plays a major, although not necessarily dominant, role in many situations (Table 4). Obviously a solid surface is not required for protein adsorption at the solution/air interface.

TABLE 4
Solid Surface Variables and Surface Characterization[a]

Dry free surface in air or vacuum:
- Topography and heterogeneity—optical and scanning electron microscopy and profilometry
- Surface composition—X-ray photoelectron spectroscopy and reflection infrared spectroscopy
- Surface potential—electrostatic probe
- Advancing contact angle—critical surface tension

Wet solid-aqueous solution interface:
- Stability in water—leaching and dissolution characteristics
- Solid-water-air contact angles—receding angles and contact angle hysteresis
- Electrical double layer potential
- Ionization tendencies and proton exchange of surface groups?
- Water binding
- Interface entropic effects
- Interface dynamics—does the surface itself reorient or "denature"?
- Interface hydrodynamics

[a] Refer to Volume 1 for further information.

The solid surface and the solid–solution interface must be well characterized and understood if we hope to be able to understand protein adsorption. The various chapters in Volume 1 provide methods which enable the characterization of solid surfaces and solid–water interfaces.

Table 5 summarizes the information desired from a protein adsorption experiment. We wish to know the amounts adsorbed as a function of bulk solution concentration—the classical adsorption isotherm. We wish to know the amount adsorbed as a function of time—the adsorption kinetics. We now know that it is not which protein is on the surface or even its amount which is important, but what shape it is in; i.e., what is its *orientation* and three-dimensional *conformation* on the surface? This is illustrated schematically in Figure 2, where immunoglobulin G molecules are shown interacting

TABLE 5
Information Desired in a Complete Protein Adsorption Experiment and Some Techniques Which Provide Such Information[a]

Amount adsorbed: adsorption isotherms, ex situ[b]
Solution depletion
Surface concentration—radio—or fluorescent-labeled proteins
Surface concentration—reflection infrared or X-ray photoelectron spectroscopy
Amount adsorbed: f (time), in situ[b]
Solution depletion
Reflection fluorescence spectroscopy
Reflection infrared spectroscopy
Reflection Raman spectroscopy?
Ellipsometry
Desorption and/or exchange?
Conformation of adsorbed proteins, bound segments, in situ
Reflection or transmission infrared spectroscopy
Reflection Raman spectroscopy?
Reflection CD?, transmission CD
Fluorescent probes
Calorimetry
Nuclear magnetic resonance spectroscopy
Thickness or extension of adsorbed layer
Ellipsometry
Viscometry?
Heterogeneity of adsorption
Fluorescent microscopy?
Microautoradiography
Electron microscopy

[a] Refer to other chapters in this volume which provide additional information.
[b] *Ex situ* means the sample must be *removed* fom the solution for the surface concentration measurement to be made—as opposed to *in situ*, where the measurements are made directly in the protein solution.

with a solid–liquid interface. Note that there are a variety of orientations present, only one of which would lead to optimum Fab interaction with an antigen in the surrounding solution. If the molecule is adsorbed through the Fab arms, then the Fc or effector region of the molecule (see also Figure 5) is exposed to the solution for subsequent interaction, such as complement activation. If the protein is adsorbed such that its three-dimensional shape or secondary structure is changed, then we say that the protein has undergone a change in conformation. A conformational change is often called "denaturation," because many of the "normal" characteristics of the protein can be expected to be altered.

In addition to these basic pieces of information desired from an adsorption experiment with only one protein species present, we wish to know how proteins in a mixture interact with the surface. How do they compete with one another?

We also wish to know the protein's desorption properties. When the protein solution is removed and replaced with buffer, does the adsorbed protein come off the surface? Does it desorb? Will it exchange with other proteins? All of these pieces of information are essential in order to fully understand protein interactions with solid surfaces.

The techniques which are available to study protein adsorption are limited and generally do not provide all the information desired. Table 6 summarizes the available methods for studying proteins at interfaces. The chapters in this volume summarize a number of the more modern methods which are useful for investigating protein adsorption, including spectroscopic tools which provide information on rates of adsorption and amounts adsorbed, as well as conformation.

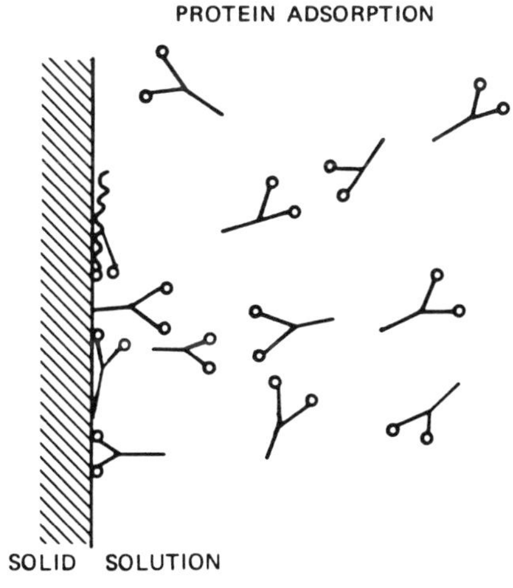

FIGURE 2. The adsorption of IgG molecule (shown for convenience as Y-shaped) may result in several orientations. The molecule may adsorb through one or both of the Fab ends or through the Fc tail or lie on its "side." In addition to such *orientation* effects, the molecule may be conformationally altered or *denatured.*

TABLE 6
Methods Used for Study of Protein Interfacial Behavior

Method/Technique	Principle	Information	Reference
Langmuir trough	Adsorption and spreading at solution/air interface.	Area/molecule, surface pressure, surface tension, surface potential, changes with time.	9–11, 13, 16–18
Pendant drop	Adsorption and spreading at L/L interface.	Interfacial tension and pressure, changes with time.	27–31
Solute depletion	Measure change in concentration of bulk solution.	Adsorption as f(time); high surface area required; no labels required.	15
Ellipsometry	Polarized light externally reflected at solid–solution interface; change in polarization upon reflection.	Thickness and optical constants of adsorbed film as f(time). Optically flat, reflective surfaces needed.	14
Radiotracers	Radiolabeled proteins, measure surface radioactivity directly.	Simplest, most sensitive, most widely used method. Assumes no effect of label on adsorption process. Most quantitative method.	12, 14, 15, 143, 145
Infrared spectroscopy	Direct measure of IR vibrations in transmission or attenuated total reflection (ATR).	Mainly amide absorption bands—water interferes. Band shifts provide information on bound fractions and conformation. Quantitation difficult.	5, 14, 35
Fluorescence spectroscopy	Intrinsic (Trp) or extrinsic (label) fluorescence—transmission or total internal reflection.	Local microenvironment influences wavelength and intensity of fluorescence. Some conformation information. Difficult to quantitate. Adsorbed layer concentration profile may be deducible via variable angle TIRF.	89–92, 116, 118, 146, Chapter 2, Chapter 3
Substrate turnover	Immobilized enzyme functions as interfacial indicator.	Interfacial mass transport information.	36

Contact angles	Liquid droplet on dry protein film: air or oil drop at protein film/water interface.	Surface or interfacial tension. Method plagued with problems.	32–34
Column chromatography	Solute retention on high surface area particle bed.	Similar to solute depletion but a dynamic, pseudo-equilibrium experiment.	127, Chapter 6
Electrophoretic mobility—surface charge pH titration	Methods to probe surface or accessible charged groups.	Surface charge nature; requires individual particles or high surface area.	Chapter 7
Microcalorimetry	Direct measure of enthalpic changes.	Direct thermodynamic information, requires high surface area particles, basic but difficult to interpret.	15, Chapter 8
Raman spectroscopy	Frequency-shifted light scattering; molecular vibrations.	Similar to IR, but with different selection rules; needs enhancement tricks to be practical for interface studies.	Chapter 4
X-ray photoelectron spectroscopy (ESCA)	X-ray induced surface photoelectrons.	Elemental analysis and chemical bonding nature of surface. Need dried surface or cryogenically cooled interface. Not feasible for direct solid-water studies. Variable-angle studies useful for determining homogeneity of adsorbed layers.	Chapter 5
Circular dichroism	Optical activity changes.	Conformational changes—only applied via transmission through many adsorbed layers. Difficult to apply.	117
Microautoradiography	Beta particle emission detected by photographic film	Distribution of adsorbed protein on surface.	1
Capillary viscometry	Apparent viscosity change due to change in capillary radius.	*In situ* measure of adsorbed film thickness	Chapter 10

3. PROTEIN CHARACTERISTICS AND PRINCIPLES

3.1. INTRODUCTION

Proteins are biological macromolecules constructed for specific and unique functions. They are high molecular weight polyamides produced by the specific copolymerization of up to 20 different amino acids. The amino acid composition, called the primary structure, is generally unique and specific to each particular protein. The hydrogen bonding characteristics of the polyamide bond in the backbone of proteins result in various secondary structures, such as the well-known α-helix and β-sheet. Intramolecular associations, including ionic interactions, salt bridges, hydrophobic interactions, hydrogen bonding, and covalent disulfide bonds, result in a unique teritiary structure for each polypeptide chain. Finally, two or more polypeptide chains, each with its own primary, secondary, and tertiary structure, can associate to form a multi-chain quaternary structure. Most proteins contain short, carbohydrate sequences off the main polypeptide chain and are, therefore, called glycoproteins.

The fundamental principles of protein structure and function are available in modern biochemistry textbooks.(37–46) The current edition of Stryer is particularly recommended.(37) The books by Dickerson and Geis(38,39) are excellent tutorials on the fundamental principles of protein structure. The recent modern text by Schulz and Schirmer,(40) *Principles of Protein Structure*, is outstanding. In addition, modern physical biochemistry texts, in particular Freifelder(41) and Cantor and Schimmel,(42) provide an essential basis for the understanding of protein structure, function, and interfacial behavior. Bull's *An Introduction to Physical Biochemistry* is still an outstanding introduction to the subject.(43) Walton's books on biopolymers are also highly recommended,(44,45) as is Tanford's book.(46)

Table 7 lists those characteristics of a protein which are normally considered essential for a minimal characterization. Most of the information desired is available for practically all proteins which may be of interest to the biomaterials investigator. Generally, all of this information is available in the literature, and the investigator need only check for purity, homogeneity, and perhaps activity of the protein preparation.

For a complete adsorption analysis, the amino acid sequence and secondary, tertiary, and quaternary structure of the protein should be known (Table 8). This is usually only available if a complete X-ray diffraction analysis has been done. Some information on ligand-binding characteristics of the protein should be available: Does it bind low molecular weight solutes? Does it have specific macromolecular binding characteristics, etc.? For example, the physical properties of albumin are different depending on its fatty acid and bilirubin composition. The properties of hemoglobin

TABLE 7
Minimal Protein Information Required for Adsorption Studies

Quantity	Method	Significance/Information
Molecular weight	Light scattering, osmometry, gel permeation chromatography	Size, shape, mass of molecule.
Size and shape	Light scattering, sedimentation	Size, shape, mass of molecule.
Amino acid composition	Amino acid analysis	Basic.
Electrophoretic mobility	Gel electrophoresis	Related to net charge.
Isoelectric point	Isoelectric focusing	Number of charged groups.
UV-visible absorption spectrum	Absorption spectroscopy	Tyrosine–tryptophan environments, presence of absorbing ligands or impurities.[a]
Fluorescence spectrum	Fluorescence spectroscopy	Tryptophan environment, presence of fluorescent ligands or impurities.[b]
CD spectrum	Circular dichroism	Secondary structure, α-helix, β-sheet, etc.
Subunits	SDS-gel electrophoresis	Quaternary structure components and their molecular weight.
Subunits and —S—S—	Gel electrophoresis in mercaptoethanol	Tertiary structure—breaks —S—S— bands.
Solubility—salting out	Salting-out fractionation	Solubility under different solvent conditions, pH, ionic strength, hydrophobic interactions.

[a] For example, heme in myoglobin and hemoglobin.
[b] For example, bilirubin in albumin.

TABLE 8
Additional Protein Information Desired for Adsorption Studies

Quantity	Method	Information/Significance
Molecular vibrations	Raman and infrared spectroscopy	Secondary and tertiary structure.
Thermal denaturation	Raman, CD, or fluorescence	Conformation (secondary/tertiary) and structure as f(temperature).
Monolayer properties	Langmuir trough or pendant drop	Air/water interfacial properties; Π–A curves; can deduce close-packed monolayer dimensions.
Binding properties	Equilibrium dialysis for lower molecular weight ligands, affinity chromatography for others	Ligand binding characteristics, such as Ca^{++} binding to fibrinogen and prothrombin, fatty acid and bilirubin binding to albumin, heparin binding to antithrombin III and fibronectin.
Amino acid sequence	Usually cleavage fragments plus sequencing	Sequence important in structure and function.
3-D structure	X-ray diffraction	Complete stereochemistry of tertiary or quaternary structure.
Enzymatic or specific biochemical activity	Substrate turnover, antibody binding, ligand binding	Is protein (enzyme) functional? Does it bind its specific ligands?

are different if its heme ligand is in the deoxy or oxy form or in the met (Fe^{+3}) form.[39] The activity of many proteins is enhanced or potentiated when bound to appropriate ligands. For example, binding of heparin to antithrombin III greatly increases the binding of thrombin. All such information is useful in predicting and interpreting protein interactions with surfaces.

The general solubility characteristics of the protein, and its behavior in different pH and ionic strength enviroments, its behavior in urea solutions, its behavior in solutions containing small amounts of methanol, ethanol, or glycerol, and related information on solution characteristics are all important in interpreting and predicting interactions at interfaces.

3.2. *FORCES AND INTERACTIONS*[21,47]

The fundamental classes of forces and interactions important in protein structure and function are summarized in Table 9. The four general classes are: (1) ionic or electrostatic interaction, due to the attraction or repulsion of two or more groups carrying a net charge; (2) hydrophobic interaction, a largely entropically driven process due primarily to water-structure effects adjacent to hydrophobic interfaces; (3) hydrogen bonding, a unique type of dipole/dipole interaction which results in interaction energies comparable to very weak covalent bonds; and (4) other interactions, primarily those of a charge-transfer or partial electron donor/acceptor type, which in aqueous solutions are often dominated by π–π interactions. It is important to point out that all of these interactions take place in an aqueous medium and that the hydrogen bonding and dielectric properties of water play a very major role.

3.2.1. *Ionic or Electrostatic* (Table 9)

In addition to the terminal amino and carboxylic acid groups of each protein chain, a number of amino acid side chains are ionized at normal conditions (Table 10). The side chain carboxyl groups of aspartic and glutamic acid have pK's of 3.9 and 4.3, respectively, and are negatively charged at pH 7.4. The side chain amino groups of lysine and arginine have pK's of 10.5 and 12.5, respectively, and are positively charged at physiological pH. Histidine, with a pK of 6.0, has some positive charge character at physiological pH. These amino acid side chains are normally located on the surface of proteins where they can be fully solvated by the aqueous environment.

Although the net charge on a protein, which is determined by the sum of these respective amino acid charges, results in electrophoretic migration of a protein in an electric field, the electrostatic interactions of proteins

TABLE 9

Major Noncovalent Forces and Interactions Important in the Organization and Stabilization of Protein Structure in Aqueous Solutions[a]

Interaction	Description	Directional?	Effect of increasing: Temperature	Effect of increasing: Ionic strength or dielectric constant	Binding energy (kcal/mol)	Free energy change $H_2O \rightarrow EtOH$ (kcal/mol)
Ionic	Coulombic interaction due to opposite charges.	No—spherical symmetry	Decreases	Decreases	——COO··· N^+H_3—— −5 kcal/mol	−1
H-bonding	A donor–acceptor effect involving the H-atom of the donor—essentially an ionic interaction. Water participates in H-bonds.	Yes, linear bond lengths ~3 Å, bond energies ~3–6 kcal/mol	Decreases	Decreases	——NH··· O=C—— −3 kcal/mol	
Charge-transfer (donor–acceptor)	In water this is mainly due to π–π interactions. Trp interaction strongest as it is strong electron donor.	Yes—analogous to H-bonding	Decreases	Increases		
Hydrophobic or entropic	Water cannot interact with apolar groups via H-bonding, resulting in water adjacent to such apolar groups being more organized or more ordered. If such apolar groups come together, some ordered water is released, increasing entropy—hence apolar groups tend to "interact." Hydrophobic interactions are particularly strong in aqueous solutions.	No	Increases	Increases; effect is ion-specific		−2.4 (phe)

[a] References (21), (40), and (47).

TABLE 10
The Twenty Standard Amino Acid Residues of Proteins[a]

Amino acid residue	Three-letter symbol	One-letter symbol	Mnemonic help for one-letter symbol	MW of residue at pH 7.0 (daltons)[b]	pK value of side chain	ΔG values for transfer of side chain from water to ethanol at 25°C (kcal/mol)
Alanine	Ala	A	Alanine	71		−0.5
Glutamate	Glu	E	gluEtamic acid	128	4.3	
Glutamine	Gln	Q	Q-tamine	128		
Aspartate	Asp	D	asparDic acid	114	3.9	
Asparagine	Asn	N	asparagiNe	114		
Leucine	Leu	L	Leucine	113		−1.8
Glycine	Gly	G	Glycine	57		
Lysine	Lys	K	before L	129	10.5	
Serine	Ser	S	Serine	87		+0.3
Valine	Val	V	Valine	99		−1.5
Arginine	Arg	R	aRginine	157	12.5	
Threonine	Thr	T	Threonine	101		−0.4
Proline	Pro	P	Proline	97		
Isoleucine	Ile	I	Isoleucine	113		
Methionine	Met	M	Methionine	131		−1.3
Phenylalanine	Phe	F	Fenylalanine	147		−2.5
Tyrosine	Tyr	Y	tYrosine	163	10.1	−2.3
Cysteine	Cys	C	Cysteine	103		
Tryptophan	Trp	W	tWo rings	186		−3.4
Histidine	His	H	Histidine	137	6.0	−0.5

[a] From Reference (40), p. 2.
[b] Weighted mean: 108.7.

appear to be relatively short-range because of the high dielectric constant of water. As one applies Coulomb's Law to charge-charge interactions, the high dielectric constant of water means that the electrostatic interaction in water is roughly two orders of magnitude less than it would be in air or vacuum. It is only when the unlike charges are close enough that very few water molecules are in between that the electrostatic interactions become very strong. Thus, it is unlikely that electrostatic interactions, *per se*, would be responsible for directing a protein to a charged interface. However, once the protein is close to the charged interface, then the appropriate electrostatic coupling would become significant and even dominant.

Those charges which are not fully solvated by water are generally involved in unlike charge interactions or salt bridge effects.[48] Salt bridges can be on the external surface of the protein or internal, functioning essentially as strong cross-links.

Recent studies suggest that, even on the surface of a protein, charges tend to be surrounded by unlike charges[49] rather than being only solvated by water. Charges in the protein are on an average surrounded by charges of opposite sign, providing an electrostatic interaction component to the overall stability of a protein.[48] Obviously if such a charge now has the opportunity to interact with a charged surface, not only is this a mechanism for attaching protein to the surface, but the interaction disturbs the electrostatic component of the structural cohesion of the protein, making it more likely that a conformational change might occur.

Ionic interactions are particularly important in regions where water can be at least partially excluded, as in the case of a charge surrounded by a hydrophobic environment, such as in some enzyme active sites.[50]

In addition to the net charge interactions, the partial ionic character of many covalent bonds results in partial charges on atoms present in proteins.

3.2.2. Hydrogen Bonding (Table 9)

Hydrogen bonds are predominantly electrostatic interactions. Due to the very small size of the hydrogen atom and its electropositive character in covalent bonds, it is easily attracted to electronegative atoms. Because hydrogen is so small this attraction results in a very short binding distance, which leads to a significant electrostatic interaction. This interaction is strong and unique enough that it is given its own name—the hydrogen bond. Hydrogen bonds are very common in proteins and are partly responsible for the stabilization of α-helix and β-sheet structures. Although we normally think of hydroxyl–carbonyl and amide–carbonyl hydrogen bonds as being the most significant, hydroxyl–hydroxyl and amide–hydroxyl are also present and of importance. The common amide–carbonyl bond energy is about 3 kcal/mol. Hydrogen bonds are generally linear and one can think of the hydrogen as a bridge between two electronegative atoms [see Reference (40), p. 34].

Hydrogen bonding is not expected to have a strong stabilizing effect in the aqueous environment. Water is at least as good at forming hydrogen bonds as any of the groups in a polypeptide. Thus, during a folding and association process, peptide and amino acid groups exchange hydrogen bonds with water, which does not result in any significant net change in hydrogen bond energy.[(51)]

3.2.3. Hydrophobic Interactions[(21–23,47,52,53)]

A very significant component to protein interactions and structure, and particularly to protein adsorption, is the hydrophobic interaction. One normally thinks of intermolecular forces in an attractive or repulsive sense and primarily enthalpic in nature. However, the stability and interactions of a protein are dependent on the overall free energy of the system, and thus one must also consider entropic effects. Hydrophobic interactions are basically entropic interactions largely due to order/disorder phenomena in the water solvent.

The fact that water orders at air or apolar interfaces, and that this ordering is entropically undesirable, means that there is a tendency to minimize such interfacial areas. This free energy driving force is responsible for minimizing the surface area of water droplets, air bubbles, or oil droplets in water. This is the main reason why the hydrophobic amino acid side chains tend to be excluded from the surface of a protein, thereby minimizing their contact with water. Most of the hydrophobic amino acids are oriented towards the interior. One normally hears protein structure discussed in terms of an oily inner phase and an outermost hydrophilic component interacting with water.

Although the majority of the nonpolar side chains are removed from the water and are located in the hydrophobic core, the protein interior also contains polar groups involved in strong hydrogen bonds and salt bridge interactions. The charged and other hydrophilic groups present on the surface result in most proteins being soluble in aqueous solutions.

One can quantitate these tendencies by considering the free energy change in moving a hydrophobic, amino acid side chain from the protein surface in contact with water to the protein interior, by measuring the free energy of transfer of the amino acid from water into a nonaqueous solvent (Table 9). This was first done by Tanford[(52)] and coworkers to quantitate the nature of the hydrophobic interaction, using ethanol to approximate the polarity of the protein interior. Typical results are given in Figure 3 where the surface area of amino acid side chains is plotted as a function

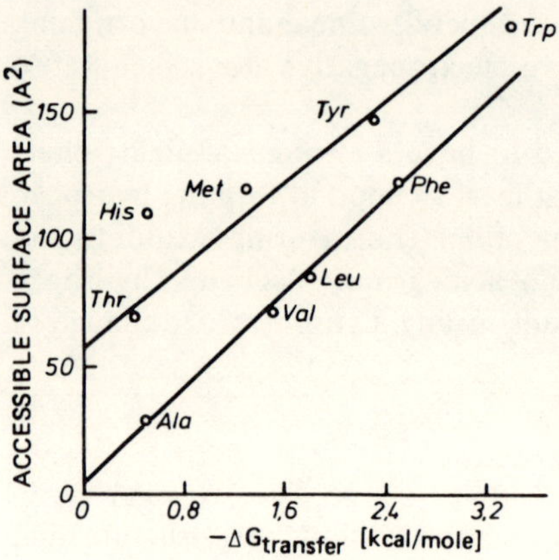

FIGURE 3. Water-accessible surface areas of amino acid side chains as a function of their hydrophobicity, that is, the free energy of transfer from water to ethanol and dioxane. These solvents are believed to resemble the protein interior. The accessible surface area is defined in Figure 4 [from Reference (40)].

of the free energy of transfer. Tryptophan, phenylalanine, and tyrosine are the most hydrophobic of the amino acids with leucine, valine, and methionine playing an intermediate role. The accessible surface is defined as the area described by the center of a water molecule that rolls over the protein (Figure 4). In all cases, the free energy of transfer is proportional to the interface area. Hydrophobic interactions are primarily dependent on the alkyl group and involve a reorientation of water molecules to create a cavity capable of accommodating the apolar residue. Much of the development

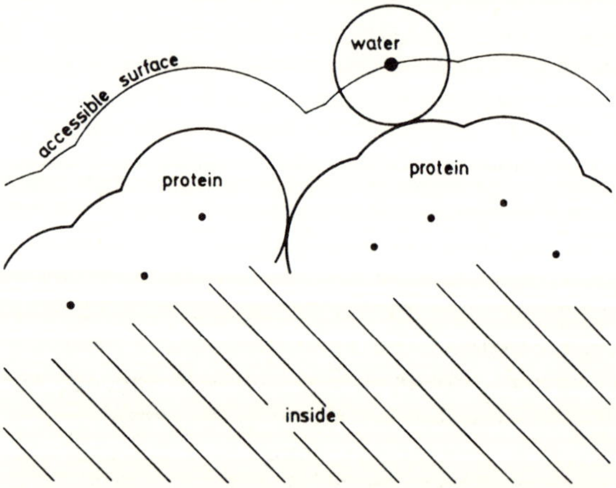

FIGURE 4. The water-accessible surface of a protein (or of a side chain). The protein surface is described by van der Waals envelopes of surface atoms (dots). The protein interior is hatched. Water molecules are considered as spheres of 1.4-Å radius. The "accessible surface" is defined as the area described by the center of a water molecule that rolls over the van der Waals envelope of the protein (or of the side chain). Note that there exist inaccessible protein (or side chain) surface areas in clefts [from Reference (40); see also Reference (80)].

of the hydrophobic interaction concept has been related to the size of the hydrophobic species, and therefore the size of the water cavity required to accommodate it. This immediately requires the concept of surface area and the interfacial free energy associated with such areas. It is not surprising, therefore, that hydrophobic interactions can also be effectively treated by interfacial free energy concepts. One can think of the driving force for such interactions as being the minimization of surface area and, therefore, of total interfacial free energy. The effects of ions on hydrophobic processes can be directly related to the effects of the same ions on the surface tension of water. Thus the common lyotropic or Hofmeister series can be equally well interpreted as a surface tension effect or as a water structure, entropic effect.[(53)]

Although the hydrophobic interaction is generally accepted in physical biochemistry, there is some controversy. There are a number of authors who argue that it should be treated as a simple case of a London dispersion interaction or in terms of interfacial tension arguments.[(52–55)]

3.2.4. Charge-Transfer Interactions[(21,24–26)]

Charge-transfer interactions are also important in protein stabilization and surface interaction. Hydrogen bonding can be thought of as a particularly strong and special case of a charge-transfer interaction in which a positive charge, in this case the hydrogen proton, is donated to a suitable acceptor. In general donor–acceptor processes one can think of excess electron density being present which can be donated to an electrophilic species; or electron density can serve as an acceptor for positive charge. In water these solute interactions are primarily due to π–π electron effects. Of all the amino acid side chains, tryptophan is the strongest electron donor. Charge-transfer chromatography has been developed recently to facilitate separations on columms of electron-donating or electron-accepting character.[(24–26)]

3.3. THREE-DIMENSIONAL AND SURFACE STRUCTURES

3.3.1. Introduction

The primary amino acid composition of a protein is normally known, including the sequence data. Secondary structures, such as alpha helices, beta sheets, beta turns, and related components are generally known for various fragments as well as often for the intact molecule (Table 11). This information is generally obtained by circular dichroism measurements and by infrared or Raman spectroscopy. In cases where the protein can be crystallized in a form suitable for X-ray diffraction, the complete tertiary

TABLE 11
Common Secondary Structure Features in Proteins[a]

Name	Description	Comments
α-helix	3.6 residues/turn, 1.5 Å rise/residue, 2.3 Å helix radius.	Stable, abundant, α-helix. Lengths range from ~5–20 segments. Average length ~17 Å (11 residues or 3 turns). Helix is a stable rod in the structure. Kinks in rods occur at each proline residue.
Reverse turns	Hydrophilic residues on surface.	~$\frac{1}{4}$ of *all* residues in globular proteins are involved.
β-pleated sheet	A H-bonded sheet structure which is not usually planar—has a left-handed twist.	Globular proteins have ~15% sheet structures. Average length of sheet is 6 residues (range 3–10), less than 6 strands per sheet.

[a] From Reference (40).

structure and even quaternary structure may be known. Proteins which have been thoroughly worked up via X-ray crystallography are the most useful for protein adsorption studies.

In many cases the various properties or characteristics of a protein are located in distinct functional domains.[(40)] These can often be deduced from enzymatic fragments and by careful specific binding studies. Knowing the binding characteristics of the various fragments and domains in the molecule allows the investigator to begin to deduce certain types of adsorption behavior. For example, it is known that human plasma fibronectin contains several heparin binding sites, with pK's in the 8–9 range and with a high, positive charge density. It would be reasonable to suspect that when fibronectin interacts with negatively charged surfaces, it may do so through such binding sites, just as it does with heparin.[(56)]

There has been considerable effort directed toward the prediction of secondary and tertiary structures of protein from the amino acid sequence using computer-aided minimal potential energy calculations.[(40,57)] The question as to how a primary amino acid sequence begins to produce secondary and super-secondary structures and folds into its equilibrium tertiary structure and functional domains is a very active field of structural biochemistry.[(40,58)] A related problem is the mechanism by which a protein unfolds or denatures,[(59–63)] which is of fundamental interest in the protein adsorption process.

3.3.2. Amino Acids

Proteins are condensation polymers of some twenty different amino acids with a wide range of side chain characteristics (Table 10). Each of the amino acids confers unique and distinct properties on the polypeptide chain. These are discussed in detail by Schulz and Schirmer.[40] Glycine, for example, has no side chain and therefore confers flexibility upon the chain at that point. The branched amino acids such as valine, isoleucine, and leucine, make the chain stiffer where they are located, as do the aromatics and proline which has a very rigid side chain. The polar amino acids aid in hydrogen bonding interactions; tyrosine in particular forms very strong hydrogen bonds. Histidine with a pK of 6 is particularly sensitive to its local microenvironment. It is found in most of the active centers of enzymes, probably because charged and uncharged states are readily available, and it is therefore suitable for catalyzing chemical reactions. The charged amino acids (Lys, Arg, Asp, and Glu) are generally found on the external surface of the protein; they are long and flexible and tend to wobble in the surrounding solution, thereby increasing the solubility of the protein globule.

3.3.3. The Chain in Water

A polypeptide chain in water has a specific pattern of polar and nonpolar groups. There is a tendency to minimize the surface area between nonpolar groups and the aqueous phase. This can be treated in terms of a surface free energy argument or in terms of the classical hydrophobic interaction argument as described earlier. In any event, the nonpolar groups tend to be excluded from the surface of the polypeptide globule with the highly polar and charged amino acids on the surface interacting strongly with water. There are some regions on the surface which are nonpolar and there are some regions in the interior which contain polar groups, involved in either hydrogen bonding or salt bridge associations. About 90% of all internal polar groups form hydrogen bonds.

The solvent entropy (hydrophobic interaction) effect contributes significantly to overall protein structural stability. The intramolecular interactions are highly efficient and effective.

The interior of the protein has a very high packing density, comparable to that in the protein crystal. The average packing density of a protein is about 0.75 which is incredibly high when one considers that a close-packed structure of spheres has a packing density of 0.74. Thus the packing density of proteins in terms of volume utilization is even more efficient than most close-packed metals.[40] One often hears the criticism that the structure of a protein in the crystal must be very different than in solution. This is not the case, as the solution packing of a protein is essentially identical to that in the crystal. It is important to note that active centers in reactive sites of proteins tend to be much more loosely packed than the rest of the molecule, thereby providing the increased flexibility required for chemical reactions and recognition phenomena.

3.3.4. Domains[40]

Careful study of the structure of available proteins suggests regions of amino acid homology and structural homology in very different protein molecules. This has led to the concept of the structural and functional domain. One can consider large globular proteins as being made up of a set of functional domains, a more important structural concept that the chain composition of the protein. Active sites are often located at the interface between two structural domains. One can, therefore, consider proteins as being modular structures with the structural or functional domains as the modules. A good example is the immunoglobulin molecule shown in Figure 5 (an IgG).

There are both functional and structural domains. A functional domain may consist of one or more structural domains. A functional domain is a functionally autonomous region of the molecule. In the case of the IgG, the Fab portion is the functional antigen recognition domain while the Fc region is an effector domain. Structural domains are geometrically separate.

The large globular proteins of blood plasma are monomers, that is they consist of only one chain or a set of covalently linked chains. An oligomer is an aggregate of monomers whereby the aggregation is held together by noncovalent bonds. Most intracellular proteins are oligomers; proteins in blood plasma are large monomers, generally consisting of several functional domains.

Disulfide bonds confer additional stability onto a protein. A frequently encountered structural component is the sequence —Cys—Cys— with both residues forming disulfide bonds with other cystines. This is a useful architectural unit and forms the basis for linking three different chain segments in close proximity. This structure is found in serum albumin.

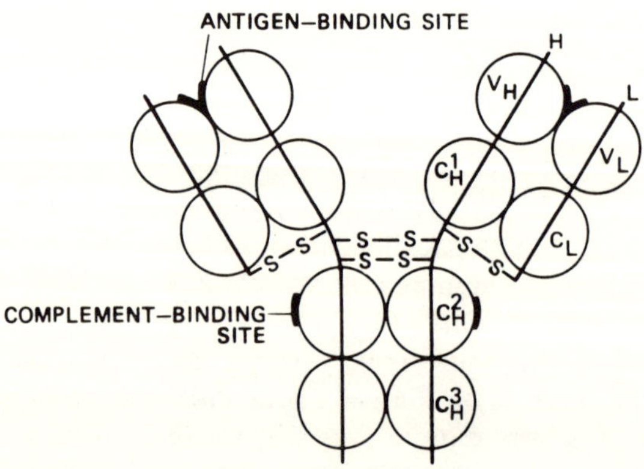

FIGURE 5. Structure of an IgG molecule showing the light (L) and heavy (H) chains, the antigen-binding sites (in the Fab regions), the complement-binding site (in the Fc region), and the constant (C) and variable (V) amino acid regions [from Reference (40)].

A structural domain can be defined as a region of locally high electron density separated by regions of low electron density or clefts in the electron density map. Amino acid residues that are far apart along the chain tend to be far apart in the three-dimensional structure. The concept of neighborhood correlation suggests that, perhaps based on kinetic folding considerations, those amino acids which are close together in a sequence tend to stay close together in the three-dimensional structure. Such neighborhood correlation is probably a consequence of the chain-folding process. The domains are probably those pieces of a chain which fold independently. Most globular proteins have domains of 100–150 residues and roughly 25 Å in diameter.

Five different classes of structural domains have been identified based on the secondary structural makeup. This is indicated in Table 12. A much more complete and thorough table is found in Schulz and Schirmer.[(40)] Structural symmetry is often evident, particularly in aggregate or oligomeric proteins.[(40)] The more contacts, the more stable the interaction.

Depending on the fit, the number and strength of contacts, and other factors, a monomer–oligomer equilibrium will exist. For example, in the case of concanavalin A, the tetramer is in equilibrium with dimer and

TABLE 12
Classes of Structural Domains in Globular Proteins[a]

Class	Secondary structure	Examples		
		Protein	Number of domains	Residues per domain
1	α-Helix only	Myoglobin	1	153
		T-4 lysozyme	2	90
		Insulin	1	51
		Hemoglobin	1	145
		Papain	2	111
2	β-Sheet only	Concanavalin *A*	1	237
		Papain	2	110
		IgG	2, 4	110
		Chymotrypsin	2	128, 114
3	Segregated α-helix and β-sheet	Hen lysozyme	1	129
		T-4 lysozyme	2	74
		Carbonic anhydrase	1	258
		Lactic dehydrogenase	2	150
4	Alternating α-helix and β-sheet	Adenylate kinase	1	194
		Lactic dehydrogenase	2	181
		Alcohol dehydrogenase	2	199
5	No α-helix or β-sheet	Phospholipase	1	124

[a] From Reference (40), pp. 88–91.

TABLE 13

Free Energy of Protein Association[a]

	Total accessible surface areas buried on association (Å^2)	Percentage of nonpolar area (%)	$\Delta G_{transfer}$ (kcal/mol)	$-T\Delta S_{assoc}$ (kcal/mol)	ΔG_{assoc} (calc) (kcal/mol)	ΔG_{assoc} (obs) (kcal/mol)
Horse oxyhemoglobin, contact $\alpha_1\beta_1$	1720	84	−43	+27	−16	<-11
Insulin, contact OP	1130	78	−28	+23	−5	−7
Trypsin–trypsin inhibitor	1390	68	−35	+27	−8	−18

[a] From Reference (40), p. 104.

monomer under normal conditions. In the case of insulin, the monomer, dimer, and hexamer are all in equilibrium. In hemoglobin, the dimer and tetramer are in equilibrium. In the case of the adsorption of such proteins, one must not only know what is the aggregation state of the protein in solution, but must be able to deduce the adsorbed state. Do the molecules adsorb as dimers?[(64)] Do they adsorb as dimers and then associate as tetramers or hexamers on the surface, etc.?

Some of the free energy considerations used to probe protein oligimerization and aggregation are also applicable to adsorption (Table 13). Protein–protein interactions usually involve surface complementarity, which can provide significant specificity. The specificity of a surface is determined by its shape, its pattern of hydrogen bond donors and acceptors, its pattern of charges, and even its pattern of hydrophobic patches. Based on immunoglobulin studies, a 100-Å^2 surface can form on the order of a thousand patterns showing different binding specificities.[(40)]

3.3.5. Folding and Denaturation[(58–63)]

An important question is how the linear, amino acid sequence spontaneously folds into the three-dimensional structure of the native form. There has been considerable work on the folding process as well as on the unfolding or denaturation process. It is possible to take a protein in its native form, and, by subtly changing its microenvironment (pH, ionic strength, temperature, addition of denaturants such as urea, addition of other ions, etc.), induce the protein to undergo a conformational change. The energetics of that change can be measured. In a number of such studies, summarized briefly in Table 14, one can conclude that the native state is only marginally stable and is in equilibrium with a variety of other states under normal conditions. One can write the equilibrium equation $N \rightleftharpoons D$. The total free energy change in going from the native (N) state to one of

TABLE 14
Estimates of the Stability of Globular Proteins under Physiological Conditions[a]

Protein	Denaturant	$\Delta G_D^{H_2O}$ (kcal/mol)
Lysozyme	GdnHCl or Urea	9.5
	T-pH	10.0
Ribonuclease	GdnHCl or Urea	8.5
	T-pH	6.0
Myoglobin	pH	14.0
α-Chymotrypsin	T-pH	14.0
Immunoglobulin, light chain, (Wes) domain	GdnHCl	5.5

[a] From Reference (63), p. 38.

the available denatured (D) states is only in the vicinity of 10 kcal/mol for an average globular protein of 150 residues.[63] This is an energy corresponding to only two to five hydrogen bonds per molecule. It is clear that protein conformations, therefore, are very fragile and truly only marginally stable. For example, the burying of just one surface tryptophan residue provides an energy of 3.4 kcal/mol. Thus, when a protein finds itself in the vicinity of an interface, its microenvironment is altered, and clearly its equilibrium will be altered. Thus the protein conformation may be altered in the adsorbed state.

In vitro folding and denaturation studies have established that these processes occur in the time span of 0.1 to 100 seconds. It is commonly assumed that there exist folding nuclei—that is, local regions of alpha helices which are stable and which function as nuclei for additional folding or organization.[40] This provides a domain, therefore, with strong local neighborhood correlation. It is expected that there are various folding pathways and intermediate structures which are used during the folding process. Some of the theoretical methods which are now being applied to the simulation of folding processes may, in modified form, have application to the simulation of the adsorption process. Stabilization of the native conformation of proteins appears to be dominated by entropic effects arising mainly from solvent interactions.

3.3.6. Dynamics and Fluctuations[56,65–69]

A protein and its associated water molecules constitute an extremely small thermodynamic system subjected to the continual Brownian motion or buffeting of the solvent molecules. Such transient changes result in fluctuations in the local thermodynamic properties of the system, and therefore fluctuations in energy. A protein of molecular weight of 25,000 experiences internal energy fluctuations of about 40 kcal/mol. This is much larger than the energy changes involved in ligand-induced conformational transitions. These thermal fluctuations can result in volume changes which provide transitory cavities or channels in the protein. These dynamic fluctuations are a consequence of the thermodynamic uncertainty of small systems. They exist within a single macromolecule and are not correlated with similar fluctuations in other molecules. It is clear from these thermodynamic fluctuations that conformational mobility must be expected. It is now known from NMR studies that even such large and rigid structures as the planar aromatic amino acid side chains can flip in and out of the protein surface.[66]

Proteins are therefore dynamic, flexible objects whose physical and chemical properties are dominated, not only by their conformation, but also by the continual changes in conformation which are a consequence of their microscopic size.[51] The marginal stability of most protein conformations suggests that processes at one point in the protein might well have an effect on a portion of the molecule far removed.

Although we have argued that proteins are dynamic structures with transitory fluctuations and conformations, it is important to know that the cooperative nature of the forces stabilizing protein conformations permits only very limited flexibility of the protein. Although dynamic processes do occur, the folded protein rarely strays very far from the compact globule observed crystallographically.[68]

The process of ligand binding to a protein also may involve conformational change. The process might be viewed as follows:[51]

> Freely diffusing ligand approaches the protein. Assume that the ligand binding side is not a rigid structural feature, but involves transitory conformations. If the protein is in a transitory "open" conformation, the ligand may diffuse in and bind. The binding interactions may now stabilize a conformational state which is optimum for ligand–protein interaction, a state which previously, and in the absence of ligands, would be unstable.

One can envision a conformational fluctuation spectrum with the native state being a mean conformation. Since that mean conformation is only marginally stable, it is clear that any local microenvironmental change, such as contact with a solid surface, may well stabilize other conformations.

3.3.7. Molecular Models and Computer Graphics[40,70–80]

In order to begin to understand and appreciate how a protein may interact with a surface, it is important to be able to "see" the protein in three dimensions. This usually requires the use of models which, in the case of proteins, are terribly cumbersome and expensive. In many institutions molecular models have been replaced by three-dimensional molecular computer graphics. There are some 70 major molecular graphics installations throughout the world which have the capabilities of imaging large macromolecular structures in three dimensions.[78]

Ideally one should be able to predict a three-dimensional structure from the amino acid sequence. There has been considerable activity along these lines. Current methods can predict the structural class of a protein or domain. Although about two-thirds of all residues in a polypeptide chain can be assigned to the correct secondary structure, the three-dimensional structure cannot be predicted. If one independently knows the three-dimensional structure, for example from X-ray crystallography, i.e., all of the atoms are identified and their atomic coordinates known, then this information can be fed into a computer and the three-dimensional structure displayed. Such coordinates and related information are available for over a hundred proteins from the Protein Data Bank.[79] The structure can be displayed as a wire or stick model or as a space-filled model. With space-filled models, one sees only the outer surface of the protein. Using one of a number of algorithms for defining a protein surface,[80] such as the van der Waals surface (Figure 4), and by color-coding the different amino acid residues or atom types present on the surface, one can quickly "see" which regions of the molecule are hydrophobic, positively or negatively charged, noncharged but polar, and other features.

This approach was developed by Richard Feldmann and coworkers at NIH.[71-74] Feldmann and others have developed a series of teaching tools for macromolecular structure using color-coded molecular graphics-derived images.[70] Connolly and others have improved upon this approach by clearly showing which portions of the protein surface are indeed accessible to a water molecule, as opposed to those portions which are inaccessible, such as in clefts, etc.[77] Images based on the Connolly methods directly show those regions of the protein surface which can be expected to interact with other molecules. It provides immediate, comprehensible information about steric complementarity.

Another major advantage of molecular graphics is that it allows real time manipulation and interaction of several interacting molecules while quantitatively monitoring the stereochemistry and even the interaction energies. These methods are only beginning to be used for studies of interactions between molecules and—to my knowledge—no one has yet used such tools for the simulation of protein adsorption.

It is recommended that any reader seriously interested in protein adsorption obtain *Teaching Aids for Macromolecular Structures*[70] which is commercially available for about $20.00. These aids clearly show the dramatic potential of surface protein structural visualization for the development of hypotheses of protein–surface interactions.

3.3.8. So? (Outlook)

How can all this protein information help our understanding of the adsorption process? Recall Figure 1, a schematic and idealized view of a single protein interacting with a single, well-characterized solid surface. Assume that the solid surface is characterized by most, if not all, of the tools noted in Volume 1. Presume that we know precisely the location of hydrophobic regions on the surface, the location of positive and negatively charged regions, neutral hydrophilic regions, etc.; i.e., the solid surface properties are thoroughly mapped on a submicroscopic level. We also presume that we have comparable information for the protein; we know its three-dimensional surface structure, the distribution of functional groups, and related properties. We now begin to allow the protein to approach the solid surface (by molecular graphics?) and sample the variety of potential interaction orientations. Clearly, one or more of these interaction orientations will tend to dominate on certain areas of the surface, depending on the interactions present.

It is clear, given the protein and solid surface in Figure 1, that the protein can interact with the solid surface in a variety of different ways, depending on the particular orientation by which it approaches and the overall binding energetics. Such mental gymnastics, aided by 3-D graphics and at least crude estimates of interaction free energies, should aid in the formulation of definitive hypotheses and in the explanation of data.

4. MASS TRANSPORT CONSIDERATIONS

4.1. INTRODUCTION

All adsorption and desorption processes depend on transport of solute to and from the interface. Many adsorption situations are dominated by mass transport processes. There are basically four major transport mechanisms (Figure 6):

(1) diffusion
(2) thermal convection
(3) flow (commonly called convective transport)
(4) coupled transport, such as convective-diffusion processes.

Turbulent or stirred solutions may incorporate all of the processes noted.

4.2. DIFFUSION[13,81,82]

Assume that a freshly prepared interface, such as an air/solution interface in a Langmuir trough, is optimally thermostated to minimize temperature gradients and thus minimize thermal convection. Assuming there is no energy barrier to adsorption during the initial stages, every molecule which hits the interface sticks and adsorbs. All molecules near the interface will be rapidly adsorbed, resulting in a depletion of solute in the volume adjacent to the interface, called the sublayer. This concentration gradient drives diffusion from the bulk of the solution towards the depleted

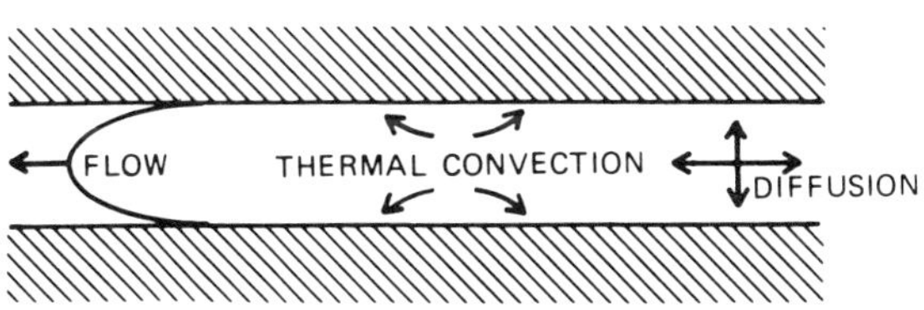

FIGURE 6. The parabolic flow profile in a thin wall channel. In addition to flow, mass transport can occur by molecular diffusion and by thermal convection.

sublayer. Under these conditions, the rate of adsorption is equal to the rate of diffusion, dn/dt:

$$\frac{dn}{dt} = C_0\left(\frac{D}{\pi t}\right)^{1/2} \tag{1}$$

where C_0 = bulk solution concentration, D = diffusion coefficient, and t = time. Thus the *rate* of adsorption is proportional to $D^{1/2}$ and $t^{-1/2}$ (Figure 7).

Integrating gives the total number of molecules, n, adsorbed at the elapsed time (t):

$$n = 2C_0\left(\frac{Dt}{\pi}\right)^{1/2} \tag{2}$$

The total surface concentration at time, t, is proportional to $D^{1/2}$ and $t^{1/2}$ (Figure 7). D for proteins is in the range of 10^{-6} to 10^{-7} cm^2/s.

All fluid interfaces contain an undisturbed layer of solution adjacent to the interface. Mass transport in this boundary layer occurs only by diffusion. The thickness of the boundary layer depends on temperature, stirring, and the interface itself. It is up to 0.1 cm thick in unstirred systems and approaches 10^{-3} cm in vigorously stirred systems.[13,81]

Table 15 shows measured and calculated times for bovine serum albumin films to reach a surface pressure of 0.1 nN/m (a surface concentration of about 70 ng/cm^2) at a fresh air/water interface.[13] Note that there is good agreement for solution concentrations greater than 0.01 g/liter. As the bulk concentration decreases, the width of the stationary boundary layer grows until thermal convection begins to play a major role in bringing material to the interface.

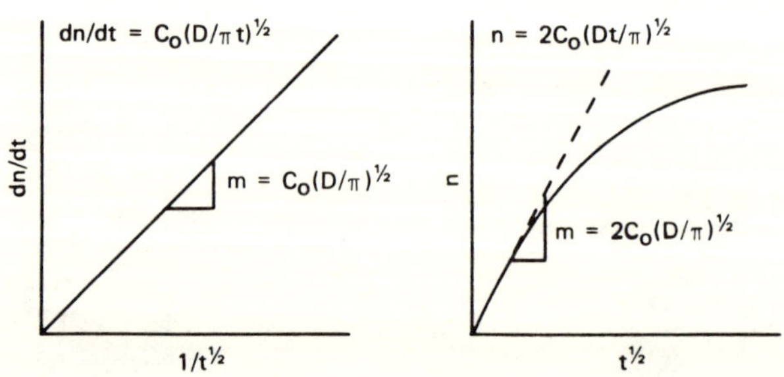

FIGURE 7. Plotting of diffusion-limited adsorption data [see Reference (13)]. (a) Rate of adsorption, dn/dt, is proportional to $t^{-1/2}$; (b) the total amount adsorbed, n, in the initial stages is proportional to $t^{1/2}$.

TABLE 15
Measured and Calculated Times for Bovine Serum Albumin Films To Reach 0.1 mN/m at a Fresh Interface[a]

Concentration (g liter^{-1})	t measured	t calculated	Width of stationary layer at t calculated (cm)
0.03	6 ± 1 s	7	0.004
0.02	15 ± 2 s	16 s	0.005
0.01	55 ± 5 s	64 s	0.011
0.005	3.5 ± 0.2 min	4.2 min	0.022
0.003	9.0 ± 0.2 min	11.9 min	0.037
0.002	19 ± 1 min	26.7 min	0.055
0.001	45 ± 5 min	107 min	0.11

[a] From Reference (13).

Once the interface is partially saturated with adsorbed solute molecules, then the rate of adsorption falls below the rate of diffusion, suggesting an energy barrier to adsorption (to be discussed later).

Molecules at or near the interface may diffuse back into the bulk solution, particularly if the free energy of adsorption is not very high. Mass transport equations which account for back-diffusion are available.[(13,81,82)] In most situations the back-diffusion of protein is probably negligible and will not be considered here.

Adsorbed molecules may diffuse laterally at the interface. Although surface diffusion is well-known in classical surface chemistry,[(83)] data on adsorbed macromolecules are sparse. Burghardt and Axelrod[(84)] and Michaeli *et al.*[(85)] have both demonstrated rapid interface diffusion of adsorbed albumin.

4.3. CONVECTION (FLOW)[(86–88)]

In many situations the system is primed with a buffer solution which is displaced by the protein solution of interest (Figure 8b). Assuming constant, laminar established flow, the velocity (V) in a rectangular flow channel of width (w), thickness (b), and length (l), where $b \ll w$, has a characteristic parabolic profile, given by [(86)]

$$V = 6\bar{V}\left[\frac{y}{b} - \left(\frac{y}{b}\right)^2\right] \tag{3}$$

where $\bar{V}$ is the average velocity and y is the y coordinate (see Figure 8a). The volumetric flow rate, Q, is equal to $A\bar{V}$, where A is the cross sectional

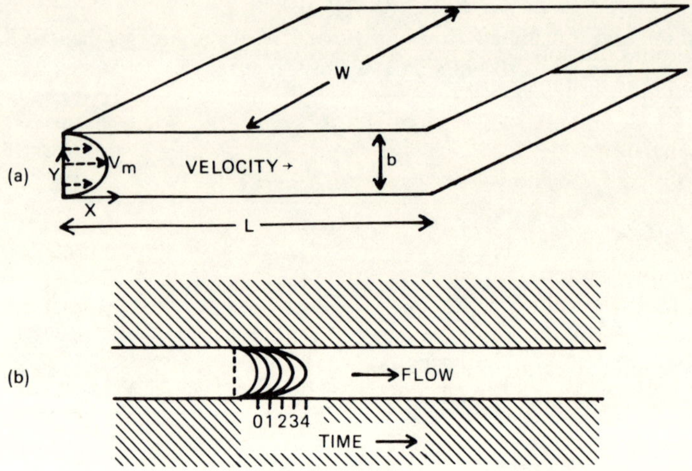

FIGURE 8. (a) The geometry and coordinate system used in the convection–diffusion treatment; (b) the development of the concentration profile due to a plug of protein solution entering a buffer-primed, thin plate flow channel. Note that a "bullet-shaped" concentration profile develops with time (assuming no diffusion).

area of the channel, i.e., $A = b \cdot w$; therefore,

$$V = \frac{6Q}{b^2 w} y \left(1 - \frac{y}{b}\right) \tag{4}$$

It can be shown that the wall shear rate, γ, is

$$\gamma = \frac{6Q}{b^2 w} \tag{5}$$

Thus

$$V = \gamma y \left(1 - \frac{y}{b}\right), \tag{6}$$

which is the equation for the parabola shown in Figure 8.

Physically, the shear rate is the change in fluid velocity over a change in distance, dV/dy, or the velocity gradient. The units of γ are cm s^{-1}/cm or s^{-1}. The shear stress is also of interest. Shear stress (τ) for Newtonian fluids is linearly related to γ,

$$\tau = \mu\gamma \tag{7}$$

where μ is the viscosity.

Note that the velocity at the wall ($y = 0$ and $y = b$) is zero, meaning nothing moves at the wall, according to classical fluid mechanics.

Now imagine a protein solution enters the flow chamber at time $t = 0$. The protein solution (of uniform concentration and flow velocity) begins displacing the buffer solution (Figure 8b). Given a parabolic velocity profile, Figure 8b shows the development of the concentration profile in the cell at various times after entrance. A "bullet-shaped" concentration profile develops. This is easily observed using a dye solution or blood.

Clearly, no protein ever reaches the surface by convective flow alone. Protein is transported to the interface by diffusion.

The flow system can be designed to produce mixing which minimizes concentration profiles. Such processes are discussed in any basic fluid mechanics or flow injection analysis textbook.[86,87]

A complete treatment of mass transport to interfaces requires combining convective (flow) and diffusion processes.[88]

4.4. CONVECTIVE DIFFUSION[88–92]

The molecules present in a flowing stream are transported by flow (convection) and by diffusion simultaneously. Fortunately, this mass transport problem is well-treated,[88] particularly in the chemical engineering literature. Robertson[89,90] and Leonard[91] are chemical engineers who have been very productive in the area of protein adsorption—they and their students have presented and used convective diffusion treatments extensively (see also Chapters 3 and 6).

For the flow direction and coordinate system given in Figure 8a, and assuming diffusion only perpendicular to the solid–liquid interface, the general convective diffusion equation in Cartesian coordinates is[88–90]:

$$\frac{\partial C}{\partial t} + V(y)\frac{\partial C}{\partial x} = D\frac{\partial^2 C}{\partial y^2} \tag{8}$$

where

$$V(y) = \gamma y\left(1 - \frac{y}{b}\right) \tag{6}$$

and C = concentration, t = time, and D = diffusion coefficient.

The origin of the coordinate system is the entrance to the flow chamber (far left of Figure 8a). Assuming there is no adsorption at the interface, the boundary conditions are[90,92]

- at $t = 0$, $C = 0$ for all y, $x > 0$,
- at $x = 0$, $C = C_0$ for all y, $t > 0$,
- at $y = b/z$, $C = 0$ for $t < x/V_m$
  $C = C_0$ for $t \geq x/V_m$
- at $y = 0$, $\partial C/\partial y = 0$ for all $x > 0$, $t > 0$.

The equation can be solved using various numerical methods.

As the protein solution moves into and through the rectangular flow channel, a parabolic concentration profile develops (Figure 8b). Diffusion washes out the profile, eventually equalizing the concentration at C_0.

In the absence of diffusion the injected material would never reach the wall due to the laminar nature of the flow. In the presence of diffusion, the concentration at the wall rises with time in an S-curve fashion. Lok *et al.* generated a dimensionless equation which defines the time course of diffusion:[90]

$$\tau = t\gamma^{2/3}\frac{D^{1/3}}{l^{2/3}} \tag{9}$$

where l = the distance from the entrance to the midpoint of the flow chamber and when $\tau = 3$ a plateau value is reached (i.e., the concentration at the wall is C_0). Solving for t, we find that

$$t = \tau\frac{l^{2/3}}{(\gamma^{2/3}D^{1/3})} \tag{10}$$

Using the typical values $D = 4.0 \times 10^{-7}$ cm^2/sec for gamma-globulin, l = 2.5 cm, $\gamma = 47.4$ s^{-1} for a volume flow rate of 2.0 ml/min, and $\tau = 3$ to reach equilibrium, Eq. (10) gives a diffusion time of 57 s from the time the newly injected fluid enters the cell (see also Chapter 2).

The study of adsorption or desorption kinetics in the first minute or so must take full account of convection–diffusion processes. Fortunately this has been well modeled by Lok *et al.*[90] and Watkins.[89,92]

Limiting cases of the general convective-diffusion equation are often helpful. If the time dependence is ignored, i.e., $\partial C/\partial t \simeq 0$, then we have

$$V(y)\frac{\partial C}{\partial x} = D\frac{\partial^2 C}{\partial y^2} \tag{11}$$

This simpler expression can be solved numerically with the appropriate boundary conditions. This is the so-called transport limited or Lévêque case.[88,91]

Another limiting case is when $V = 0$ (no flow); then we get the common diffusion expression

$$\frac{\partial C}{\partial t} = D\frac{\partial^2 C}{\partial y^2} \tag{12}$$

Techniques which allow one to monitor the boundary layer as a function of time, such as total internal reflection fluorescence (TIRF) spectroscopy (see Chapters 2 and 3), permit a quantitative evaluation of the interfacial mass transport processes using, for example, fluorescently tagged macromolecules which do not adsorb, such as fluorescein-labeled dextran.[90]

In any adsorption situation, the adsorption process itself is involved in the boundary conditions, thus affecting the solution of the equations.[89-91]

Vroman[93] has recently demonstrated the great importance of volume and concentration in limiting interfacial transport and thus influencing the adsorption process. By using a convex lens on a flat slide, Adams *et al.*[94] studied the adsorption of proteins from plasma and blood as a function of volume/surface area ratio.

5. MULTIPLE EQUILIBRIA MODEL

5.1. INTRODUCTION

It is important to carefully analyze adsorption data to extract the optimal amount of *reliable* information. Most of the data available in the literature are minimally analyzed. Usually the data are plotted as amount of protein per unit area as a function of time and/or as a function of equilibrium bulk protein concentration.

This section addresses the question of presentation and analysis of protein adsorption data, utilizing the concepts of protein–ligand multiple equilibria.[95-103] Since this theory assumes that the ligand of interest is a rigid, point molecule, it is often called the rigid binding model. The rigid binding models of multiple equilibria will be presented and transposed to the situation of protein adsorption at solid–liquid interfaces.

5.2. PROTEIN–LIGAND INTERACTIONS[96-98]

Protein–small ligand interactions can be treated assuming the protein is a surface containing different binding sites in equilibrium with one or more solutes or ligands in the solution. The solutes are assumed to be rigid spheres.

Consider a protein (P) with one binding site of ligand A (Figure 9). We can say

$$\mathrm{P} + \mathrm{A} \rightleftharpoons \mathrm{PA} \tag{13}$$

where the association or binding constant, k, is

$$k = \frac{[\mathrm{PA}]}{[\mathrm{P}] \cdot [\mathrm{A}]} \tag{14}$$

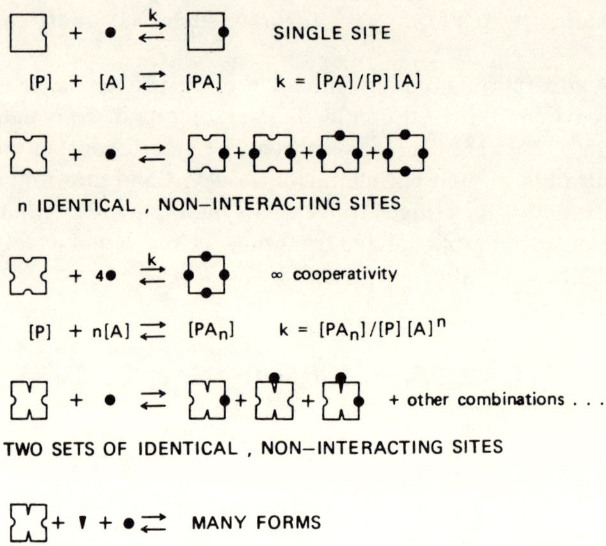

FIGURE 9. Schematic diagram of the multiple equilibria models of Table 16. From top to bottom: (a) one site/protein; (b) n identical, noninteracting sites; (c) $n = 4$ sites of infinite cooperativity; (d) two sets of identical, noninteracting sites; (e) competitive binding of two different ligands (bottom).

The experimental binding data are usually expressed as a ratio, $\bar{\nu}$, where

$$\bar{\nu} = \frac{\text{moles bound ligand}}{\text{moles total protein}} = \frac{[\mathrm{PA}]}{[\mathrm{P_T}]} \tag{15}$$

where

$$[\mathrm{P_T}] = [\mathrm{P}] + [\mathrm{PA}] \tag{16}$$

Combining Eqs. (13)-(15)

$$\bar{\nu} = \frac{k[\mathrm{A}]}{1 + k[\mathrm{A}]} \tag{17}$$

Other equivalent forms of the equation are available (Table 16), which permit the data to be plotted in various ways (Figure 10).

If the protein has *n identical* binding sites for ligand A, and *if* the sites do not interact in any way (Figure 9), then it can be shown that (Table 16)

$$\bar{\nu} = \frac{nk[\mathrm{A}]}{1 + k[\mathrm{A}]} \tag{18}$$

TABLE 16

Summary of Derived Equilibrium Binding Constants, k, and Molar Binding, $\bar{\nu}$, for Various Multiple Equilibria Models[a]

Model	Equilibria	Equilibrium constant	Isotherms $\bar{\nu} =$	Langmuir form $1/\bar{\nu} =$	Langmuir form $\frac{\bar{\nu}}{n-\bar{\nu}} =$	Hill form $\ln \frac{\bar{\nu}}{n-\bar{\nu}} =$	Scatchard form $\frac{\bar{\nu}}{[A]} =$
1 site/protein ($n = 1$)	$P + A \leftrightarrow PA$	$k = \frac{[PA]}{[P]\cdot[A]}$	$\frac{k[A]}{1 + k[A]}$	$1 + \frac{1}{k[A]}$	$k[A]$	$\ln k + \ln [A]$	$k - k\bar{\nu}$
n identical noninteracting sites	$P + iA \leftrightarrow PA_i$	$k = \frac{[PA_i]}{[P][A]^i}$	$\frac{nk[A]}{1 + k[A]}$	$\frac{1}{n} + \frac{1}{nk[A]}$	$k[A]$	$\ln k + \ln [A]$	$nk - k\mathrm{J}$
n highly cooperative identical sites	$P + nA \leftrightarrow PA_n$	$k = \frac{[PA_n]}{[P][A]^n}$	$\frac{nk[A]^n}{1 + k[A]^n}$	$\frac{1}{nk[A]^n} + \frac{1}{n}$	$k[A]^n$	$\ln k + n \ln [A]$	not useful
n identical but interacting (cooperative) sites	see text	$k = k^0 e^{-\Phi(\bar{\nu})}$	$\frac{nk^0 e^{-\Phi(\bar{\nu})}[A]}{1 + k^0 e^{-\Phi(\bar{\nu})}[A]}$				
n nonidentical but independent sites; no cooperativity	see text	k_1, k_2, k_3	$\sum_i \left(\frac{n_i k_i [A]}{1 + k_i [A]}\right)$				
n nonidentical cooperative sites	see text	$k_1 = k_1^0 e^{-\Phi_1(\bar{\nu})}$ $k_2 = k_2^0 e^{-\Phi_2(\bar{\nu})}$	$\sum_i \left(\frac{n_i k_i [A]}{1 + k_i [A]}\right)$				

[a] References (98)-(103).

FORM	TYPE	PLOT
$\bar{\nu} = (k[A]n)/(1 + k[A])$	BINDING ISOTHERM	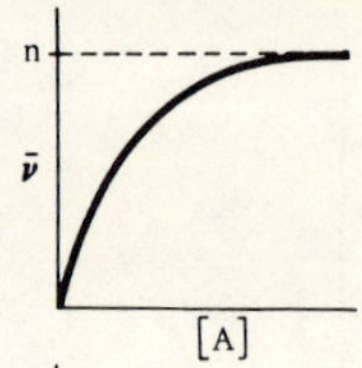
$1/\bar{\nu} = 1/n + 1/(nk[A])$	RECIPROCAL	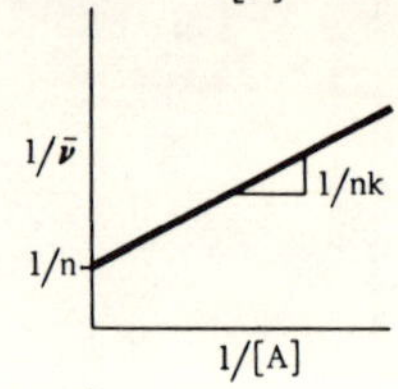
$\bar{\nu}/[A] = nk - k\bar{\nu}$	SCATCHARD	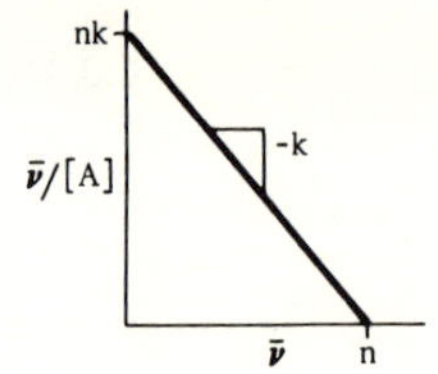
$\bar{\nu}/(n - \bar{\nu}) = k[A]^n$ or $\ln((\bar{\nu})/(n - \bar{\nu})) = \ln k + n \ln[A]$	HILL	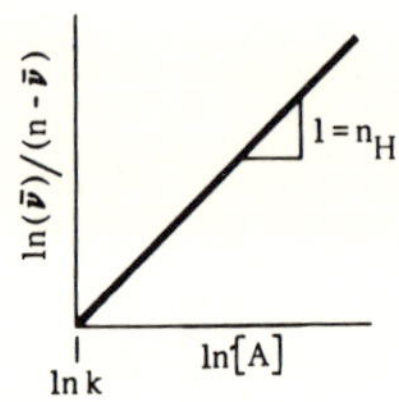

FIGURE 10. Ways of treating and plotting binding data for the case of *n* identical, noninteracting sites interacting with a spherical, rigid ligand. See text for symbol definitions.

The different forms of the equation are given in Table 16. Figure 10 presents the different standard plots.

Other models include the case of *n* identical sites where the binding of the first ligand facilitates the binding of the second, which further facilitates binding of the third, etc., until all *n* are bound [95-99] (line 3 in Table 16). The binding of oxygen to hemoglobin (where $n = 4$) is an example of such positive cooperativity (Figure 11). Negative cooperativity is also included in the general treatment (Table 16) and is readily observed.[103]

Consider

$$P + nA \rightleftharpoons PA_n \tag{19}$$

where a bound protein molecule has *n* molecules of bound A. Only unbound and PA_n species can exist. This assumed infinite cooperativity, i.e., the first

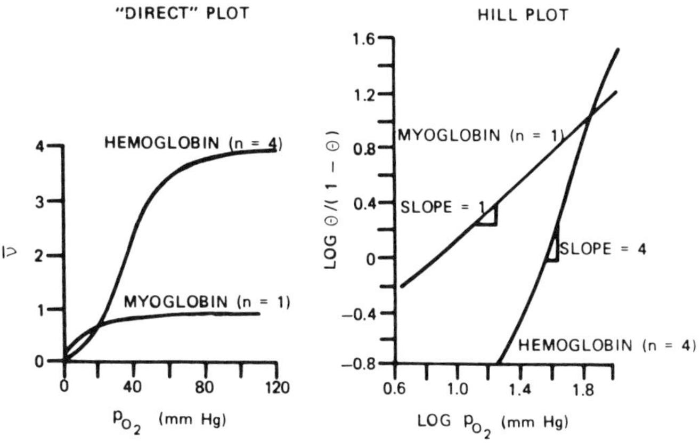

FIGURE 11. The O_2 (ligand) binding isotherms of myoglobin and hemoglobin: (a) (left) the conventional $\bar{\nu}$ vs. p_{O_2} linear plot; (b) (right) the Hill plot [from Reference (97)].

molecule bound facilitates and is immediately followed by the binding of the remaining $n - 1$ molecules (Figure 9). Then

$$k = \frac{[PA_n]}{[P] \cdot [A]^n} \tag{20}$$

$$\bar{\nu} = \frac{nk[A]^n}{1 + k[A]^n} \tag{21}$$

$$\frac{\bar{\nu}}{n - \bar{\nu}} = k[A]^n \tag{22}$$

$$\ln \frac{\bar{\nu}}{n - \bar{\nu}} = \ln k + n \ln [A] \tag{23}$$

Equation (23) is usually written as

$$\ln \frac{\bar{\nu}}{n_m - \bar{\nu}} = \ln k + n_H \ln (A) \tag{24}$$

where n_m is the maximum value of n and is usually estimated from a Scatchard plot. n_H is the Hill coefficient, obtained from a plot of $\ln (\bar{\nu}/(n_m - \bar{\nu})]$ vs. $\ln [A]$. Such a plot yields an intercept ($\ln k$) and a slope (n_H) called the Hill coefficient (Figure 10). For ideal, infinite cooperativity, $n_m = n_H$. For purely noninteracting sites, $n_H = 1$ (Figure 10); $1 < n_H \leq n$ suggests positive cooperativity and $0 < n_H < 1$ suggests negative cooperativity. The

Hill coefficient (n_H) is treated as an empirical coefficient representing the type and extent of cooperativity present in the binding interaction. Figure 11 shows the data for oxygen binding to myoglobin.[97] Myoglobin has only one binding site, $n_H = n_m = 1$. Hemoglobin has four binding sites, $n_m = 4$, which are highly cooperative. $n_H = 4$ only exists for a very narrow, binding region. The average cooperativity is closer to 2.8 for the common range of oxygen pressures normally encountered (Figure 11).

The Hill coefficient used in the literature for protein–ligand binding is usually determined at 50% saturation. It can be shown that the Hill coefficient is always a measure of the minimal number of interacting sites.[103] One can also show that the limiting value of the Hill slope at very low and very high saturation is 1. Since these regions correspond to the binding of the first and last ligand, it is clear that under these conditions there can be no cooperativity, and the slope must be 1 (Figure 12).

Scatchard ($\bar{\nu}/[A]$ vs. $\bar{\nu}$) plots ideally provide information regarding cooperativity (Figure 12). Positive cooperativity produces a concave-down plot; negative cooperativity produces a concave-up plot. Figure 12 shows the effect of cooperativity on the four common types of plots discussed.

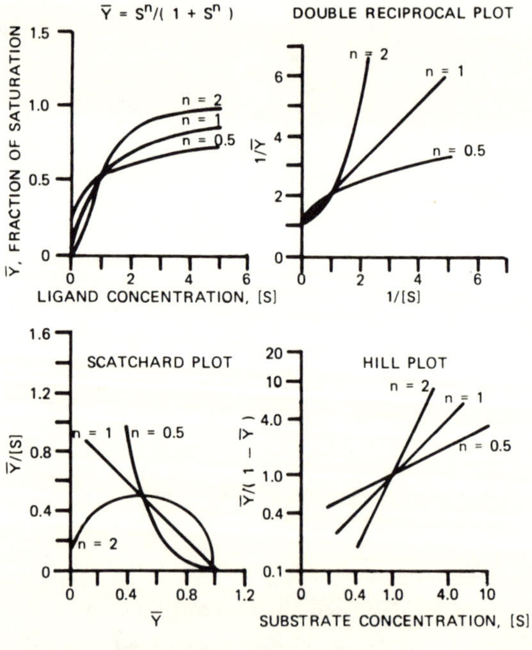

FIGURE 12. Noncooperative, positively cooperative, and negatively cooperative ligand binding. Four different ways to plot binding data are presented. The specific example used is a protein exhibiting either noncooperative binding ($n_H = 1$), positively cooperative binding ($n_H = 2$), or negatively cooperative binding ($n_H = 0.5$) [from Reference (103), p. 5].

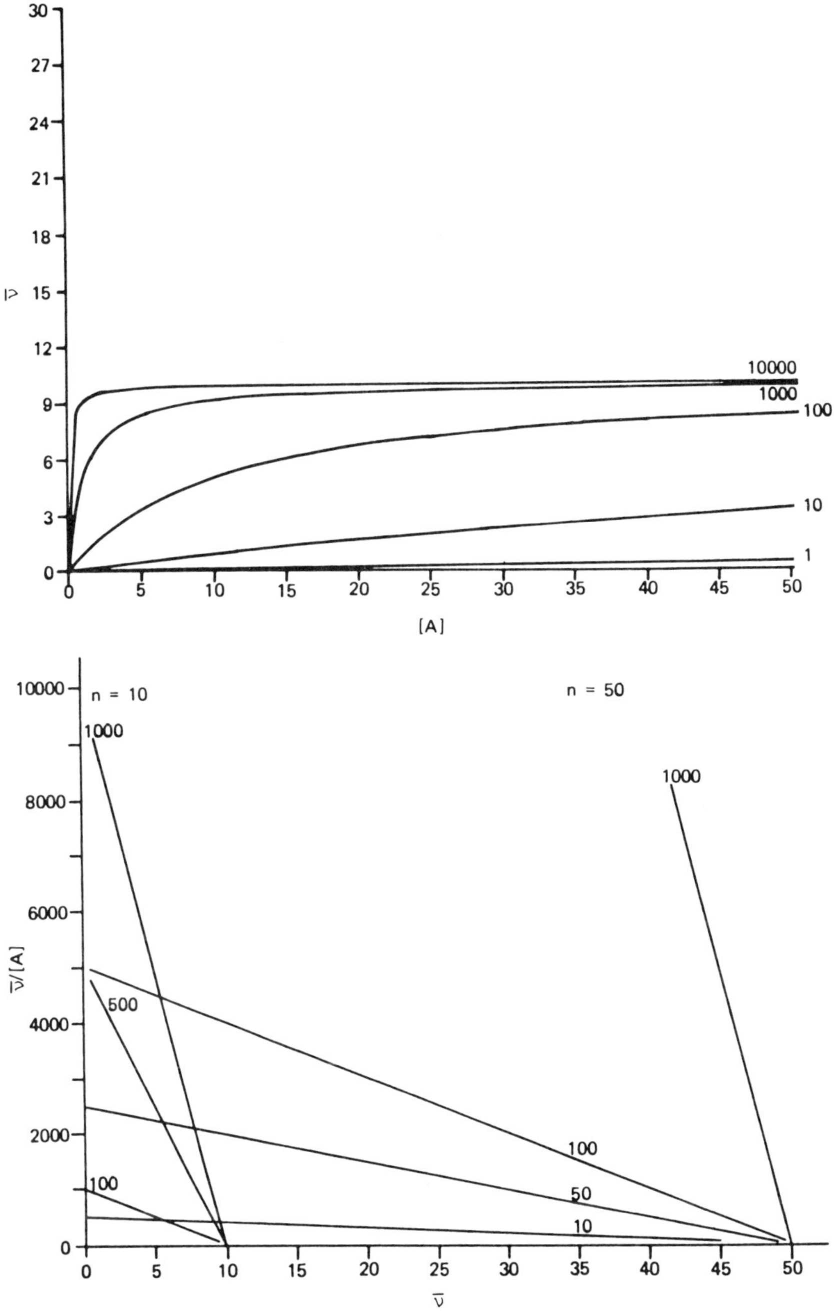

FIGURE 13. (a) (top) The conventional $\bar{\nu}$ vs. [A] (binding isotherm) plots for a hypothetical protein with ten identical, noninteracting sites binding ligand with different binding constant (k) values of 1, 10, 100, 1000, and 10,000; (b) (bottom) the Scatchard ($\bar{\nu}$/[A] vs. $\bar{\nu}$) plots for n = 10 and 50 and for k ranging from 10 to 1000.

Figure 13a presents the plots for the case of a single set of sites. Clearly, for low k values one must go to very high ligand concentrations before saturation is achieved. Conversely, for high k, saturation occurs at low ligand concentrations. The comparable Scatchard plots are given in Figure 13b.

Rarely do proteins have only one class of binding sites for a particular ligand. The case of two or more classes of binding sites can also be treated, especially if no cooperativity or site–site interaction can be assumed (Table 16, Figure 9d). In such cases, there will be a set of n_i values of k_i for *each* of the i binding sites.

The case of i different sets of identical, noninteracting sites is readily treated:

$$\bar{\nu} = \sum_i \left(\frac{n_i k_i[\mathrm{A}]}{1 + k_i[\mathrm{A}]} \right) \tag{25}$$

For two sets of sites, n_1, k_1, and n_2, k_2, we have

$$\bar{\nu} = \frac{n_1 k_1[\mathrm{A}]}{1 + k_1[\mathrm{A}]} + \frac{n_2 k_2[\mathrm{A}]}{1 + k_2[\mathrm{A}]} \tag{26}$$

Under such conditions, the Scatchard curve will appear concave-upward, even though no cooperativity is assumed (Figure 14).

Clearly the high-affinity sites dominate the behavior at low [A] and the low-affinity sites are only important at much higher [A] values (Figure 14). From the Scatchard plot one can estimate n_1, k_1, and n_2, k_2 by standard means,[99] assuming that the interactions are due to two sets of independent, noncooperative sites.

Note from Table 16 and Figures 10 and 13 the significance of the intercepts; n and k are readily deduced from the intercepts and slopes of these various plots. The Scatchard plot is generally considered to be the most informative, though strong arguments have been made for log plots, particularly in log [A].[100,101]

Comparing Figures 12 and 14, note that negative cooperativity with a single set of sites is not readily distinguishable from the case of two or more sets of sites with no cooperativity. The case of two or more sets of *cooperative* sites is a difficult one to treat. The case of competitive binding (more than one ligand) can be treated with some difficulty.[46] Cooperative behavior is often treated as a ligand-induced conformational change, and several models are available.[107,108]

When the sites interact with each other, an interaction function, $\Phi(\bar{\nu})$, can be estimated from theory or deduced from an assumed model and fit to the experimental data (see Table 16).[95–99]

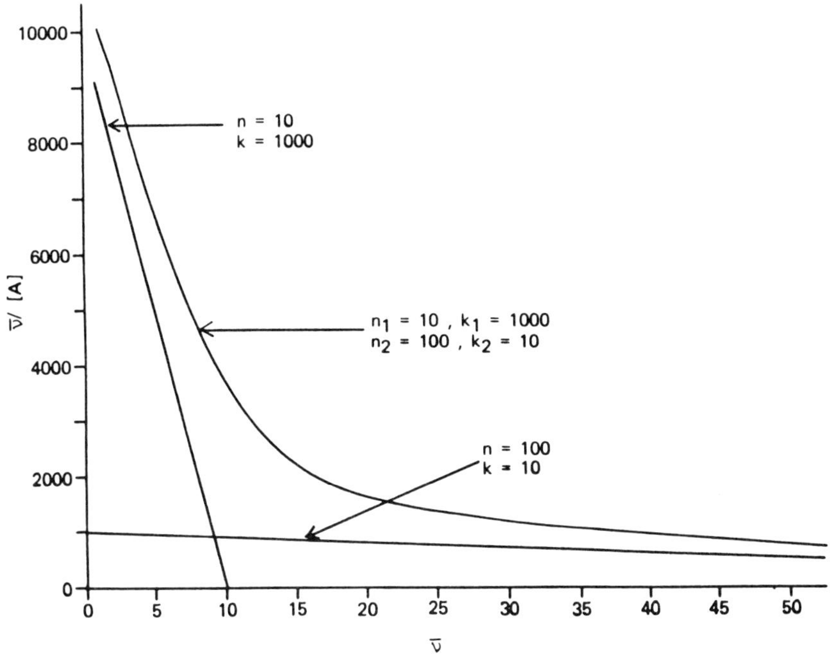

FIGURE 14. Scatchard plot for the case of two different sets of identical, noninteracting sites. Set one has $n_1 = 10$, $k_1 = 1000$; set two has $n_2 = 100$, $k_2 = 10$, i.e., a small number of high-affinity sites and a large number of low-affinity sites. Note the curvature of the plot.

Thus, by appropriate plotting and computer fitting of the experimental data to different models, one can estimate the degree of cooperativity and site heterogeneity for the protein–ligand system under study, given certain assumptions.

The units of [A] are commonly in moles/liter; units for k are thus usually liters/mole or units of $[A]^{-1}$. If the concentration of A is expressed as a mole fraction, k is, or course, unitless. $\bar{\nu}$ is best expressed as moles bound ligand/moles total protein. If a $\bar{\nu}_{max} = n_m$ can be estimated from the data, then

$$\frac{\bar{\nu}}{\bar{\nu}_{max}} = \theta$$

which is defined as fractional coverage; thus, $\bar{\nu} = n_m\theta$. $\bar{\nu}/(n - \bar{\nu})$ is accordingly often expressed as $\theta/1 - \theta$.

The theory and analysis of protein–ligand multiple equilibria are widely used for the analysis of protein–small solute equilibria and even for protein–membrane receptor interactions and related processes.[110] Extensive reviews are readily available.[95–109]

Can such treatments be useful in protein adsorption? First, let us consider the conventional treatments of adsorption from solution.

5.3. ADSORPTION FROM SOLUTION[111,112]

The classical Langmuir theory for gas adsorption can be applied to adsorption from solution, if the solution is sufficiently dilute. The surface is considered to consist of sites of about the same area as the projected area of the solute of interest (Figure 15).

Let $\bar{\nu}_s$ = number or moles of solute molecules adsorbed per area of surface. The subscript on $\bar{\nu}$ refers to the case of adsorption on a *solid surface* as contrasted with $\bar{\nu}$ used in the case of protein–ligand equilibria. Let [A] = equilibrium solute concentration. The adsorption isotherm is then [compare with Eq. (17)]

$$\bar{\nu}_s = \frac{k[\mathrm{A}]}{1 + k[\mathrm{A}]} \tag{27}$$

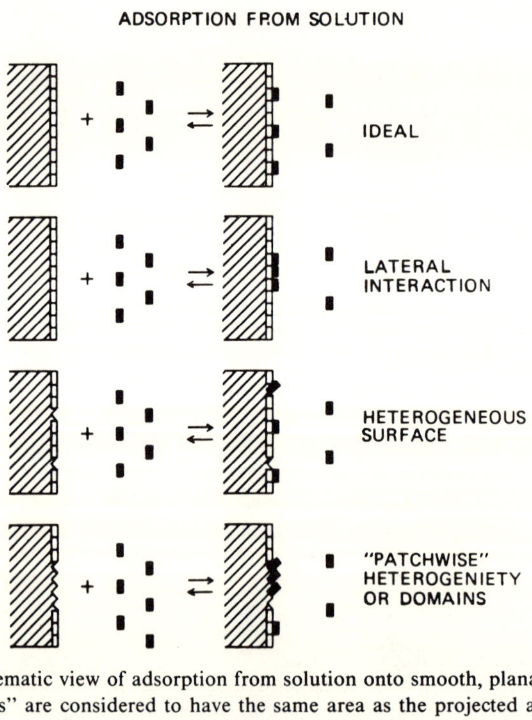

FIGURE 15. Schematic view of adsorption from solution onto smooth, planar surfaces where the surface "sites" are considered to have the same area as the projected area of the solute of interest. (a) (top). The ideal (Langmuir) case; (b) clustering of adsorbed solute due to attractive lateral interactions or positive cooperativity; (c) heterogeneous surface, i.e., two sets of binding sites; (d) "patchwise heterogeneity" or surface domains of different adsorptive properties.

The classical Langmuir treatment of adsorption assumes:

(1) only one molecule can be adsorbed per site (commonly called the monolayer assumption);
(2) only one type of site is present (homogeneous surface);
(3) the adsorption of one molecule does not affect the adsorption energy of other molecules (no lateral interactions or cooperativity);
(4) only one adsorbing species is present (no competitive adsorption);
(5) dilute solution; and
(6) reversible adsorption.

These assumptions are identical to those in the ideal theory of multiple equilibria just discussed.

Figure 16 shows how these assumptions can be considered in more refined treatments of adsorption from solution.[111]

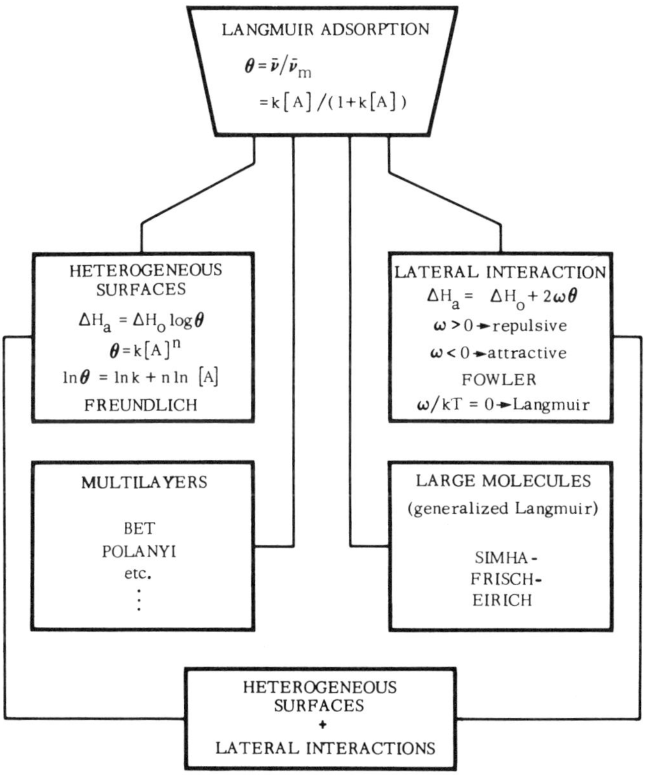

FIGURE 16. The relationships between the different models of adsorption. The various assumptions involved in the classical Langmuir treatment can be considered explicitly in other, more complex models [see Reference (111) for further details].

The case of the heterogeneous surface (Figure 15) is treated as

$$\bar{\nu}_s = \sum_i \left(\frac{k_i[\mathrm{A}]}{1 + k_i[\mathrm{A}]} \right) \tag{28}$$

The case of patchwise heterogeneity versus "homogeneous" heterogeneity cannot be distinguished by adsorption data using low molecular weight solutes.[111] The case of lateral interactions can also be treated using assumed functions for the lateral interaction (Figure 15 and 16).[111]

Clearly, the adsorption theories and protein–ligand equilibria theories are formally identical. If one simply treats the protein molecule as a heterogeneous surface (two or more classes of binding sites), interacting with identical low molecular weight ligands, the equations are identical. The assumptions involved are generally identical—for example, one ligand per site, which is the same as one adsorbed molecule/site or the monolayer assumption of Langmuir adsorption. Lateral interaction in the Langmuir treatment becomes cooperativity in the ligand equilibria treatment; heterogeneous surfaces are the same as two or more classes of binding sites; etc.

5.4. PROTEIN CHROMATOGRAPHY

There is a class of adsorption or ligand equilibria data which is of particular importance and relevance to our discussion—the literature of protein chromatography. Clearly, proteins can be separated by ion exchange, hydrophobic, and charge-transfer processes on suitable chromatographic supports. Generally, a support is used which has very low interactions with protein (k is very low). A common one is cross-linked agarose. The support is then derivatized with the proper type and density of binding site (immobilized ligand). The ligand may be a carboxylic acid group, a quaternary ammonium group, a charge-transfer (electron donor or electron acceptor) group, or a hydrophobic group. We shall only consider the hydrophobic case as an example here. The underivatized agarose is a high water content gel consisting mainly of —OH groups, which are considered to have a negligible interaction with most proteins in water.

Consider Figure 17 where the binding equilibrium is expressed diagrammatically. Clearly the binding groups on the agarose bead surface now play the same role as the soluble ligand in our previous treatments. Jennissen[113] (see also Chapter 9) has used this approach to study the interaction of protein with alkyl-substituted agaroses of different degrees of substitution (alkyl group densities or surface concentrations).

Figure 18 from Jennissen[114] is a Scatchard plot of phosphorylase kinase on butylamino Sepharose,* where C refers to the alkyl group density

* Sepharose is the commercial cross-linked agarose produced by Pharmacia, Inc.

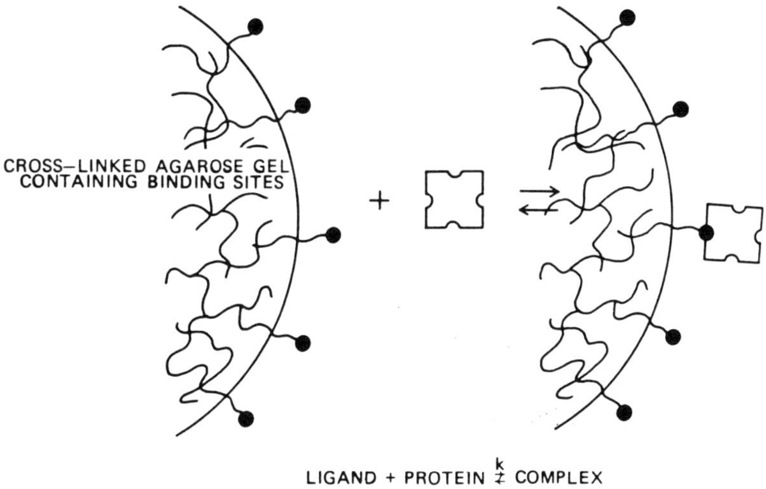

FIGURE 17. Schematic of a hypothetical protein interacting with a cross-linked agarose gel containing certain binding sites. The binding sites on the gel are considered to be immobilized ligands.

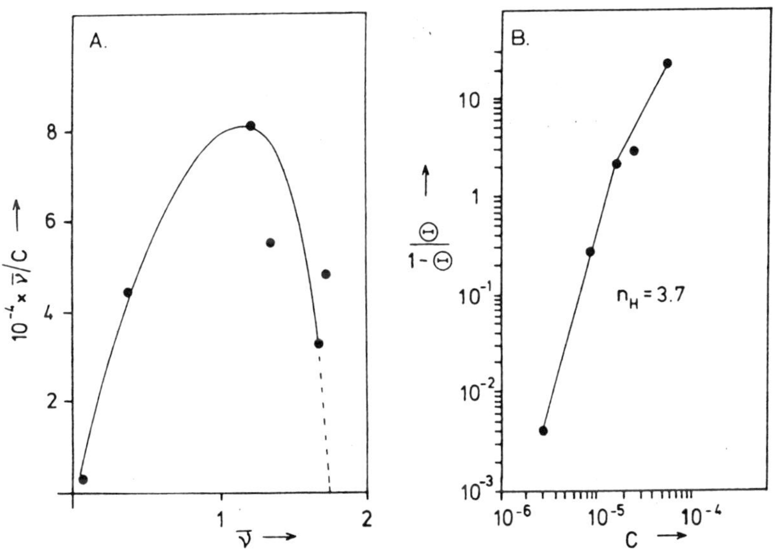

FIGURE 18. Scatchard plot (left) and Hill plot (right) of adsorption of phosphorylase kinase on butylamine-Sepharose of increasing alkyl group density. In the plots, $\bar{\nu}$ denotes the amount of phosphorylase kinase adsorbed per mg/ml packed Sepharose; Θ symbolizes fractional saturation (saturation being extrapolated as 1.75 mg enzyme/ml packed Sepharose); and C corresponds to the density of butyl residues in mol/ml packed Sepharose [from Reference (114)].

of the modified Sepharose. The concave-down character of the plot suggests a cooperative binding process. The Hill plot slope (Hill coefficient is about 3.7) suggests that about four butyl residues (binding sites) are involved in the adsorption. The Hill coefficient for the corresponding methyl derivative of Sepharose is about 6.7, suggesting that six to seven methyl groups are required for adsorption.[114]

However, the usual adsorption experiment consists of a fixed surface and a variable protein concentration (next section). Thus, the "ligand" concentration is constant and the protein concentration is varied. Returning to the basic binding equilibria presented earlier and in Table 16, the expressions can be derived in terms of [P] rather than [A], where [P] is the equilibrium protein concentration. Indeed, a strong argument can be made for varying both [A] and [P] in protein–ligand equilibria studies.[100,101] The expressions in terms of [P] are identical to those presented earlier, except for the definition of $\bar{\nu}$ [Eq. (15)]:

$$\bar{\nu} = \frac{\text{moles bound protein}}{\text{total moles ligand}} = \frac{[\text{PA}]}{[\text{A}_\text{T}] = [\text{PA}] + [\text{A}]} \tag{29}$$

For n identical, noninteracting sites

$$\bar{\nu} = \frac{nk[\text{P}]}{1 + k[\text{P}]} \tag{30}$$

Due to steric considerations, n will generally be a lower number than in the free ligand case. For the adsorption of proteins on a solid surface, the total ligand concentration is usually represented as the actual surface area of the adsorbent. In studies utilizing cross-linked gel chromatographic adsorbents, ligand concentration is often given as moles/ml packed gel. Units of k are, as before, reciprocal concentration units.

Figure 19 presents Jennissen's data and analysis for phosphorylase *b* on butyl-Sepharose.[113,114] In Figure 19a, the phosphorylase *b* initial concentration was fixed at 0.05 mg/ml and the butyl group density (in μmol/ml) was varied. The Hill coefficient is about 3, indicating positive cooperativity involving three butyl groups. Figure 19b is at a fixed butyl concentration (44 μmol/ml packed Sepharose) with varying protein concentration. The Hill coefficient is now 0.44, obviously less than 1, indicating negative cooperativity.

Thus in terms of ligands bound (butyl groups), the process is positively cooperative. Binding of protein to the first butyl group increases the probability of interaction with a second butyl group, etc. However, in terms of protein binding or protein adsorption, binding of one protein to the butyl groups decreases the probability that a second protein can interact, or the

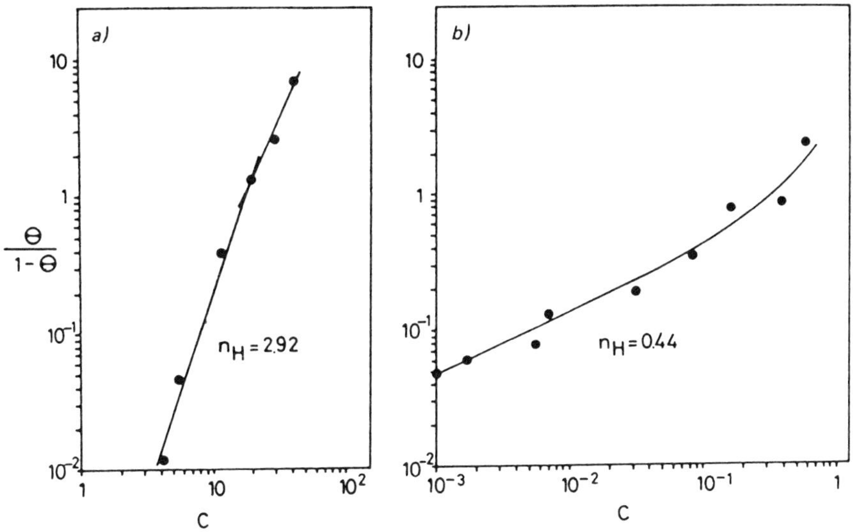

FIGURE 19. Hill plots of cooperative binding on hydrophobic agaroses. (a) Hill plot of the adsorption of phosphorylase *b* on butyl-Sepharoses of increasing butyl residue density at an equilibrium concentration of phosphorylase *b* of 0.05 mg/ml. Ordinate: (Θ) symbolizes fractional saturation [saturation being extrapolated as 12.5 mg phosphorylase *b* per ml packed Sepharose on a Scatchard plot (not shown)]. *C* corresponds to the density of butyl residues in μmol/ml packed Sepharose. (b) Hill plot of the adsorption of phosphorylase *b* on butyl-Sepharose (44.3 μmol/ml packed Sepharose) as a function of the free phosphorylase *b* equilibrium concentration (*C*) in mg/ml. (Θ) symbolizes fractional saturation being extrapolated as 49 mg phosphorylase *b* per ml packed Sepharose on a Scatchard plot (not shown). n_H denotes the Hill coefficient calculated by the least-squares method. All experiments were performed at 34°C in the presence of 1.1 *M* ammonium sulfate. For further details see Reference (125); reprinted with permission.

system exhibits negative cooperativity. Figure 20 summarizes these concepts in simple, diagrammatic form.

5.5. REAL BIOMATERIAL SURFACES—OR SURFACES OF VERY HIGH DENSITIES

The adsorption or interaction of proteins with chromatographic supports discussed in the previous section made one overriding assumption: the underivatized support does not significantly interact with proteins; i.e., the underivatized agarose has such a low interaction energy per site with proteins that for all practical purposes that interaction is negligible. This is not the case with most other surfaces with which proteins come in contact. Using our earlier definition of surface site from Figure 15, Figure 21 schematically shows what happens when a typical surface of very high site density interacts with a protein. This is the case with hydrophobic surfaces

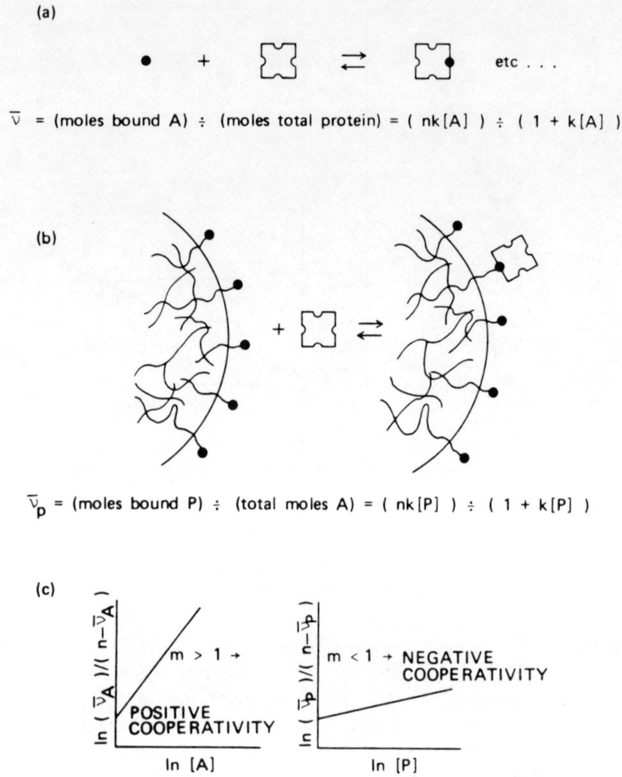

FIGURE 20. (a) Protein–soluble ligand classical binding equilibria and definition of $\bar{\nu}$ as moles bound ligand/moles total protein. (b) Protein adsorption: binding equilibria now treated in terms of soluble protein concentration; $\bar{\nu}$ now defined as moles bound protein/moles total immobilized ligand. (c) Hill plots for each case, demonstrating that positive cooperativity with respect to binding of *ligand* is negative cooperativity with respect to binding (adsorption) of protein.

such as, for example, polyethylene, polypropylene, polystyrene, and related biomedical polymers. One can consider every monomer repeat unit exposed at the surface in such polymers to be a surface site interacting with the hydrophobic domains or patches on a protein by strong hydrophobic interactions. A large number of such interactions per individual protein molecule results in irreversible adsorption. This is, of course, also experienced in conventional protein adsorption chromatography, whether by hydrophobic or ionic (electrostatic) mechanisms. When the site density on the chromatographic support is too high, the protein is irreversibly bound (see Chapter 9). Sometimes this condition can be overcome by changing the elution medium. In the case of ion exchange chromatography, one would go to a higher ionic strength eluant, which would decrease the magnitude of the

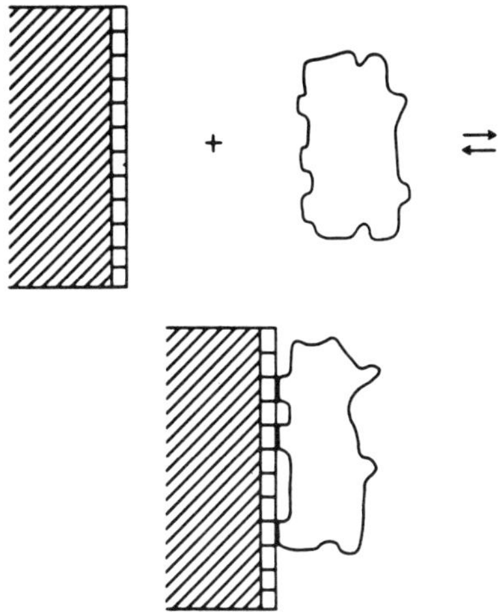

FIGURE 21. Typical adsorbent surfaces can be considered to have very high binding site ("ligand") densities, resulting in multivalent interactions with adsorbed protein. If the multivalent interactions are of sufficient number and energy, the adsorptive interaction is "irreversible."

electrostatic interactions. Under these new ionic strength conditions, the protein would then be reversibly bound and could be eluted under the proper conditions. The same is true of irreversible hydrophobic adsorption. If one elutes with an eluant which destroys or decreases the magnitude of the hydrophobic interaction, then what was an irreversible process is reversible in the new solution environment.[23] Most studies of protein adsorption onto solid surfaces in the literature are under conditions of irreversible adsorption or at best semi-reversible adsorption, which are difficult to treat with the models presented in this section.

Ideally, protein adsorption experiments should be conducted on surfaces of low binding site density so that the interaction remains reversible and can be analyzed by multiple equilibria and related models. This will then allow us not only to determine estimates of numbers of interaction sites, but to obtain measures of binding constants and thereby the interaction energies. This is not only useful for the characterization of the surface, but for the characterization of the protein itself, i.e., the accessible regions on the protein surface which are involved in the adsorption process. There are cases and conditions in which proteins can be studied on hydrophobic surfaces in a reversible fashion—for example, proteins which appear to be very weakly hydrophobic or at the bottom of the scale in hydrophobic chromatography experiments.[23] Albumin is highly hydrophobic and near the top of the hydrophobic chromatography scale and would be expected to adsorb irreversibly to most low-energy surfaces. On highly hydrophilic surfaces, such as clean, wettable quartz, it is known that some proteins adsorb in a highly reversible fashion, including albumin.[14] Here, the hydrophobic interactions are probably minimal, and the major adsorptive interaction is probably ionic.

The most useful information, however, on the interactions of proteins with surfaces will come from studies analogous to those of protein chromatography, where well-characterized and understood proteins are studied with well-characterized surfaces of known functional group type and density. The information obtained is then analyzable in such a way as to deduce interaction site densities and interaction energies. Only with such data in hand will we be able to begin to treat and understand protein adsorption.

5.6. CONCLUSIONS

Given the assumptions and limitations outlined in this section, conventional adsorption isotherm data can by analyzed in terms of the multiple equilibria models. The general scheme is outlined in Figure 22. First, the adsorption data are taken with sufficient care to get a wide range of points over a wide range of concentrations, preferably as both a function of protein solution concentration and immobilized ligand concentration, as outlined by Weber.(100,101) In many cases, the ligand surface concentration will be kept constant or will be largely unknown. The adsorption isotherm is then plotted. It is then most convenient to go immediately to a Scatchard plot; if the plot is approximately linear, an estimate of n and k can be derived. If the plot has a strong concavity, then one can use the approximation techniques to derive a set of n and k for two different assumed site populations. The shape and concavity of the Scatchard plot immediately indicate several things. If it is concave-down, one can assume that this is due primarily to a positive cooperativity process. For example, the papers by Jennissen make such conclusions. If it is concave-up, one can assume two or more populations, i.e., heterogeneous sites. One can also assume negative cooperativity. The concavity may be due to the fact that one is using a flexible ligand, and a flexible ligand model is needed to analyze such data.(105) However, from the Scatchard plot or from the original isotherm, one can generally obtain an estimate of the upper value for n. Using this value for n, one can then go to a Hill plot and plot $\log \bar{\nu}/(n - \bar{\nu})$ vs. $\log [P]$. From the slope of the Hill plot one can deduce some measure of cooperativity.

For many proteins and many surfaces as discussed earlier, the adsorption will be essentially irreversible, which will translate into a Scatchard plot with a shape which is not easily analyzed. The models and methods presented here are not adequate for treating such data. It is important under such conditions to redo the adsorption experiment, preferably on a modified surface of lower immobilized ligand density. Figure 23 outlines in schematic

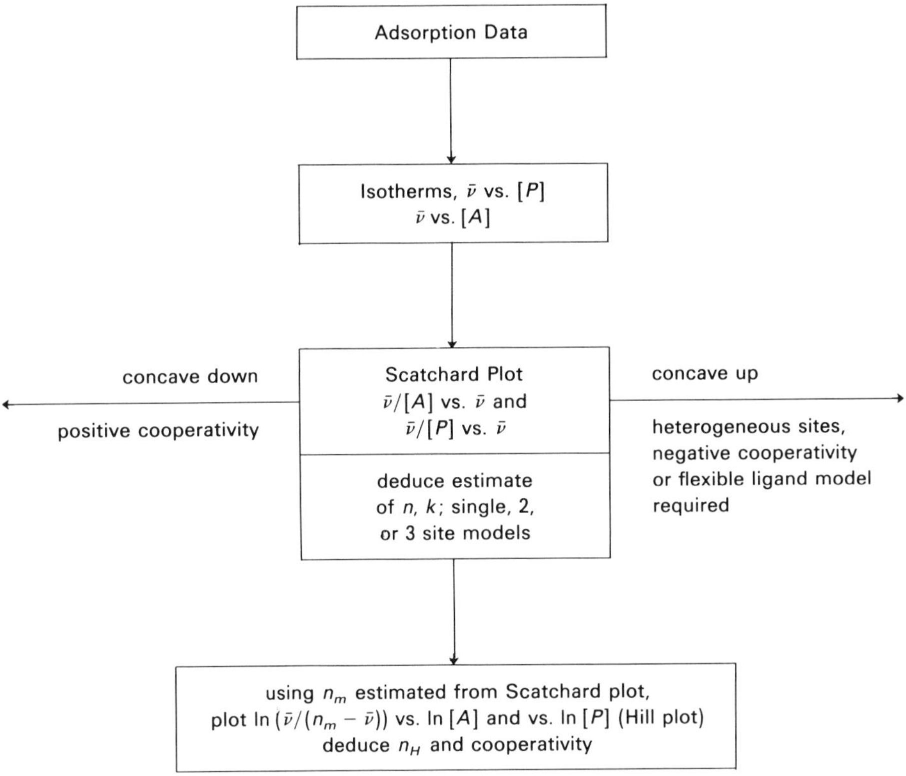

FIGURE 22. Scheme for the analysis of protein adsorption data using the methods of multiple equilibria of proteins.

form what one might expect if this is done, although no such data are at present available in the literature. Let us assume that we have a surface—for example, a substituted agarose analogous to the chromatography supports discussed earlier—in which the ligand surface density is allowed to vary. Let us further assume that each of these different supports of different ligand group density is studied with respect to different protein solution concentrations and the amount adsorbed is determined. One then has two sets of isotherms, basically $\bar{\nu}$ vs. [P] as a function of ligand group density or $\bar{\nu}$ vs. [A], i.e., ligand group density or concentration, as a function of [P]. These can be plotted in a three-dimensional form as indicated in Figure 23 and analyzed by the methods described herein. This has been described briefly by Weber.[101] Note from Figure 23 that we are assuming that as the ligand group density [A] increases, the adsorption isotherm goes into an irreversible form, which is no longer within the domain of analysis of the models and methods presented here. However, the adsorption behavior can

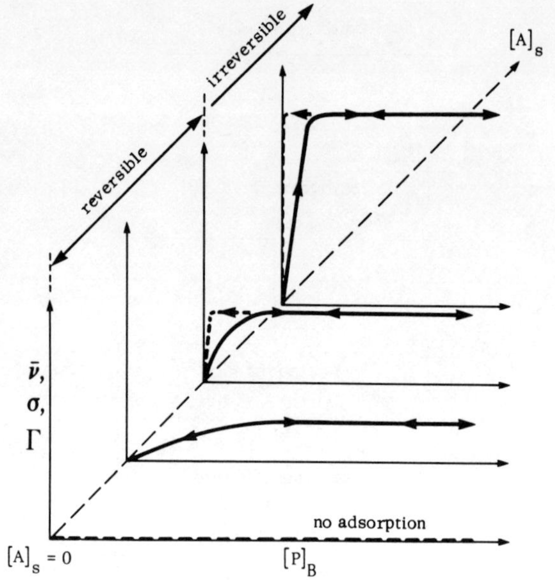

FIGURE 23. Hypothetical 3-D plot of protein adsorption isotherm ($\bar{\nu}$ plotted against protein bulk concentration $[P]_B$) as a function of surface ligand concentration, $[A]_s$. Note that the system is reversible only up to a critical $[A]_s$ and then behaves irreversibly for higher ligand surface concentrations. The right-facing arrows (→) denote adsorption; the left-facing ones (←), desorption.

be shifted to a reversible domain if the solution conditions are changed as discussed earlier.

There are some inherent limitations and assumptions in the multiple equilibria treatment. One is the fact that the treatment was devised primarily for ion interactions with proteins—that is, small, spherically symmetric rigid solutes. There are models available for nonrigid solutes,[104–107] which are more appropriate to fatty acid binding to proteins, particularly through the hydrophobic alkyl chains. Such models would be more appropriate to the study of protein interactions with highly mobile surfaces, such as those of elastomers and low glass transition materials, and probably even the substituted agaroses. Finally, both the rigid solute and the soft solute models[104–107] assume that the protein is conformationally rigid, although the rigid model has been treated in terms of allosteric phenomena or conformational coupling of the binding process.[108] In most cases of particular interest, the protein cannot be assumed to be conformationally rigid, and further methods, models, and means of obtaining data will be required in order to interpret and understand such processes. The allosteric treatments of conventional rigid ligand–protein equilibria,[108] coupled with the treatments of mobile ligand–rigid protein equilibria,[104,105] should allow such problems to be approached. The methodologies and discussion in this section, therefore, are but a first step in the analysis of protein interactions at solid–liquid interfaces.

6. CONFORMATION EFFECTS

6.1. INTRODUCTION

We have already established that the protein adsorption process may result in significant conformational changes (Sections 1.–3.). In addition to adsorbed amounts and rates, the orientation and conformation of the adsorbed protein are critical (Figure 2). Conformation refers to the secondary (α-helix, β-sheet, etc.), tertiary, and quaternary structures.

Certain orientations may make a specific site on the protein inaccessible to ligand, substrate, or antigen. For example, consider the adsorptive immobilization of a specific IgG for a solid-phase immunoassay. The procedure will be optimal if the Fab domains are free to bind antigen (Figures 2 and 5).

If the protein's structure is changed due to the adsorption process and/or the new local microenvironment, then it is said to be fully or partially "denatured," meaning that its properties are no longer those of the native protein. Conformational changes can occur due to ligand binding (for example, hemoglobin), substrate binding (many enzymes), heparin binding (antithrombin III), and surface binding (Hageman Factor, fibronectin, albumin, etc.). Although there have been many hypotheses relating to protein adsorption conformational effects, there are few direct data available (see next section and Chapter 2).

The surface itself may also be conformationally altered or "denatured" due to its interaction with the protein, particularly polymer surfaces which tend to relax or change in response to new environments (see Volume 1).

We have already established (Section 3.3.6.) that proteins are highly dynamic structures, which are constantly "sampling" different conformations due to local thermodynamic fluctuations. Some such transient conformations can be stabilized in new microenvironments, such as at a solid–liquid interface.

6.2. ADSORPTION AND CONFORMATION

Many studies of proteins at air–solution interfaces have indirectly established that the adsorbed proteins undergo detectable conformational changes.[13] Similar studies at solid–liquid interfaces are few. We review here only several key studies.

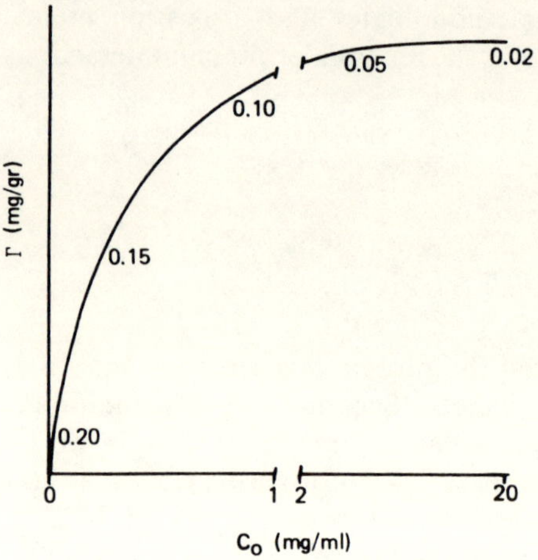

FIGURE 24. Morrissey's bound fraction data obtained by transmission IR measurements for the adsorption of bovine gamma-globulin in D_2O, pD 7.4, onto silica particles. Note that at low bulk protein concentration, the bound fraction is maximal while at high solution concentration the bound fraction is very low [from Reference (14)].

Morrissey[14] used transmission infrared spectroscopy to study protein adsorption onto silica particles in a heavy water (D_2O) buffer. By observing the shift in the amide I absorption band, he could deduce the fraction of protein carbonyl groups involved in bonding to the silica surface. He found that bovine gamma-globulin had a bound fraction of 0.20 at low bulk solution concentrations, but only about 0.02 at high solution concentrations (Figure 24). However, neither prothrombin nor bovine serum albumin exhibited a change in bound fraction with concentration.

Parallel experiments with flat silica plates using ellipsometry showed that the gamma-globulin adsorbed layers had an optical thickness of 140 Å and a surface concentration of 1.7 mg/m^2 at low bulk solution concentration—in concentrated solutions, the surface amount was 3.4 mg/m^2 with a thickness of 320 Å (Figure 25).

Morrissey suggested that at low solution concentrations the adsorbed protein has sufficient time and "elbow room" to accommodate to its new

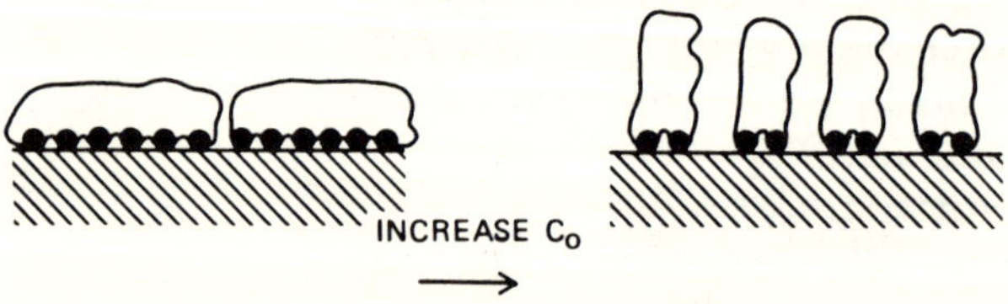

FIGURE 25. Schematic interpretation of Morrissey's results (Figure 24). At low solution concentration, the protein has no neighbors on the surface and thus can optimally adapt to the surface, maximizing the number of binding interactions. At high solution concentration, any one adsorbed protein is immediately surrounded by neighbors, minimizing the probability that it can conformationally adapt to the interface. This behavior leads to differences in adsorbed amount and adsorbed protein thickness (determined by ellipsometry), as indicated.

microenvironment by conformational change resulting in significant hydrogen bonding to the silica surface. At high solution concentrations, the collision frequency with the surface is so high that an adsorbed protein has neither the time nor the room to optimize its interaction with the surface. This general interpretation is common in the air/solution field as well.[(13)] Air/water studies show that the protein only needs to get a sufficient "foothold" on the surface to minimize the probability for desorption (see Section 8.1.). Once attached to the surface, the rest of the protein is "dragged" to the interface, thereby optimizing the interaction. The foothold need only be roughly 100–200 Å^2 whereas the cross sectional area of typical proteins is 1000–10,000 Å^2. Morrissey's high bulk concentration case may represent a minimum foothold.

Walton and his students have also pioneered the study of protein conformational changes using fluorescence and circular dichroic spectroscopies.[(116–118)] Transmission CD spectra of Hageman Factor, adsorbed on a stack of quartz plates, suggested conformational changes upon adsorption. CD spectra of proteins eluted after different contact times with the surface show that the degree of conformational change is directly related to the contact time for periods of up to 10 days.[(117)] It is now generally accepted that protein conformational change can be a rather slow event.

Brash and Chan have also used CD to study eluted proteins and find large changes in the α-helix content of fibrinogen, perhaps due to enzymatic fragmentation produced by the surface-induced activation of plasminogen to plasmin.[(119,120)]

Transmission fluorescence studies of adsorbed Hageman Factor show fluorescence changes which can be interpreted in terms of conformational

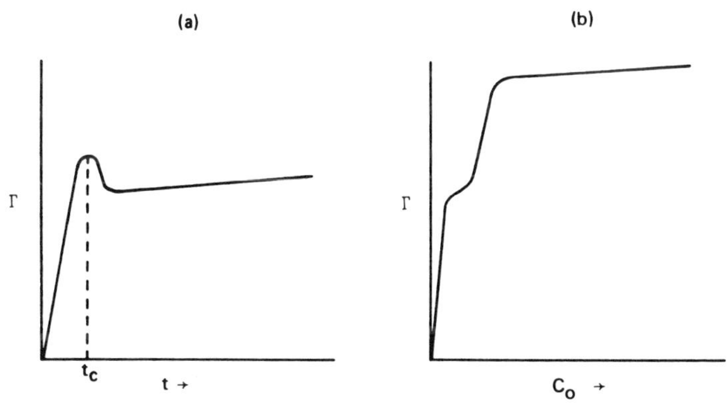

FIGURE 26. (a) A schematic of kinetic "overshoot" adsorption data, as discussed by Soderquist and Walton,[(118)] Van Dulm and Norde,[(121)] and others. See text for explanations. (b) The "kink" at about half-saturation often observed in protein adsorption isotherms. See text and Reference (118) for discussion.

and other microenvironmental effects (Chapter 6),[116] although such studies must consider substrate effects on fluorescence which may not directly involve conformational changes.[116]

Adsorption may lead to an increase or decrease in titrable groups. Titration data can thus be interpreted in terms of conformational changes (see Chapter 7).

Soderquist and Walton[118] showed an "overshoot" in adsorbed amount as a function of time and a kink or inflection in the isotherm at about half-saturation (Figure 26). They proposed:

(a) Within the first minute of contact, adsorption is rapid and reversible—a pseudo-equilibrium is present. The protein is adsorbed in a random arrangement at coverages less than 50%.
(b) At coverages greater than 50%, surface transitions and ordering may develop which lead to more efficient packing and an increase in adsorbed amount, hence the kink in the isotherm.
(c) Given sufficient time, adsorbed proteins undergo conformational changes which lead to increased surface interaction. During this process, proteins less optimally adsorbed undergo desorption, hence the overshoot in the time curve.
(d) The desorption rate decreases with increasing residence time.
(e) Desorbed protein may be permanently denatured, suggested by CD results.[117,119]

Although this last point may not be general, the others are now accepted as rough "rules of thumb" for protein adsorption, though clearly each protein–surface–solution system is unique and may not exhibit all of the features noted.

Van Dulm and Norde,[121] in a study of human plasma albumin on negatively charged polystyrene lattices, showed a fast initial adsorption followed by desorption, probably due to conformational change of the adsorbed albumin, which induces the release of less well-bound protein. This result was observed at pH 4, where the albumin has a net positive charge, and was not observed at pH 7.4, where it is highly negative.

The attenuated total reflection (ATR) Fourier transform infrared spectroscopic (FTIR) studies of protein adsorption by Gendreau, Jakobsen, and others[35] have the potential for direct determination of conformational changes during the adsorption process due to shifts in infrared absorption bands.[14] Sakurai *et al.*[147,148] have used ATR–FTIR as well as CD to probe conformational changes upon adsorption.

Total internal reflection fluorescence (TIRF) spectroscopy has recently been applied by several groups[89–92] and complete reviews are now available (Chapters 2 and 3).[122] The method can easily follow the kinetics of adsorption using proteins labeled with extrinsic fluors, such as fluoroscein or rhodamine. The intrinsic UV fluorescence of tryptophan (Trp) can be used to follow adsorption. The UV approach has the advantage that the tryptophan fluorescence is sensitive to the local microenvironment and no label is required. The major disadvantage of the UV method is the photochemical changes which occur, although such changes can be minimized by working at low radiation dose levels (Chapter 2).

The intrinsic UV fluorescence of proteins is dominated by the tryptophan indole rings. The absorption maximum is 280–290 nm with the fluorescence maximum ranging from 315–355 nm,[41] depending on the local environment of the indole side chains. Quantum yields range from 0.04 to 0.50; 0.10 is a common value. As the local environment polarity or dielectric constant increases, the maximum shifts up to 355 nm, such as for an indole ring in water or buffer. Trp moieties in highly hydrophobic environments fluoresce at 315–320 nm. Thus the fluorescence emission maximum (and the quantum yield) provides indirect information as to the local environment of the Trp fluors.

Although a number of proteins of interest (human serum albumin, for example) contain a single Trp, most contain two or more (Chapter 2). Thus the spectrum observed is the sum of all active Trp fluors, making it difficult to deduce the local environment of each fluor. Nevertheless, the UV fluorescence emission spectrum is useful in deducing orientation and/or conformation changes upon adsorption (Chapter 2).

Fibronectin (Fn) adsorption in the static mode from 0.05 mg/ml solution showed very different adsorption kinetics on hydrophobic and hydrophilic surfaces. The data for static and flow adsorption and desorption have been reported.[56] The quantity of interest here is the fluorescence maximum, which for Fn on hydrophilic silica is identical to that in bulk solution, suggesting no major conformational change upon adsorption (Table 17).

The Fn–surface interactions probably involve the charged groups on the surface of the molecule, most likely the highly positively charged heparin binding regions of the molecule (pK = 8–9). Fn adsorbed on hydrophobic

TABLE 17
Fluorescence Maximum of Fn in Different Environments[a]

Environment	Fluorescence maximum, nm
Bulk–PBS buffer	321
Bulk–PBS/8 M urea	330
Bulk–PBS/3 M guanidine Cl	350
Adsorbed on hydrophobic silica	326
Adsorbed on hydrophilic silica	321

[a] Reference (56).

silica, however, fluoresces at 326 nm suggesting a slight denaturation of the molecule. Fn interactions with the hydrophobic surface may involve some of the apolar residues in the protein interior, suggesting a partial denaturation. Clearly, studies on surfaces of a range of charge, polar, and apolar character would be of interest.

Bovine albumin has been studied via intrinsic TIRF only on hydrophilic quartz to date (Chapter 2). The fluorescence maximum (1 mg/ml BSA in PBS) is 342 nm and shifts to 333 nm upon adsorption. These results suggest that the adsorption of BSA onto silica changes the conformation of the molecule such that the two Trp fluors are in a more hydrophobic environment.

Studies using radio-iodinated proteins in the TIRF apparatus permit a direct measurement of the amount adsorbed, thus allowing one to deduce if the quantum yield is also changed on adsorption. Such studies are in progress.[(123)]

6.3. CONCLUSIONS

Intrinsic UV TIRF of the Trp residues in proteins permits one to monitor the adsorption and desorption of single-component protein solutions at solid-liquid interfaces. The measurement is made *in situ*, in a controlled hydrodynamic environment, continuously, using native, unlabeled proteins. More importantly, the fluorescence emission maximum of the adsorbed protein, when compared against fluorescence maxima in different bulk solutions, permits one to deduce microenvironmental changes upon adsorption, which are related—though not straightforwardly—to the orientation and/or conformation of the adsorbed proteins.

Other spectroscopic methods have also been used to monitor conformation and orientation of adsorbed proteins, in particular circular dichroism, attenuated total reflection infrared spectroscopy, and possibly surface Raman spectroscopy (Chapter 4).

Other less direct methods include calorimetry, hydrogen ion titration, and extrinsic fluorescence probes.

Clearly, adsorbed proteins often exhibit different conformations than their bulk solution counterparts. Methods for studying the conformation of adsorbed proteins are becoming available. Considerable progress in the very near future can be expected.

7. HYSTERESIS AND TIME DEPENDENCE

7.1. INTRODUCTION

The adsorption of macromolecules is rarely an equilibrium process. Just as the properties of synthetic polymers are often dependent on nonequilibrium processes and relaxation phenomena (Volume 1, Chapter 2), so do the properties of adsorbed proteins depend on time, metastable states, and hysteresis processes.

Although many of the authors already cited have discussed the reversibility, "semi-reversibility," and irreversibility of protein adsorption, it will be briefly discussed here for completeness. I am indebted to Jennissen[124-126] for stimulating my interest in adsorption hysteresis. Jennissen's Chapter 9 in this volume presents the subject in greater depth.

Section 5.5 dealt briefly with surfaces of high adsorption site densities (refer to Figure 23). As the adsorption site density or total free energy of adsorption increases, one moves from the realm of "reversibility" to that of "irreversibility." As proteins can undergo conformational and orientational changes on a surface, they can optimize their interfacial interactions so as to produce the maximum free energy of adsorption. Such conformational alterations are relatively slow (Section 3.3.6.) and hence very time-dependent.

7.2. TIME AS A VARIABLE

The fact that the surface-induced conformational changes of an adsorbed protein are time-dependent means that time is a critical variable. Soderquist and Walton[118] showed that the desorption rate is a function of residence time. The adsorption process can be characterized in three stages:

(1) Short times where adsorption is reversible; presumably little or no time is available for conformational changes.
(2) At longer times, where slow conformational changes occur, the process is semi-reversible, and desorption occurs very slowly.
(3) Long times where the conformational change is completed: adsorption is now irreversible, and desorption is improbable.

Those proteins which do not undergo any significant conformational change at a particular interface may not show any significant time dependence.

Soderquist and Walton proposed that the kinetic "overshoot" (Figure 26a) and the kink at around half-saturation (Figure 26b) often observed in protein adsorption studies may be due to time-dependent, surface-induced transitions. In the case of the time effect (Figure 26a), the conformational changes probably result in some unfolding and an increased number of protein sites contacting the surface; a conformationally changed protein requires greater surface area. Those proteins further along in the conformational change process "consume" surface at the expense of their neighbors which are less tightly adhered to the surface. Therefore, the amount of protein adsorbed can go through a maximum as noted in Figure 26a. Van Dulm and Norde have made similar observations.[121]

The isotherm ("equilibrium") case follows a similar argument (Figure 26b). At low C_0, the surface is not saturated, and the adsorbed proteins are in random orientations on the surface. At greater than 50% coverage, lateral interactions among neighbors become important. These may produce an ordering of the adsorbed proteins, which can be thought of as a surface phase transition. The more ordered adsorbed layer requires less area per molecule, hence the surface has "room" for more proteins to adsorb—thus the kink in the isotherm (Figure 26b).

Soderquist and Walton[118] developed a mathematical model of the protein adsorption process which reduced to classical Langmuir (small-molecule) adsorption as $t \rightarrow 0$. As $t \rightarrow \infty$, their model showed that adsorption was irreversible. Their model and conclusions are complicated by the observation that at long times the now fully denatured adsorbed protein actually desorbs. This conclusion is controversial and may not be general.

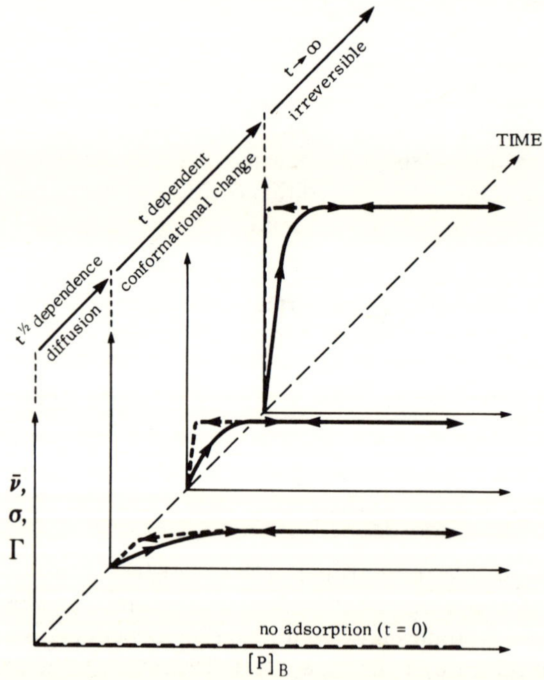

FIGURE 27. Schematic adsorption "isotherms" with a constant surface site concentration ($[A]_s$ in Figure 23 is here constant), but with adsorption time as a variable. At very short times, adsorption is diffusion-controlled. At short times, the protein has insufficient time to conformationally adjust to the interface; thus adsorption can be reversible and of the Langmuir type. At longer times, conformational adjustments begin, leading to the commonly observed semi-reversible or irreversible behavior of protein adsorption. Other nomenclature same as Figure 23.

Nevertheless, a model which reduces to Langmuir adsorption at short contact time and leads to irreversible adsorption at very long contact times is certainly consistent with much of the data in the literature. Walton and Koltisko [Reference (127) and Chapter 6 in this volume] have examined both protein stability and surface-induced, time-dependent conformational changes. Such treatments deserve to be more fully developed and investigated.

The tendancy for an adsorbed protein to undergo conformational change is protein- and interface-specific, as well as time-dependent. Chapter 2 presents a total internal reflection fluorescence (TIRF) study of IgG adsorption on hydrophobic and hydrophilic surfaces and presents desorption data as a function of residence time, showing clearly the time dependence as well as the surface dependence of desorption.

Figure 27 summarizes the situation in schematic form. Assuming a constant surface binding site concentration (fixed surface properties), we examine the amount bound as a function of time and bulk solution concentration. At very short times and particularly at low bulk concentrations, adsorption is diffusion-controlled and shows a $t^{1/2}$ dependence. At longer times, the sluggish conformational changes begin to become important. A significant hysteresis now begins to appear on the adsorption–desorption isotherm. At long contact times, the conformational adjustments are complete, adsorption free energy is maximized, and adsorption is irreversible, producing maximum hysteresis.

Clearly, proteins with a high concentration of binding sites on their surfaces may adsorb in the proper orientation, resulting in a multi-point attachment and high adsorption free energy, even without any conformational adjustment. Thus, hysteresis can be present in some systems even at very short contact times. Proteins with very stable tertiary or quaternary structures may not show significant time dependence due to the low probability for conformational change [see Reference (127)].

7.3. EXCHANGE VS. DESORPTION

It is commonly observed that protein desorption can be very slow or even nonexistent, but protein exchange can be rapid. This "anomally" has been pointed out many times.[129,130] Fortunately Jennissen's studies[124–126] and the arguments and work discussed in this chapter lead to a reasonable explanation.

Consider the cartoon in Figure 28. Imagine some adsorbed proteins with multiple binding sites. Call each protein binding site a "foot" (see also Section 8.1.). Statistically a protein foot may lift off every now and then. Clearly, all feet must be lifted off in order for the protein to desorb. This is statistically improbable, thus desorption cannot occur. We say the protein

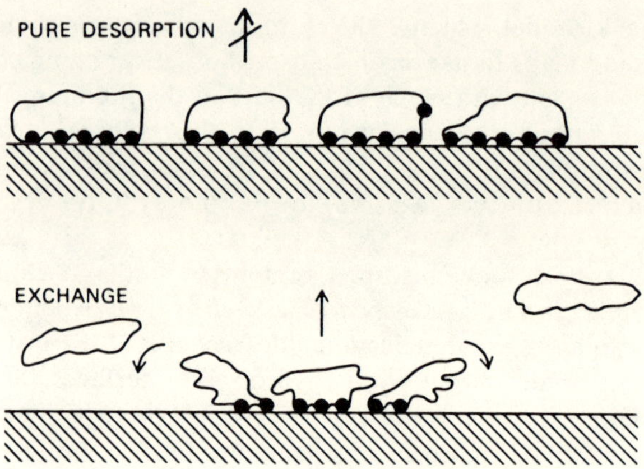

FIGURE 28. (a) Cartoon showing the improbability of desorption for a protein adsorbed by four or five binding sites. (b) Exchange of adsorbed protein with protein in the bulk solution is probable, however, due to the fact that as one foot releases, a foot from a different protein may attach. See text for further discussion. Explanation due to Jennissen [Chapter 9, and References (124)–(126)].

is irreversibly bound. Exchange is a different matter. Now we have proteins present in the local environment, diffusing to and colliding with the surface. If one of these proteins should statistically place a foot on the space vacated by a "desorbing" foot of an adsorbed protein, then a competition develops. If the new protein then puts a couple of feet down and gets itself anchored, it may indeed induce the removal of the first protein. Therefore, one protein can, in essence, be lifted off by a number of proteins adsorbing and attempting to accommodate with the surface. This is a very schematic and simplistic explanation, but it appears to at least qualitatively explain the situation. Figure 28b (bottom) shows that such a "dance" can result in the adsorbed protein being desorbed by exchange with proteins originally in the bulk solution.

This is probably the basis of the complex protein exchange processes observed by Vroman and coworkers.[(128)] The more feet and energy per foot (total adsorption free energy), the lower the probability that exchange can occur. Hence in a competitive adsorption process, one finds that those molecules present in the highest amounts and of smaller size bind first (due to diffusion and collision arguments) and are then exchanged by the more strongly adsorbing components. Vroman and Adams have shown[(128)] for plasma proteins adsorbing on glass and metal oxides that the exchange hierarchy is: albumin < gamma-globulin < fibrinogen < fibronectin < Factor XII (Hageman Factor) < high molecular weight kininogen (HMWK). (Albumin adsorbs first.) Although competitive adsorption was excluded from discussion here (Section 1.), due to its complexity, it is clear that the principles deduced from the study of simpler, single-component systems are leading to the understanding of complex systems.

A more complete treatment of protein adsorption hysteresis is given in Chapter 9 by Jennissen; see also Chapter 6 by Koltisko and Walton.

8. PRINCIPLES AND "RULES OF THUMB"

8.1. AREA REQUIRED FOR INITIAL ADSORPTION[13]

At an air/water interface the two-dimensional interfacial pressure (Π) can be easily monitored using an instrumented Langmuir trough.

We have already seen (Section 4.2.) that the adsorption rate at a clean surface is simply the rate of diffusion

$$\frac{dn}{dt} = K_a C_0 \tag{31}$$

where K_a is the adsorption rate constant. Assume that the interface is partially occupied and the interfacial pressure is Π. "In order for a molecule to adsorb, it must compress molecules already adsorbed against the interfacial pressure Π to create an area of interface ΔA, equal to that required for the molecule to move into" [Reference (13), p. 289]. The work required is $\int_0^{\Delta A} \Pi dA$ or $\Pi \Delta A$ if Π is approximately constant. Then

$$\frac{dn}{dt} = K_a C_0 \exp\left(\frac{-\Pi \Delta A}{kT}\right) \tag{32}$$

where k is Boltzmann's constant, T is absolute temperature, and no desorption has been assumed (valid for $\Pi < 15\ \text{mN/m}$). Equation (32) can be expressed as:

$$\ln\left(\frac{dn}{dt}\right) = \ln(K_a C_0) - \frac{\Pi \Delta A}{kT} \tag{33}$$

A simple linear plot of the data (Figure 29) allows ΔA to be obtained. The results of a set of experiments (Table 18) are surprising. ΔA is independent of the size or MW of the protein. Although the cross sections of the proteins studied range from ~1000 to 10,000 Å², ΔA is nearly constant at 100 to 200 Å². Conclusion: "*... only a small portion of the protein molecule needs to enter the interface in order for adsorption to then proceed spontaneously.*" [Reference (13), p. 290]. It is as if only a small "foothold" or "handhold" is required to stabilize the molecule against desorption. Now firmly planted

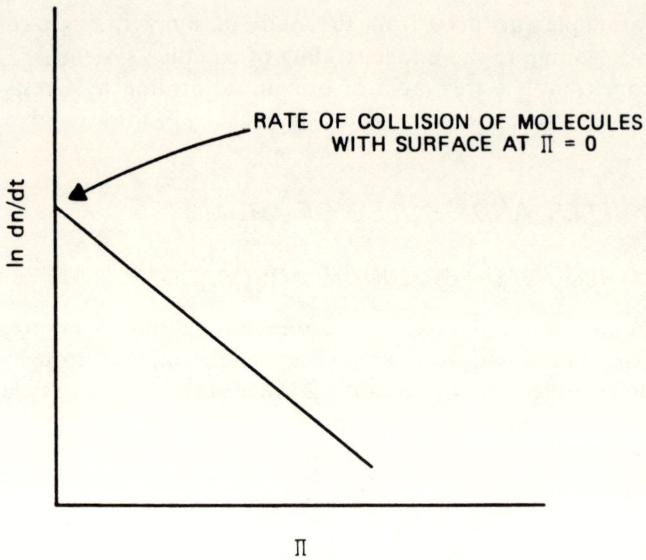

FIGURE 29. A plot of Eq. (33) as $\ln(dn/dt)$ against Π produces a straight line of slope $-\Delta A/kT$, where ΔA is interpreted as the "foot" or "anchor" size needed in order for adsorption to proceed spontaneously [see References (13) and (131)].

at the interface, the molecule can optimize its interfacial interactions by time-dependent orientation and perhaps conformational changes. The size of the "foot" is obviously relevant to the exchange discussion in Section 7.3.

An alternative interpretation of the data is that ΔA is proportional to the number of water molecules per protein participating in the adsorption and denaturation process and is related to water activity at the interfaces.[131]

TABLE 18
Values of ΔA for Various Proteins[a]

Protein	Concentration (g liter^{-1})	ΔA (Å^2)	Molecular weight ($\times 10^{-3}$)
Myosin	0.03	145	600
Human γ-globulin	0.01	130	180
Human albumin	0.02	100	70
Ovalbumin	0.03	175	44
Lysozyme	0.01	100	15

[a] From Reference (13).

8.2. ELECTRICAL POTENTIAL BARRIER[13]

A similar experiment to that described in Section 8.1. above can be performed, but now let the interface be populated by a molecular layer at constant Π and known interface electrical potential or zeta-potential. A molecule adsorbing at such an interface must do work against the electrical potential barrier as well as against the interfacial pressure. We get[(13)]

$$\frac{dn}{dt} = K_a C_0 \exp - \left[\frac{(\Pi \Delta A + q\psi)}{kT} \right] \tag{34}$$

where q is the charge on the adsorbing molecule and ψ is the interface electrical potential. Table 19 and Figure 30 present data on lysozyme adsorption into different monolayers. Clearly the *electrostatic effects are very significant.*

Because of the large size of the protein and the ΔA required for initial adsorption, only a small region of the molecule need participate in the initial stage of the process.

It is well known that *protein adsorption tends to be at a maximum at the isoelectric point* because the protein has zero net charge. Under such conditions, electrostatic barriers to adsorption are minimized.

Protein adsorption is dependent on the nature and concentration of the electrolyte.[(15)] Norde and Lyklema (see Chapters 7 and 8) have shown that adsorption generally increases with decreasing charge on the protein and on the surface, although counterion effects can often override this generalization (Chapter 7).[(15)]

TABLE 19
Relative Initial Rates of Adsorption of Lysozyme into Different Monolayers[a,b]

Monolayer	Rate	ζ-Potential (mV)
Cephalin	34.5	−68.1
Polyglutamic acid	13.2	−51.5
Pepsin	5.4	−16.7
Serum albumin	4.5	−14.1
Octadecanol	5.0	Not measured
Trypsin	3.0	+3.2
Lysozyme	(1.0)	+12.2
Polylysine	0.3	+38.5

[a] pH 6.5, I.S. = 0.01, charge on lysozyme ≃ +9 units.
[b] From Reference (13).

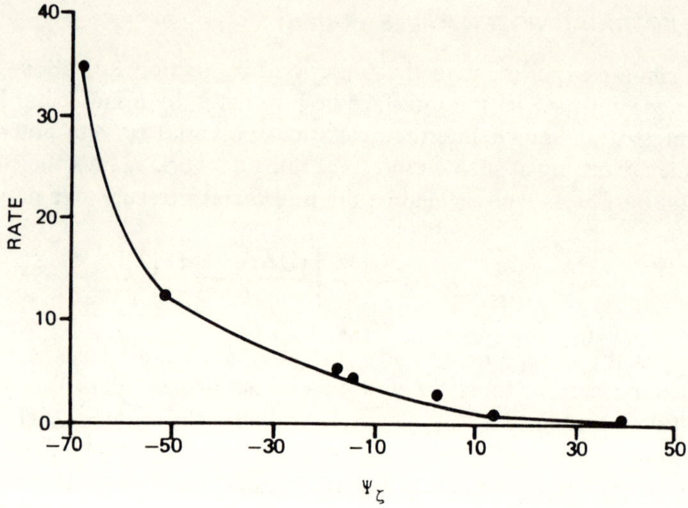

FIGURE 30. Plot of the data of Table 19[13] showing rate of lysozyme adsorption at air/water monolayers of different zeta-potentials. Refer to Eq. (34) and text for details.

8.3. HYDROPHOBICITY

The now classic studies by Norde, and Lyklema, and others have established that the *interaction between a protein and a surface increases with increasing hydrophobicity of the surface and increases with increasing hydrophobicity of the protein.* "Desorption from hydrophobic surface usually does not occur whereas proteins can often be removed from hydrophilic surfaces by exposure to extreme pH, high ionic strength, or by extensive rinsing."[15]

8.4. OTHER PRINCIPLES

Diffusion and mass transport: Proteins have a low diffusivity with $D \sim 2$ to $6 \times 10^{-7}\ \text{cm}^2/\text{sec}$. Therefore, the *initial stages of adsorption are generally diffusion-limited* and the amount adsorbed is proportional to $t^{1/2}$. If the adsorption rate, dn/dt is:

(a) $\propto 1/t^{1/2}$, diffusion is rate limiting:
(b) faster, then convection or other mass transport effect is present;
(c) slower, the adsorption is reaction limited; i.e., there are energy barriers to the adsorption process.

Time: As *contact or residence time increases, the protein tends to orientationally and conformationally adjust to the interface,* leading to stronger bonding, and greater irreversibility of adsorption.

Surface: Surface adsorption site energy and density are very important. Most *biomaterial surfaces have very high site densities*, making it difficult to study the mechanisms governing adsorption. Low site density surfaces are available (see also Section 9.2.). Heterogeneous surfaces, such as block copolymers and polymer blends, may have very unique adsorption properties. If one of the phases or domains tends to dominate the surface, it may act as a homogeneous surface. If both phases are present on the surface, then two or more very different classes of adsorption sites will be present. Protein adsorption on each of the two phases can be very different.[147,148]

Protein: The protein's intrinsic properties (size, molecular weight, three-dimensional structure, surface site density, conformational stability) are all very important and must be fully characterized and understood in order to interpret adsorption data.

9. OTHER TOPICS

9.1. PROTEIN SENSORS

This book presents many methods for detecting proteins at interfaces. Some of these are applicable to the development of protein sensors for diagnostic and industrial applications. The total internal reflection fluorescence (TIRF) method is potentially well-suited for such applications. The original work on TIRF applied to proteins at interfaces by Kronick and Little[132] was used to develop an immunoassay technique based on the detection of interfacial fluorescence. Specific antibodies can be used to confer binding specificity at an interface. Using an immobilized antibody, one can detect the binding of antigen and hence develop a sensor for antigen;[132,133] conversely, the antigen can be bound and its specific antibodies detected—thus a sensor for antibody levels.[133]

Recently Newby *et al.*[134] have shown that the TIRF experiment can be performed with a single optical fiber. They used the evanescent field (see Chapters 2 and 3) created at the interface between the fiber core and surrounding medium as the excitation source. They exposed the fiber core by stripping the cladding from the last few centimeters and capping the exposed tip with an opaque material. By coupling light into the fiber, fluorescence was excited in the sample medium adjacent to the exposed surface of the fiber. An important attribute of the sensor design is that it produces a localized, evanescent excitation field in the proximity of the core surface so that spectroscopic data are obtained preferentially from the adsorbed layer rather than from the bulk solution. The fiber end is terminated with an opaque or highly reflecting material to prevent excitation of the bulk medium.

Collection of the spectroscopic signal from the adsorbed layer occurs "automatically" using the single fiber configuration. The evanescent excitation field at the exposed core excites dipoles in the adsorbed layer. Because the near-field components of these dipoles are evanescent in nature,[135] they couple back into the fiber core and are guided back to a spectrometer. As the spectroscopic signal (e.g., fluorescence or Raman scattering) is typically shifted in frequency from the excitation radiation, it is relatively easy to separate the signal from the incident light. While optimized for monitoring chemical adsorption, the applications for the remote sensor are not limited to adsorbing species, but also include studies of almost any medium in which a fiber can be immersed.

Figure 31 shows spectra obtained after exposing the sensor tip to rhodamine isothiocyanate (RITC)-labeled IgG in phosphate-buffered saline (PBS) solution. The top spectrum in Figure 31 was obtained in air after the sensor had been exposed to RITC–IgG solution for 20 min. The measured protein fluorescence spectrum resulted only from adsorbed protein. Because spectra obtained with the sensor are evanescently excited, molecules adsorbed onto the sensor tip should be the major contributor to the observed fluorescence since the evanescent field intensity decays exponentially away from the sensor tip surface. The second spectrum in Figure 31 was obtained with the sensor immersed in PBS and after 30 min of exposure to the PBS. Exposure of the sensor tip to the buffer solution reduced the fluorescence intensity as proteins desorbed from the probe surface. However, a residual spectrum remained from the irreversibly adsorbed IgG molecules. The bottom spectrum in Figure 31 was the initial signal from the PBS prior to introduction of the protein. The spectrum is flat except for the —CH Raman bands from the fiber coating. The data of Figure 31 demonstrate that the sensor is capable of detecting a monolayer of adsorbed proteins. Figure 31 clearly shows that a single fiber optic can be used as a remote sensor for the detection of protein.

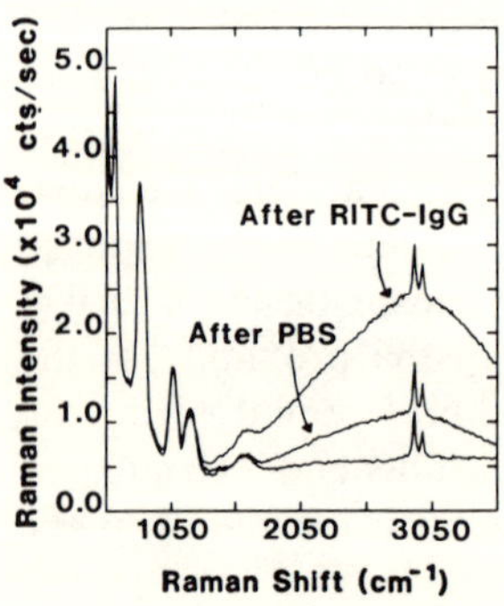

FIGURE 31. Protein adsorption detected remotely using an unclad optical fiber as the sensor. Conditions: Rhodamine isothiocyanate (RITC)-labeled IgG, pH 7.4, phosphate-buffered saline, 0.8 μg/ml protein, excitation wavelength = 488 nm. The four peaks at the right are the characteristic Raman peaks of the silica fiber optic. The general feature at the left is the characteristic rhodamine fluorescence. The fine features on the fluorescence spectrum are the ——CH Raman modes of the fiber coating. RITC–IgG was detected evanescently at the unclad tip of the fiber [from Reference (134)].

It is expected that such methods can be developed into practical sensors for diagnostic and industrial applications.[133]

9.2. SURFACES WHICH MINIMIZE PROTEIN ADSORPTION

Surfaces which show minimal protein adsorption are important in many areas, particularly for blood-contacting devices (to minimize the activation of coagulation), membranes for separation processes (to minimize polarization and fouling), sensors (to minimize nonspecific binding), chromatography (to minimize nonspecific binding), etc.

The chromatography field has examined many such surfaces, in particular, uncharged polysaccharides (such as dextran and agarose) and synthetic polymers (such as polyacrylamide). Such materials are used in chromatography and electrophoresis because protein binding is minimized. The aqueous gel–water interface has a very low interfacial free energy[136] and hence a very low driving force for protein adsorption.

Recent studies by Gregonis *et al.*[137] have shown that poly(ethylene oxide) (PEO) surfaces greatly minimize protein adsorption. These results confirm the work of Merrill and Salzman,[138] Mori and Nagaoka,[139] and others which show that PEO-rich surfaces tend to minimize cell adhesion and show good blood compatibility.

The mechanism is apparently related to interface excluded volumes and surface mobilities and the inability of PEO to significantly enter into hydrogen bonding or hydrophobic interactions.

9.3. LABELING EFFECTS

Most protein adsorption studies use radio-labeled proteins, generally radio-iodinated. Although this is one of the simplest and most sensitive experimental methods, it is not without potential artifact. The presence of the label may change the physicochemical properties of the protein sufficiently to affect its adsorption properties.[140] Furthermore, if the protein–radionuclide bond is labile and the label is released with time,[140] the experimental data can be in serious error.

Iodine labeling results in iodine being linked to the aromatic ring of tyrosine, probably affecting the hydrogen-bonding and charge-transfer properties of the labeled tyrosine residues. Side reactions of iodination include oxidation of free sulfhydryl groups,[144] iodination of tryptophan,[141] and others.[141]

A solution of radiolabeled protein must be considered to be a two-component system: native, unlabeled protein and labeled protein.[142] In addition, the labeled material can have 1, 2, 3, . . . , *n* labels present, thereby producing a heterogeneous protein population. Studies with ^{125}I-labeled

TABLE 20
General Principles of Protein Adsorption

Subject	Principle	Reason	Key source
Foothold	A minimum contact area is required for initial adsorption.	A minimum contact area (~100 to 200 Å^2) results in sufficient stabilization against desorption to permit the adsorbed protein to optimize its adsorbed orientation and conformation.	Section 8.1, References (13), (131)
Adsorption rates	$\frac{dn}{dt} \alpha t^{-1/2}$	Adsorption at short times is usually transport-limited. Faster rates suggest augmented mass transport (convection), slower rates suggest energy barriers to adsorption.	Section 4, Reference (13)
Electric potentials	Electrical potentials can lead to enhanced (unlike charges or fields) or decreased (like charges or fields) adsorption.	Interface potential energy barrier can be positive or negative depending on nature of charges and their distribution.	Section 8.2, References (13), (15)
Isoelectric point	Adsorption is generally maximal at a pH equal to the isoelectric point of the protein.	As net charge on protein goes to zero, interface electrical energy barriers go to zero. Also adsorbed protein at IEP does not have any lateral interactions within the adsorbed layer.	Reference (13)

Hydrophobicity	Adsorption increases with increased hydrophobicity of the surface and/or the protein.	Hydrophobic interaction is very significant and appears to be a major driving force for protein adsorption.	Sections 8.3, 3.2.3, Reference (15)
Reversibility	Protein adsorption on uncharged hydrophilic surfaces is often at least partly reversible. Adsorption on hydrophobic surfaces is often irreversible.	Interaction site densities and energies.	Chap. 9, Sections 5 and 7
Hysteresis	Protein adsorption generally shows adsorption–desorption hysteresis.	Due to metastability and irreversibility of protein adsorption.	Chapter 9, Section 7
Time	The longer the residence time, the lower the probability for desorption or exchange.	Proteins enhance their interaction with the surface by orientation and conformation adjustments, which are time-dependent.	References (19), (118), Section 7
Denaturability	The more stable a protein (more difficult to denature), the more reversible will be its adsorption behavior.	High intrinsic stability makes surface-induced conformational changes unlikely.	Sections 3.3.5, 3.3.6, Section 7

hormones have shown that the labeled and unlabeled hormones bind differently to the receptors, resulting in nonlinear Scatchard plots, which are basically due to a competitive adsorption process.[110,142] It must also be noted that it is particularly important to have data at very low saturation, particularly for Scatchard plot analysis.[98–102] This is specifically not the case for most of the available protein adsorption data because of the difficulty in obtaining data in the low concentration range.

A number of studies have demonstrated that labeled proteins may adsorb very differently from unlabeled ones.[140,143,145] Clearly, the application of labeled proteins to adsorption studies must be done carefully and cautiously.

9.4. LIQUID/WATER INTERFACE STUDIES

Protein adsorption at the oil/water interface has been studied in detail.[27–31] The oil/water interface is a good model system which may have relevance to the interface between water and soft, mobile hydrophobic polymers, such as silicones, polyhexene, etc. (see Chapter 2 in Volume 1).

Bagnall and coworkers have applied and extensively developed the pendant drop technique,[28,29] which permits one to study adsorption *in situ*, in real time, using native, unlabeled proteins, and to obtain information on interfacial pressures, tensions, and denaturation processes. One of Bagnall's conclusions is that the spreading pressure of rapidly unfolding fibrinogen is sufficient to desorb other plasma proteins,[29] substantiating Vroman's observations at rigid solid interfaces.[93,94]

Contact angle methods provide information on the interfacial tension at adsorbed protein–water–air (or octane) interfaces.[32–34] Such measurements can be helpful, but are very difficult to interpret due to the effects caused by the presence of the air or octane phase, such as changes in adsorbed protein conformation.

10. SUMMARY/CONCLUSIONS

The general principles of protein adsorption are beginning to be identified and understood (Table 20). New techniques and methods are now available (Chapter 2–5) which, together with well-established methods (Chapters 6–9), allow one to thoroughly probe the adsorption process.

In the past, the solid surface and the solid/solution interface were often poorly characterized and poorly understood. A range of new tools and concepts are now available for the study of solid/solution interfaces (Volume 1).

Protein surface structure and conformational dynamics are now much better understood. In the near future, we can expect improved computer molecular graphics to enable us to better visualize and comprehend potential protein–surface interactions.

Very qualitatively, we have a rough idea of what is going on. The challenge now is to take what we know about proteins, surfaces, and adsorption and to begin to quantitatively model (see Chapter 10) the protein adsorption process.

The future looks bright!

ACKNOWLEDGMENTS

Discussions with Mr. Jon Pierce and Ms. Barbara Zdasiuk (deceased) on Section 5 were most helpful. Portions of Section 5 were developed in the Department of Chemistry at the University of Coimbra, Portugal during a Fulbright visit and presented at the 3rd International Conference on Plastics in Medicine and Surgery, June 22, 1979, at Twente University, Enschede, The Netherlands.

This work was supported by NIH grants HL-16921, HL-18519, and HL-29209.

I am particularly indebted to R. Van Wagenen, D. E. Gregonis, V. Hlady, J. Lyklema, A. Silberberg, H. Jennissen, L. Vroman, B. Morrissey, and others for discussions on these topics over the years.

Ms. Anne Grossenbach expertly typed the manuscript.

REFERENCES

1. L. Vroman and E. F. Leonard, eds., *Behavior of Blood and Its Components at Interfaces, Ann. N. Y. Acad. Sci.* **283** (1977).
2. R. E. Baier, ed., *Applied Chemistry at Protein Interfaces, Adv. Chem. Ser.* **145** (1975).
3. S. L. Cooper and N. A. Peppas, *Biomaterials, Adv. Chem. Ser.* **199** (1982).
4. J. L. Brash, Mechanism of adsorption of proteins to solid surfaces, in: *Biocompatible Polymers* (M. Szycher, ed.), pp. 35–52, Technomic Publ., Lancaster, PA (1983).
5. J. L. Brash and D. J. Lyman, Adsorption of proteins and lipids to nonbiological surfaces, in: *Chemistry of Biosurfaces*, Vol. 1 (M. L. Hair, ed.), pp. 177–232, Dekker, New York (1971).
6. D. F. Chessman and J. T. Davies, Physicochemical and biological aspects of proteins at interfaces, *Adv. Protein Chem.* **9**, 439–501 (1954).
7. C. W. N. Cumper and A. E. Alexander, Proteins at interfaces, *Rev. Pure Appl. Chem.* **1**, 121–151 (1951).
8. F. R. Eirich, Conformational states of macromolecules adsorbed at S–L interfaces, in: *Colloid and Interface Sci.*, Vol. 1 (M. Kerker, ed.), pp. 447–460, Academic Press, New York (1977).

9. M. J. Fraser, Behaviour of proteins at interfaces, *J. Pharm. Pharmacol.* **9**, 497–521 (1957).
10. E. Gorter, Properties of protein monolayers, *Ann. Rev. Biochem.* **10**, 619–636 (1941).
11. L. K. James and L. G. Augenstein, Adsorption of enzymes at interfaces, *Adv. Enzymology* **28**, 1–40 (1964).
12. T. A. Horbett, Protein adsorption on biomaterials, in: *Biomaterials* (S. L. Cooper and N. A. Peppas, eds.), *Adv. Chem. Ser.* **199**, 233–244 (1982).
13. F. MacRitchie, Proteins at interfaces, *Adv. Protein Chem.* **32**, 283–326 (1978).
14. B. W. Morrissey, Adsorption and conformation of proteins, *Ann. N.Y. Acad. Sci.* **288**, 50–64 (1977).
15. W. Norde, Adsorption of proteins at solid surfaces, in: *Adhesion and Adsorption of Polymer* (L-H Y. Lee, ed.), Vol. 2, pp. 801–826, Plenum Press, New York (1980).
16. M. C. Phillips, Conformation and properties of proteins at liquid interfaces, *Chem. Ind.*, 170–176 (5 March, 1977).
17. A. Rothen, Films of proteins in biological processes, *Adv. Protein Chem.* **3**, 123–137 (1947).
18. L. Ter-Minassian-Saraga, Spread monolayers, adsorption and desorption, *J. Colloid Sci.* **11**, 398–418 (1956).
19. A. G. Walton and B. Koltisko, Protein structure and the kinetics of interaction with surfaces, in: Reference (3), pp. 245–264.
20. C. A. Zittle, Adsorption studies of enzymes and other proteins, *Adv. Enzymology* **14**, 319–374 (1953).
21. J. Lyklema, Molecular interactions in affinity chromatography, in: *Affinity Chromatography* (T. C. J. Gribnau, ed.), pp. 11–27, Elsevier, Amsterdam (1982).
22. S. Hjerten, Hydrophobic interaction chromatography, *Adv. Chromatogr.* **19**, 111–123 (1980).
23. B. H. J. Hofstee and N. F. Otillio, Non-ionic adsorption chromatography of proteins, *J. Chromatogr.* **159**, 57–69 (1978).
24. J. Porath, Charge-transfer adsorption, *J. Chromatogr., Chromatogr. Rev.* **159**, 13–24 (1978).
25. J. Porath, Charge-transfer adsorption in aqueous media, *Pure Appl. Chem.* **51**, 1549–59 (1979).
26. W. Holstein and H. Hemetsberger, Donor-acceptor complex chromatography, *Chromatographia* **15**, 186–190, 251–258 (1982).
27. S. Ghosh and H. B. Bull, Adsorbed films of bovine serum albumin, *Biochim. Biophys. Acta* **66**, 150–157 (1963).
28. R. D. Bagnall, J. A. D. Annis, and P. A. Arundel, Adsorption of plasma proteins, *J. Biomed. Materials Res.* **12**, 653–663 (1978).
29. R. D. Bagnall, Adsorption of plasma proteins, I and II, *J. Biomed. Materials Res.* **11**, 947–978 (1977) and **12**, 203–217 (1978).
30. E. Tornberg, Drop volume technique, *J. Colloid Interface Sci.* **64**, 391–402 (1978).
31. A. J. Ward and L. H. Regan, Pendant drop studies, *J. Colloid Interface Sci.* **78**, 389–394 (1980).
32. A. Van der Scheer and C. A. Smolders, Contact angle measurements, *J. Colloid Interface Sci.* **63**, 7–15 (1978).
33. D. R. Absolom, C. J. Van Oss, W. Zingg, and A. W. Neumann, Surface tension of albumin altered at protein–air interface, *Biochim. Biophys. Acta* **670**, 74–78 (1981).
34. A. Baszkin and D. J. Lyman, Adsorption/desorption studies of plasma proteins on polymer surfaces, in: *Biomaterials/1980* (G. D. Winters, ed.), pp. 393–398, Wiley, New York (1982).
35. R. M. Gendreau, R. I. Leininger, S. Winters, and R. J. Jakobsen, FT-IR for protein–surface studies, in: Reference (3), pp. 371–394.
36. J. L. Dimicoli, M. Nakache, and P. Peronneau, Visualization of diffusion–convection at liquid–solid interfaces, *Biorheology* **19**, 281–300 (1982).

37. L. Stryer, *Biochemistry*, 2nd edition, W. H. Freeman Co., San Francisco (1981).
38. R. E. Dickerson and I. Geis, *Structure and Action of Proteins*, Harper & Row, New York (1969).
39. R. E. Dickerson and I. Geis, *Hemoglobin*, Benjamin/Cummings, Menlo Park, CA (1983).
40. G. E. Schulz and R. H. Schirmer, *Principles of Protein Structure*, Springer-Verlag, New York (1979).
41. D. Friefelder, *Physical Biochemistry*, 2nd edition, W. H. Freeman Co., San Francisco (1982).
42. C. R. Cantor and P. R. Schimmel, *Biophysical Chemistry*, W. H. Freeman Co., San Francisco (1980).
43. H. B. Bull, *An Introduction to Physical Biochemistry*, F. A. Davis Co., Philadelphia (1964).
44. A. G. Walton and J. Blackwell, *Biopolymers*, Academic Press, New York (1973).
45. A. G. Walton, *Polypeptides and Protein Structure*, Elsevier, New York (1981).
46. C. Tanford, *Physical Chemistry of Macromolecules*, Wiley, New York (1961).
47. F. Franks and D. Eagland, Solvent interactions in protein conformation, *CRC Crit. Rev. Biochem.* **3**, 165–219 (1975).
48. M. F. Perutz, Electrostatic effects in proteins, *Science* **201**, 1187–1191 (1978).
49. A. Wada and N. Nakamura, Nature of the charge distribution in proteins, *Science* **293**, 757–758 (1981).
50. A. Warshel, Energetics of enzyme catalysis, *Proc. nat. Acad. Sci. U.S.A.* **75**, 5250–5254 (1978).
51. A. Cooper, Conformational fluctuation and change in biological macromolecules, *Sci. Prog.* **66**, 473–497 (1980).
52 C. Tanford, *The Hydrophobic Effect*, 2nd edition, Wiley, New York (1981).
53. W. Melander and C. Horvath, Salt effects on hydrophobic interactions, *Arch. Biochem. Biophys.*, **183**, 200–215 (1977).
54. R. Srinivasan and E. Ruckenstein, Physical forces in hydrophobic interaction chromatography, *Sep. Purif. Meth.* **9**, 267–370 (1980).
55. C. J. Van Oss, D. R. Absolom, and A. W. Neumann, Mechanism of hydrophobic chromatography, *Sep. Sci. Tech.* **14**, 305–317 (1979).
56. G. K. Iwamoto, L. Winterton, R. A. Van Wagenen, J. D. Andrade, and D. Mosher, Fibronectin adsorption via TIRF, *J. Colloid Interface Sci.*, in press (1985).
57. M. J. E. Sternberg and J. M. Thornton, Prediction of protein structure from amino acid sequence, *Nature* **271**, 15–19 (1978).
58. M. J. E. Sternberg, F. E. Cohen, and W. R. Taylor, Prediction of the tertiary fold, *Biochem. Soc. Trans.* **10**, 299–301 (1982).
59. C. B. Anfinsen, Principles that govern the folding of protein chains, *Science* **181**, 223–230 (1973).
60. M. LaBrecque, Protein folding, *Mosaic* (Nat. Sci. Found. Publ.), 3–9 (May/June, 1983).
61. C. Tanford, Protein denaturation, *Adv. Protein Chem.* **23**, 121–277 (1968).
62. S. Lapanje, *Physiochemical Aspects of Protein Denaturation*, Wiley, New York (1978).
63. C. N. Pace, Protein denaturation, *CRC Crit. Rev. Biochem.* **3**, 1–43 (1975).
64. J. Chen and J. D. Andrade, Hemoglobin adsorption onto glass and polymer surfaces, *Biomaterials*, in press (1985).
65. F. R. N. Gurd and T. M. Rothgeb, Motions in proteins, *Adv. Protein Chem.* **33**, 73–158 (1979).
66. O. Jardetzky, NMR studies of macromolecular dynamics, *Acc. Chem. Res.* **14**, 291–298 (1981).
67. J. A. McCammon and M. Karplus, Simulation of protein dynamics, *Ann. Rev. Phys. Chem.* **31**, 29–45 (1980).
68. T. E. Creighton, Conformational flexibility in proteins, in: *Structural Aspects of Recognition* (M. Balaban, ed.), pp. 57–73, Balaban ISS, Publ., Philadelphia (1981).

69. R. Huber and W. S. Bennett, Jr., Functional significance of flexibility in proteins in: *Mobility and Recognition in Cell Biology*, pp. 21–48, W. deGryter Co., New York (1983).
70. R. J. Feldmann and D. H. Bing, *Teaching Aids for Macromolecular Structure*, Taylor Merchant Corp., New York, 1979.
71. R. J. Feldmann, Computing systems for molecular modeling, *Ann. Rev. Biophys. Bioeng.* **5**, 477–490 (1976).
72. R. J. Feldmann, D. H. Bing, B. C. Furie, and B. Furie, Interactive computer graphics, *Proc. Nat. Acad. Sci. U.S.A.* **75**, 5409–5412 (1978).
73. B. Furie, D. H. Bing, R. J. Feldmann, D. J. Robinson, J. P. Burnier, and B. C. Furie, Computer-generated models of blood coagulation factors, *J. Biol. Chem.* **257**, 3875–3882 (1982).
74. J. B. Tucker, Designing molecules by computer, *High Technology*, 52–60 (January 1984).
75. A. M. Lesk and K. D. Hardman, Computer-generated schematic diagrams of protein structures, *Science* **216**, 539–546 (1982).
76. R. Langridge, T. E. Ferrin, I. D. Kuntz, and M. L. Connoly, Real-time color graphics in studies of molecular interactions, *Science* **211**, 661–666 (1981).
77. M. L. Connolly, Solvent-accessible surfaces of proteins and nucleic acids, *Science* **221**, 709–713 (1983).
78. M. L. Connolly, Molecular Biology Dept., Scripps Clinic and Research Inst., La Jolla, California, 92037 and Quantum Chem. Program Exchange, Dept. of Chemistry, Indiana University, Bloomington, Illinois, 47405.
79. Protein Data Bank, Brookhaven National Laboratory.
80. F. M. Richards, Areas, volumes, packing, and protein structure, *Ann. Rev. Biophys. Bioeng.* **6**, 151–176 (1977).
81. J. T. Davies and E. K. Rideal, *Interfacial Phenomena*, Academic Press, New York (1963).
82. F. MacRitchie, Interface effects on chemical reaction rate, in: *Interfacial Synthesis, vol. I. Fundamentals* (F. Millich and C. E. Carraher, Jr., eds.), pp. 103–138, Dekker, New York (1982).
83. R. Gomer, Surface diffusion, *Sci. Am.*, 98–109 (Aug., 1982).
84. T. P. Burghardt and D. Axelrod, TIRF photobleaching recovery of albumin adsorption dynamics, *Biophys. J.* **33**, 455–468 (1981).
85. I. Michaeli, D. R. Absolom, and C. J. Van Oss, Diffusion of Adsorbed Protein within the Plane of Adsorption, *J. Colloid Interface Sci.* **77**, 586–587 (1980).
86. R. L. Daugherty and J. B. Franzini, *Fluid Mechanics*, McGraw-Hill, New York (1977).
87. J. Ruzicka and E. H. Nansen, *Flow Injection Analysis*, Wiley, New York (1981).
88. V. G. Levich, *Physico Chemical Hydrodynamics*, Prentice-Hall, Englewood Cliffs, New Jersey (1962).
89. R. W. Watkins and C. R. Robertson, TIRF for the examination of protein adsorption, *J. Biomed. Materials Res.* **11**, 915–938 (1977).
90. B. K. Lok, Y-L Cheng, and C. R. Robertson, TIRF, *J. Colloid Interface Sci.* **91**, 87–103 (1983).
91. R. L. Beissinger and E. F. Leonard, IgG sorption and desorption rates on quartz, *Am. Soc. Art. Int. Org. J.* **3**, 160–175 (1980).
92. R. W. Watkins, TIRF for the Study of Protein Adsorption, Ph.D. Dissertation, Stanford University (Nov., 1976).
93. L. Vroman, Protein/surface interaction, in: *Biocompatible Polymers* (M. Szycher, ed.), pp. 81–88, Technomic Publ., Lancaster, PA (1982).
94. A. L. Adams, G. C. Fischer, P. C. Munoz, and L. Vroman, Convex-lens-on-slide—human plasma and blood in narrow spaces, *J. Biomed. Materials Res.* **18**, 643–654 (1984).
95. J. Steinhardt and J. A. Reynolds, *Multiple Equilibria in Proteins*, Academic Press, New York (1970).

96. K. E. Van Holde, *Physical Biochemistry*, Prentice-Hall, Englewood Cliffs, New Jersey (1971).
97. A. G. Marshall, *Biophysical Chemistry*, Wiley, New York (1978).
98. I. M. Klotz, Protein interactions, in: *The Proteins*, Vol 1B (H. Neurath and K. Bailey, eds.), pp. 729-806, Academic Press, New York (1953).
99. G. Scatchard, J. S. Coleman, and A. L. Shen, Binding of small anions to serum albumin, *J. Am. Chem. Soc.* **79**, 12-15 (1957); also **72**, 535-540 (1950).
100. G. Weber, Energetics of ligand binding to proteins, *Adv. Protein Chem.* **29**, 1-83 (1975).
101. G. Weber, Bonding of small molecules to proteins, in: *Molecular Biophysics* (B. Pullman and M. Weissbluth, eds.), pp. 364-396, Academic Press, New York (1965).
102. D. E. Koshland, Jr., Molecular basis for enzyme regulation, in: *The Enzymes*, Vol. 1 (P. D. Boyer, ed.), 3rd edition, pp. 341-396, Academic Press, New York (1970).
103. A. Levitzki and D. E. Koshland, Jr., Role of negative cooperativity, *Curr. Topics Cell Regulation* **10**, 1-40 (1976).
104. G. Nemethy, Molecular interactions and allosteric effects, in: *Subunits in Biological Systems, Part C* (S. N. Timisheff and G. D. Fasman, eds.), pp. 1-90, Dekker, New York (1975).
105. N. Laiken and G. Nemethy, New model for binding of flexible ligands to proteins, *Biochem.* **10**, 2101-2106 (1971).
106. N. Laiken and G. Nemethy, Binding of flexible ligands to the surfaces of proteins, *J. Phys. Chem.* **74**, 4421-4441 (1970).
107. D. Poland, *Cooperative Equilibria in Physical Biochemistry*, Oxford University Press, Oxford (1978).
108. A. Levitski, *Quantitative Aspects of Allosteric Mechanisms*, Springer-Verlag, Berlin (1978).
109. V. L. Hess and A. Szabo, Ligand binding to macromolecules, *J. Chem. Educ.* **56**, 289-293 (1979).
110. K-J. Chang, S. Jacobs, and P. Cuatrecasas, Hormone-receptor interactions of high affinity, *Biochim. Biophys. Acta* **406**, 294-303 (1975).
111. A. W. Adamson, *Physical Chemistry of Surfaces*, 4th edition, Wiley, New York (1983).
112. J. J. Kipling, *Adsorption from Solutions of Nonelectrolytes*, Academic Press, New York (1965).
113. H. P. Jennissen, Evidence for negative cooperativity in the adsorption of phosphorylase b on hydrophobic agaroses, *Biochem.* **15**, 5683-5692 (1976).
114. H. P. Jennissen, Multivalent adsorption of proteins on hydrophobic agaroses, *Z. Physiol. Chem.* **357**, 1201-1203 (1976).
115. J. D. Andrade, Effects of plasma protein adsorption on conformation and activity, *Pure Appl. Chem.* in press (1984).
116. A. G. Walton and F. C. Maenpa, Applications of fluorescence spectroscopy to the study of proteins at interfaces, *J. Colloid Interface Sci.* **72**, 265-278 (1979).
117. C. R. McMillin and A. G. Walton, Circular dichroism technique for study of adsorbed protein structure, *J. Colloid Interface Sci.* **48**, 345-349 (1974).
118. M. E. Soderquist and A. G. Walton, Structural changes in proteins adsorbed on polymer surfaces, *J. Colloid Interface Sci.* **75**, 386-392 (1980).
119. B. M. C. Chan and J. L. Brash, Conformational change in fibrinogen desorbed from glass surfaces, *J. Colloid Interface Sci.* **84**, 263-265 (1981).
120. J. L. Brash and B. M. C. Chan, *Abstracts, 8th Ann. Meet. Biomaterials Soc.* (April, 1982), p. 85.
121. P. Van Dulm and W. Norde, Adsorption of human-plasma albumin on solid surfaces with special attention to the kinetic aspects, *J. Colloid Interface Sci.* **91**, 248-255 (1983).
122. D. Axelrod, T. P. Burghardt, and N. L. Thompson, Total internal reflection fluorescence, *Ann. Rev. Biophys. Bioeng.* **13**, 247-268 (1984).

123. D. Reinecke, R. A. Van Wagenen, J. D. Andrade, and L. M. Smith, Probing protein adsorption, *Biomaterials 84, Trans. Soc. Biomaterials* **7**, 276 (1984).
124. H. P. Jennissen, Multivalent interaction chromatography, *J. Chromatogr.* **159**, 71–83 (1978).
125. H. P. Jennissen, Hydrophobic agaroses, *Z. Physiol. Chem.* **357**, 1727–1733 (1976).
126. H. P. Jennissen and G. Botzet, Protein binding to 2-D hydrophobic binding sites, *Int. J. Biologic Macromolecules* **1**, 171–179 (1979).
127. A. G. Walton and B. Koltisko, Protein structure and the kinetics of interaction with surfaces, in: Reference (3), pp. 245-264.
128. L. Vroman and A. L. Adams, Multiple plasma/surface interactions, *Blood*, in press (1984).
129. B. M. C. Chan and J. L. Brash, Adsorption of fibrinogen on glass: Reversibility aspects, *J. Colloid Interface Sci.* **82**, 217–225 (1981).
130. E. Brynda, M. Houska, Z. Pokorna, N. A. Cepalova, Y. V. Moiseev, and J. Kalal, Irreversible adsorption of albumin onto polyethylene, *J. Bioeng.* **2**, 411–418 (1978).
131. L. Ter-Minassian-Saraga, Protein denaturation on adsorption and water activity at interfaces, *J. Colloid Interface Sci.* **80**, 393-401 (1981).
132. M. N. Kronick and W. A. Little, New immunoassay based on fluorescence excitation by internal-reflection spectroscopy, *J. Immunol. Meth.* **8**, 235–240 (1975).
133. J. D. Andrade and R. A. Van Wagenen, Process for conducting fluorescence immunoassays, U.S. Patent, 4,368,047 (11 Jan., 1983).
134. K. Newby, M. Reichert, R. A. Van Wagenen, and J. D. Andrade, Remote interfacial chemical sensor using a single multimode optical fiber, *Appl. Optics* **23**, 1812–1815 (1984).
135. E. H. Lee, R. E. Benner, J. B. Fenn, and R. K. Chang, Angular distribution of fluorescence from liquids by evanescient excitation, *Appl. Opt.* **18**, 862–868 (1979).
136. J. D. Andrade, Interfacial phenomena and biomaterials, *Med. Inst.* **7**, 110–120 (1973).
137. D. E. Gregonis, D. E. Buerger, R. A. Van Wagenen, S. K. Hunter, and J. D. Andrade, PEG surfaces to minimize protein adsorption, *Biomaterials 84, Trans. Soc. Biomaterials* **7**, 766 (1984).
138. E. W. Merrill and E. W. Salzman, PEO as a biomaterial, *Am. Soc. Art. Int. Org.* 60–64 (1983).
139. Y. Mori and S. Nagaoka, Polyethylene oxide chains, *Trans. Am. Soc. Art. Int. Org.* **28**, 459–463 (1982).
140. R. E. Crandall, J. Janatova, and J. D. Andrade, Effects of radioiodination and fluorescent labelling in albumin, *Prep. Biochem.* **11**, 111–138 (1981).
141. G. Mourier, L. Moroder, and A. Previero, Prevention of tryptophan oxidation during iodination of tyrosyl residues, *Z. Naturforsch.* **39b**, 101–104 (1984).
142. S. I. Taylor, Binding of hormones to receptors, *Biochem.* **14**, 2357–2361 (1975).
143. A. Van der Scheer, J. Feijen, J. K. Elhorst, P. G. K. Dagneaux, and C. A. Smolders, Radiolabelling for albumin adsorption studies, *J. Colloid Interface Sci.* **66**, 136–145 (1978).
144. C. E. Hayes and I. J. Goldstein, Radioiodination of sulfhydryl -sensitive proteins, *Anal. Biochem.* **67**, 580–584 (1975).
145. H. G. W. Lensen, D. Bargeman, P. Bergueld, C. A. Smolders, and J. Feijen, HPLC to measure competitive adsorption of proteins, *J. Colloid Interface Sci.* **99**, 108 (1984).
146. P. Suci, Variable Angle Total Internal Reflection Fluorescence, M. Sc. Thesis, University of Utah (July, 1984).
147. T. Okano, S. Nichiyama, I. Shinohara, T. Akaike, and Y. Sakurai, Interaction between plasma protein and micro-phase separated structure of copolymers, *Polymer J.* **10**, 223–228 (1978).
148. Y. Sakurai, T. Akaike, K. Kataoka, and T. Okano, Interfacial phenomena in biomaterials chemistry, in: *Biomedical Polymers* (E. P. Goldberg and A. Nakajima, eds.), pp. 335-379, Adacemic Press, New York (1980).

2

Total Internal Reflection Intrinsic Fluorescence (TIRIF) Spectroscopy Applied to Protein Adsorption

V. Hlady, R. A. Van Wagenen, and J. D. Andrade

1. INTRODUCTION

The ideal technique for studying protein adsorption at solid–liquid interfaces would: (1) quantitate the amount of protein adsorbed, (2) provide essentially real-time information concerning macromolecular adsorption–desorption kinetics, (3) be capable of *in situ* sensing, (4) provide information concerning protein conformational changes occurring during adsorption, (5) be applicable to flat, low-surface area polymer, metal, and ceramic surfaces, and (6) be simultaneously applicable to any proteins competitively adsorbing from complex solutions. All popular techniques used for study of protein adsorption (solution depletion, radiolabeling, ellipsometry, ORD, FTIR) are only approximations to this ideal.

Interfacial fluorescence generated by a total internal reflection (TIR) evanescent wave has been used to study proteins at interfaces for twenty years;[1,2] however, the TIR fluorescence technique (TIRF) did not receive significant application until recently. TIRF has been utilized to study the adsorption of extrinsic fluor-labeled plasma proteins on quartz[3,4] and poly(dimethylsiloxane) films.[5,6] Intrinsic fluorescence TIRF has been applied to the study of insulin adsorption via tyrosine fluorescence.[7] Extrinsic TIRF has also been applied to the study of energy transfer in adsorbed albumin[8] and in the observation of interfacial polymer depletion layers.[9]

V. Hlady, R. A. Van Wagenen, and J. D. Andrade • Department of Bioengineering, College of Engineering, University of Utah, Salt Lake City, Utah 84112. This work was carried out while V. Hlady was on leave of absence from Insitute "Ruder Bošković," Zagreb, Yugoslavia. Correspondence should be addressed to J. D. Andrade.

The TIRF approach to protein adsorption offers a number of advantages relative to conventional techniques. For instance, TIRF is well suited to *in situ* sensing. Protein adsorption and desorption may be followed with excellent time resolution as they occur, either on amorphous silica or on thin, transparent polymer films supported by silica. Furthermore, the technique is applicable to flat, low-surface area samples which are optically uniform and transparent. This does, of course, limit its applicability to many metallic and ceramic materials. If the protein of interest is labeled with an extrinsic fluor, its adsorption may be followed even in the presence of other proteins which have been shown not to fluoresce under the appropriate excitation conditions. The greatest weakness of TIRF is the necessity to quantitate the amount of protein bound by an independent technique—generally using radioisotope-labeled proteins.

Extrinsic TIRF utilizes an extrinsic fluor bound to the protein of interest. This allows visible light excitation and fluorescence emission. However, it is necessary to prove that the presence of the fluor does not alter the physicochemical properties of the protein which in turn influence adsorption. Most proteins contain intrinsic structural amino acids which are inherently fluorescent if excited in the ultraviolet (UV) region of the spectrum (260–290 nm). Intrinsic protein fluorescence is affected by the luminescent properties of these amino acids which are in turn affected by their local microenvironment. In principle, the local microenvironment may be altered by changes in protein conformation during adsorption and denaturation. These changes should be manifest by changes in the emission spectrum of the adsorbed protein. Consequently, total internal reflection intrinsic fluorescence (TIRIF) has the potential of discriminating protein conformational alterations during adsorption; however, because many proteins in a mixture such as serum contain fluorescent amino acids, competitive adsorption of specific proteins cannot be resolved by TIRIF.

This paper discusses and illustrates advantages and disadvantages of TIRIF. In order to understand the basics of TIRIF spectroscopy applied to the study of protein adsorption, a brief outline of intrinsic protein fluorescence and the theory of an evanescently decaying surface wave are given before the results of the TIRIF studies. Another chapter of this book deals with extrinsic TIRF as well as other methods of studying protein adsorption.[10]

2. INTRINSIC PROTEIN FLUORESCENCE

Most proteins fluoresce naturally due to the intrinsic fluorescence of one or more of the aromatic amino acids comprising them: tyrosine (Tyr),

TABLE 1
Luminescence Properties of the Fluorescent Aromatic Amino Acids[a]

Luminescence property	Phenylalanine	Tyrosine	Tryptophan
Molar extinction coefficient, ε ($m^2\ mol^{-1}$) at long wavelength UV absorption maxima (nm)	20 at 258 nm, shoulder at 260 nm	145 at 275 nm, shoulder at 280 nm	560 at 280 nm, shoulder at 265 nm, and second maximum at 288 nm
Fluorescence quantum yield, ϕ_F	0.02–0.04	0.14–0.21	0.15–0.20
Fluorescence lifetime (ns)	6.4–6.8	2.6–3.6	2.5–3.3
Fluorescence peak maximum (nm)	282	304 ± 1	353 ± 2

[a] In aqueous solution at neutral pH and room temperature.

phenylalanine (Phe), or tryptophan (Trp). An understanding of the luminescence properties of these amino acids is essential for a correct interpretation of protein fluorescence. A summary of the relevant parameters is listed in Table 1. Teale and Weber[(11)] were the first to fully characterize the fluorescence properties of these moieties at neutral pH in aqueous media. They demonstrated the equivalency of the absorption spectra and the fluorescence excitation spectra, thus proving that fluorescence emission was the direct result of absorption. They also determined that quantum yield and fluorescence emission were independent of excitation wavelength, which in turn suggested that emission occurs from the same lowest singlet excited energy level regardless of the extent to which higher states are populated by absorption.

Protein fluorescence emission spectra are not simple summations of the emission from constituent amino acids. Proteins are grouped into two fluorometric classes: A and B.[(12)] Class A proteins possess tyrosine but lack tryptophan. Their fluorescence spectra are characteristic of zwitterionic tyrosine—the absorption maximum is 276 nm and the emission maximum is 304 nm. The Stokes shift for Class A proteins is invariant, and the overall fluorescence quantum yield is somewhat affected by microenvironment and typically ranges from 0.01–0.05.

There are several quenching mechanisms[13] which act on tyrosyl protein residues and would explain the generally low quantum yields of Class A proteins:

(1) Tyr localized at the protein surface may suffer collisional quenching with the carboxylate side chains of Asp and Glu residues,
(2) Tyr-carboxylate complexes in the hydrophobic protein interior quench excited states, where carboxylate anions again arise from the side chains of Glu and Asp residues,
(3) hydrophilic interior amide moieties form nonfluorescent complexes with Tyr,
(4) Tyr residues in vicinal positions relative to disulfide bonds can be quenched,
(5) polypeptide bonds markedly quench Tyr fluorescence.

Consequently, the majority of Class A protein fluorescence probably arises from surface-situated Tyr residues since Tyr localized in hydrophobic environments is quenched as a result of carboxylate–phenoxyl complexes or amide carbonyl hydrogen bonds to the phenoxyl of Tyr.

The Class B proteins contain Trp and may also contain Phe and Tyr. Their excitation spectra are characteristic of tryptophan with fluorescence quantum yields about half that of free tryptophan and a variable Stokes shift strongly dependent upon the local microenvironment surrounding the tryptophan. Overall protein Trp quantum yields vary widely and range from 0.04 to 0.50.

The variability in Stokes shift and fluorescence quantum yield exhibited by Trp-containing globular proteins is probably due to local microenvironmental differences around these residues which is a reflection of the local microenvironment of the protein. It should be stressed that the fluorescence emission spectrum reflects the emission properties of all the fluorescent amino acids and the composite of the local microenvironments surrounding them.

Protein tryptophan fluorescence can be quenched in a number of ways, i.e., by adjacent sulfhydryl groups, detergents, organic solvents, and pH variations. Alkaline pH invariably enhances quenching by optimizing energy transfer from Trp to nearby weakly fluorescent tyrosinate groups. Acid titration of albumin leads to a pH 3.5 conformational transition which exposes several tyrosyl groups, resulting in an elevated Tyr fluorescence and a reduced Trp fluorescence.

The variation in Stokes shift encountered in protein tryptophan fluorescence emission depends principally upon the dielectric constant of the environment surrounding each fluor. The absorbance of such fluors is at least a factor of ten less dependent. The wide variation in Stokes shifts is a reflection of at least four different fluor microenvironments in the native tertiary protein structure[13]:

(1) Trp buried in a completely apolar, hydrophobic protein interior,
(2) Trp buried in the hydrophobic interior but complexed with either an amide, carbonyl, or polar side chain,
(3) Trp partially exposed on the molecular surface where the majority of the environment is still hydrophobic with only slight accessibility to solvent,
(4) Trp residues on the protein surface fully exposed to polar water molecules.

However, there are often no obvious indications of fluorescence spectral heterogeneity in various proteins when one considers that a protein having a single tryptophan does not differ appreciably in its emission spectrum from one having many tryptophans. There are two possible mechanisms which may account for variable Stokes shifts. In a photophysical sense the variability in the extent of solvation due to local Trp microenvironment could cause variations in local dielectric constant. Alternatively, a photochemical effect resulting from microenvironmentally discrete dipole moments could generate lower energy centers termed exciplexes which serve as the principal sources of fluorescence emission.

Even in Class B proteins containing tyrosine, the tryptophan fluorescence will predominate. For instance, most of the fluorescence of human serum albumin arises from its single Trp residue even though there are 17 tyrosines in the protein. In many globular proteins a tyrosyl contribution to the emission spectra can be determined, but only if there is good signal/noise ratio, good spectral resolution, and the emission spectra are corrected.

Urea denaturation of Class A proteins does not affect fluorescence emission while comparable denaturation in Class B proteins drives the Stokes shift and fluorescence quantum yield toward zwitterionic Trp values. This probably occurs because the potential for electronic excited state energy transfer is greatly reduced during denaturation as the chains are farther apart in the random coil configuration, unless Trp and Tyr are vicinal in the amino acid sequence.

The actual protein fluorescence depends upon the excitation wavelength, the Trp:Tyr ratio, the potential for excited state energy transfer, and quenching. For instance, at 278 nm the fractional absorbance ratio (Trp:Tyr) is about 10:1, the molar ratio (Trp:Tyr) varies from 0.1 to 1.0, and the overall protein quantum yield ratio (Trp:Tyr) is typically 10:1. The result is that the fractional Tyr fluorescence contribution at 304 nm is usually less than 10% of the Trp fluorescence at 325 nm. Generally the tyrosine component to the emission spectrum can be greatly minimized by exciting at 295 nm or above. The tyrosine component can, however, be attined via difference spectra taken at 278 nm and 295 nm. The Tyr emission spectra are not sensitive to protein conformation while the quantum yield is generally reduced by a factor of 5–10 when the amino acid is incorporated into a polypeptide structure. Tyr quantum yields of 0.02 to 0.05 are common in proteins. This great attenuation can be ascribed to enhanced internal conversion although perhaps 30% results from excited state energy transfer to Trp. The three major energy transfer routes are: Tyr → Tyr, Tyr → Trp, and Trp → Trp. Two additional routes (Phe → Tyr and Phe → Trp) may also exist but the small fractional absorbance of Phe makes study very difficult.

Because phenylalanine has such a low molar extinction coefficient and fluorescent quantum yield, it can generally be ignored as a detectable contribution to the fluorescence of any protein. Also, it is likely that dipolar resonance electronic energy transfer from Phe to Tyr and on to Trp occurs in proteins where inter-residue distances are comparable to known values for dipolar interactions.

TIRIF spectroscopy employs the aforementioned properties of protein fluorescence in the study of protein adsorption. For instance: the proportionality between the fluorescence intensity and protein concentration at the interface can be used to determine the adsorbed amount in protein adsorption/desorption kinetic studies as well as in finding equilibrium protein distribution between the interface and solution. Differences in Stokes shifts in fluorescence emission spectra between adsorbed and nonadsorbed protein molecules can be used in qualitative descriptions of protein conformational changes in the adsorbed state. Similarly, the accessibility of intrinsic fluors in adsorbed protein molecules can be probed with fluorescence quenchers of various sizes and natures giving some information about protein conformation at interfaces. The results of TIRIF protein adsorption studies are given in Section 5. in more detail.

3. EVANESCENT WAVE THEORY

Total internal reflection intrinsic fluorescence uses an evanescently decaying surface wave of ultraviolet wavelength to excite fluorescence of adsorbed proteins. In order to understand the basic characteristics of TIRIF spectroscopy some of the evanescent wave theory will be presented here. A more elaborate approach to evanescent wave spectroscopy is also available.[14–16] In the case of electromagnetic wave reflection from an interface it is known that a standing wave is established due to superposition of incident and reflected waves.[17] The phenomenon also occurs in total internal reflection where the standing wave forms in the optically more dense medium 1 and the nature of the standing wave is a function of both

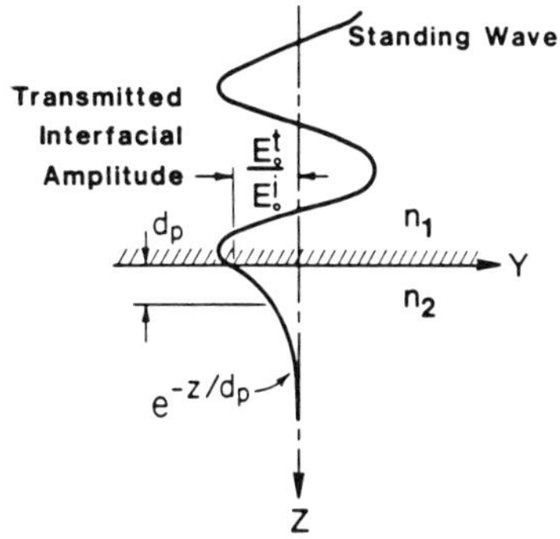

FIGURE 1. Schematics of the surface wave in the y-z plane showing the standing wave pattern and exponential decay of the transmitted electric field amplitude into the rarer medium 2.

the dense medium 1 and the less optically dense medium 2. In the case of metallic reflection, the standing wave is aligned with its node at the reflecting interface, but the electric field amplitude of the standing wave formed at the interface between two dielectric media has a nonzero value. Since the boundary condition of the electromagnetic wave reflection does not allow any sharp discontinuity of tangential field across the interface, the electric field amplitude at the interface but in medium 2 will be equal to the nonzero electric field amplitude of the standing wave at the interface in medium 1. The solution of the Maxwell wave equation for total internal reflection at the interface between two dielectric media shows the existence of a surface wave that propagates along the interface. This surface wave is described by the following equation:

$$\bar{E}^t_\perp = \hat{y} t_\perp E^i_{0\perp} \exp\left\{-j\frac{2\pi x}{\lambda_2}\left(\frac{n_1}{n_2}\right)\sin\theta_1 - \frac{2\pi z}{\lambda_2}\left[\left(\frac{n_1}{n_2}\right)^2 \sin^2\theta_1 - 1\right]^{1/2}\right\} \tag{1}$$

where $\bar{E}^t_\perp$ is the perpendicularly polarized ($\perp$) component of transmitted field intensity (given here as a vector), $\hat{y}$ is a Cartesian unit vector, $t_\perp$ is the Fresnel transmission coefficient, $E^i_{0\perp}$ is the electric field amplitude at the interface (i) in medium 2, λ_2 is the wavelength in medium 2, θ_1 is the incident light angle from the interfacial normal, and n_2 and n_1 are refractive indices of media 1 and 2, respectively.* A schematic portrayal of an evanescently decaying surface wave is presented in Figure 1. The first exponential term in Eq. (1) accounts for part of the displacement between the incident

* Note that Eq. (1) uses real parts of refractive indices only. This implies negligible absorption of the electromagnetic wave in both media. This assumption is irrelevant for medium 1 since as a rule only relative values of electric field amplitudes at the interface are considered. However, in the case of medium 2, the assumption of negligible absorption is justified only if the absorption index k_2, defined as $k_2 = \varepsilon c\lambda/4\pi n_2$, <0.001 for $\lambda_0 = 285$ nm, $\theta_1 = 70°$, $n_1 = 1.5$, and $n_2 = 1.33$. For any larger values of k_2 the complex index of refraction $\tilde{n}_2 = n_2 - jk_2$ must be used.[14,15]

and reflected beams in the x-direction. The second exponential term accounts for evanescent decay of the surface wave into medium 2 (z-direction). By setting

$$\frac{1}{d_p} = \frac{2\pi}{\lambda_2}\left[\left(\frac{n_1}{n_2}\right)^2 \sin^2\theta_1 - 1\right]^{1/2}$$

and

$$\bar{E}_{\perp}^{t'} = \hat{y}t_{\perp}E_{0\perp}^{i} \exp\left[-j\frac{2\pi x}{\lambda_2}\left(\frac{n_1}{n_2}\right)\sin\theta_1\right]$$

Eq. (1) can be simplified to:

$$\bar{E}_{\perp}^{t} = \bar{E}_{\perp}^{t'}\,\mathrm{e}^{-z/d_p} \tag{2}$$

If one arbitrarily defines d_p as the depth of penetration where the electric field amplitude decreases to e^{-1} of its original value, i.e., $\mathrm{e}^{-z/d_p} = \mathrm{e}^{-1}$, then:

$$d_p = \frac{\lambda_2}{2\pi[(n_1/n_2)^2 \sin^2\theta_1 - 1]^{1/2}} \tag{3}$$

or using the relation $\lambda_2 = (n_1/n_2)\lambda_1 = \lambda_0/n_2$:

$$d_p = \frac{\lambda_1}{2\pi[\sin^2\theta_1 - (n_2/n_1)^2]^{1/2}} = \frac{\lambda_0}{2\pi(n_1^2\sin^2\theta_1 - n_2^2)^{1/2}} \tag{3a}$$

For parallel polarized electromagnetic waves ($\|$) an equation similar in form to Eq. (2) can be derived.[14]

In order to correlate the spectroscopic phenomena with evanescent wave intensity, it is necessary to know how the electric field amplitude at the interface decays with distance z into medium 2. The exact solution for interfacial amplitudes can be derived using the Fresnel coefficient $t_{\perp}$ (or $t_{\|}$) and its complex conjugate $t_{\perp}^{*}$ (or $t_{\|}^{*}$).[17,18] Keeping x constant (by choosing any slice of surface wave parallel to the y–z plane) $\bar{E}_{\perp}^{t'}$ becomes a complex constant with magnitude equal to the electric field amplitude at the interface but in medium 2. Thus, the evanescently decaying electric field amplitude $E_{\perp}^{e}$ is

$$E_{\perp}^{e} = |\bar{E}_{\perp}^{t'}|\,\mathrm{e}^{-z/d_p} \tag{4}$$

and the magnitude of $|\bar{E}_{\perp}^{t'}|$ is given by

$$\bar{E}_{\perp}^{t'} = [t_{\perp}t_{\perp}^{*}]^{1/2}E_{0\perp}^{i} \tag{5}$$

Since

$$t_\perp = \frac{2n_1 \cos \theta_1}{n_1 \cos \theta_1 + n_2 \cos \theta_2} \tag{6}$$

which under the conditions of TIR becomes[14]:

$$t_\perp = \frac{2n_1 \cos \theta_1}{n_1 \cos \theta_1 - j[n_1^2 \sin^2 \theta_1 - n_2^2]^{1/2}} \tag{7}$$

Eq. (4) becomes

$$E_\perp^e|_{z=0} = E_{0\perp}^e = \frac{2 \cos \theta_1}{[1 - (n_2/n_1)^2]^{1/2}} E_{0\perp}^i \tag{8}$$

The corresponding equation for parallel polarized electromagnetic waves can be derived:

$$E_\parallel^e|_{z=0} = E_{0\parallel}^e = \frac{2(n_2/n_1) \cos \theta_1}{[1 - (n_2/n_1)^2]^{1/2}[(1 + (n_2/n_1)^2) \sin^2 \theta_1 - (n_2/n_1)^2]^{1/2}} E_{0\parallel}^i \tag{8a}$$

More often, the electric field amplitude as a function of distance z is expressed per unit of incident electric field amplitude at the interface, so that

$$E_\perp^e = \frac{E_{0\perp}^e}{E_{0\perp}^i} e^{-z/d_p} = \frac{2 \cos \theta_2}{[1 - (n_2/n_1)^2]^{1/2}} e^{-z/d_p} \tag{9}$$

where the ratio $E_{0\perp}^e / E_{0\perp}^i$ is often denoted as E_{y0}^t, the transmitted interfacial electric field amplitude per unit incoming amplitude (subscript y denotes the direction of the electric vector of the perpendicularly polarized electromagnetic wave). From the experimental point of view, it is important to recognize that the profile of $(E_\perp^e)^2$ vs. z, and the resolution of adsorbed layer signal from the signal of nonadsorbed protein molecules, is dependent on the choice of λ_0, θ_1, n_1, and n_2, only. Figure 2 compares $(E_\perp^e)^2$ decay as a function of λ_0; case (1) corresponds to $\lambda_0 = 488$ nm, the excitation wavelength of extrinsic fluors like fluorescein isothiocyanate (FITC); case (2) is for $\lambda_0 = 284$ nm, the wavelength often used in excitation of intrinsic protein fluorescence. It can be seen that for intrinsic fluorescence (2), the evanescent intensity decays more rapidly, making TIRIF relatively more sensitive to adsorbed proteins rather than bulk protein. Hence, the fluorescence signal coming from adsorbed proteins will dominate the overall TIRIF fluorescent signal intensity.

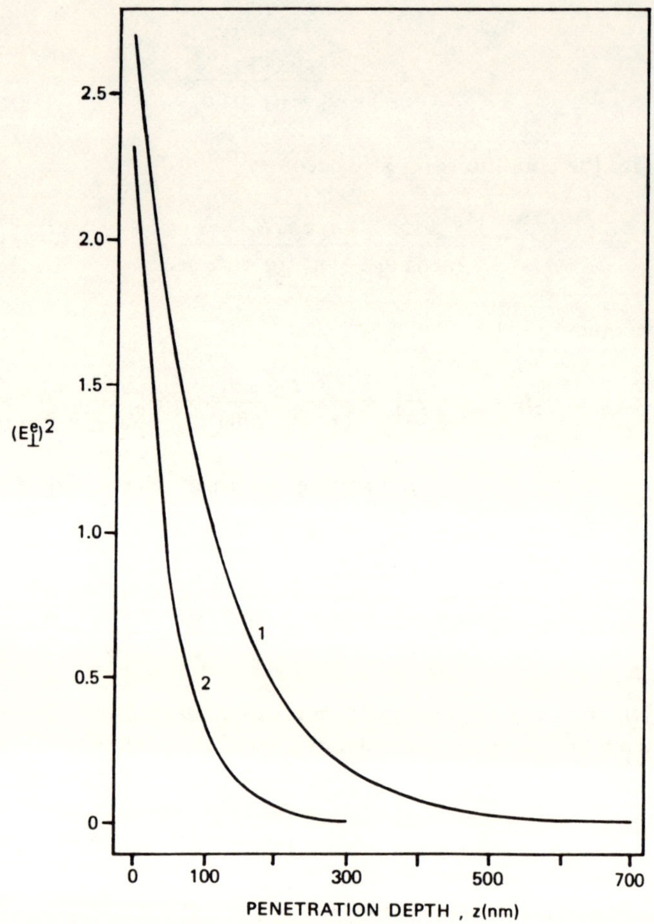

FIGURE 2. Exponential decay in $(E_{\perp}^{e})^2$ with distance z from the reflecting interface; case (1) extrinsic fluor (e.g., fluorescein isothiocyanate), $\lambda_{ex} = 488$ nm, $n_1 = 1.467$, $n_2 = 1.333$, $\theta_1 = 70°$; case (2) intrinsic protein fluor, $\lambda_{ex} = 284$ nm, $n_1 = 1.496$, $n_2 = 1.353$, $\theta_1 = 70°$.

4. METHODOLOGY

4.1. EXPERIMENTAL EQUIPMENT

The TIRIF approach to protein adsorption is schematically illustrated in Figure 3. The essential feature is an amorphous silica internal reflection element (IRE). The IRE prism (P) is optically coupled to a sample plate (Q) of the same material via glycerol. An elliptically shaped poly(dimethylsiloxane) (PDMS) gasket (M) forms a flow field (N) between the IRE

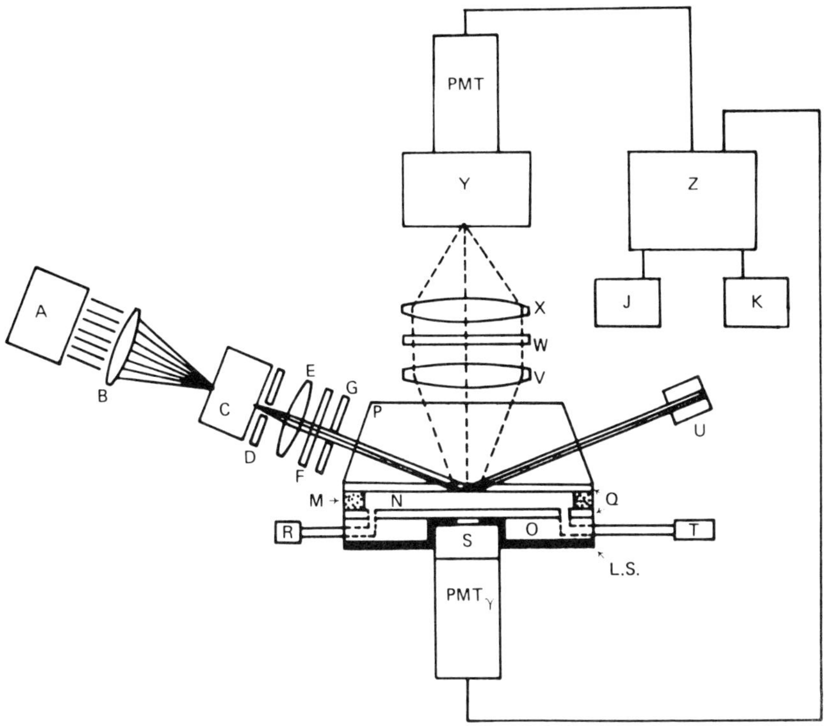

FIGURE 3. TIRIF system schematic: light source (*A*), focusing lens (*B*), excitation and emission monochromators (*C* and *Y*, respectively), electronic shutter (*D*), collimating lens (*E*), UV polarizer (*F*), iris (*G*), amorphous silica IRE (*P*), beam stop (*U*), amorphous silica sample plate (*Q*), Silastic gasket (*M*), flow channel (*N*), flow cell base (*O*), lead shield (*LS*), syringe pump (*T*), waste reservoirs (*R*), scintillator (*S*), photomultiplier tubes for fluorescence and γ-counting (*PMT* and γ-*PMT*), collection lens (*V*), focusing lens (*X*), emission polarizer (*W*), photon counter (*Z*), analog recorder (*J*), and microcomputer (*K*).

coupled sample plate and a second identical sample plate separated from the flow cell base (*O*) by a solid PDMS sheet. This plate has two holes drilled through it to allow flow from the syringe pump (*T*) to flow into the base, through the flow field and back out to a waste reservoir (*R*).

Some portion of the evanescent wave-generated interfacial protein fluorescence traverses the IRE, is collected by lens (*V*), and is focused upon the entrance slit of an emission monochromator (*Y*). Fluorescence is amplified by a photomultiplier tube (*PMT*), quantitated by photon counting (*Z*), recorded in analog form on a strip chart recorder (*J*), and recorded digitally on an Apple II microcomputer (*K*). When radioisotope-labeled proteins are used for calibrating, a solid NaI–Tl-doped scintillator crystal (*S*) detects γ-rays from labeled proteins in the flow cell. The signal is amplified by a second PMT which is also accessed by the photon counter on a time-shared basis.

The light source (*A*) is a xenon lamp (ILC Technology) or a nitrogen pumped dye laser (Molectron). An excitation monochromator (*C*) selects an appropriate UV wavelength from the xenon lamp to excite UV fluorescence. The collimating lens directs the beam into the prism through a suitable UV polarizer (*F*) and iris diaphragm (*G*). An electronic shutter (*D*) limits UV exposure to adsorbed protein. Additional aspects of the TIRIF technique have been previously described.[19]

An independent approach to TIRIF quantitation which utilizes interfacial radioisotope counting as a calibration technique is being developed. A NaI(T1) scintillator crystal, optically coupled to a PMT, interfaces to one side of the TIRIF flow. One plate is optically coupled to the TIRIF prism and the other plate having 1-mm diameter holes at the foci of the ellipse is coupled to the scintillator–PMT interface plate. Fluid enters and exits the flow cell via these holes.

The 30-keV X-ray photon from ^{125}I is currently being used. The protein is labeled with ^{125}I using chloramine-T[20] or lactoperoxidase[21] labeling procedures. Cold labeling with iodine is carried out in parallel. Bulk solution spectrofluorometry is used to check for iodine-induced changes in protein

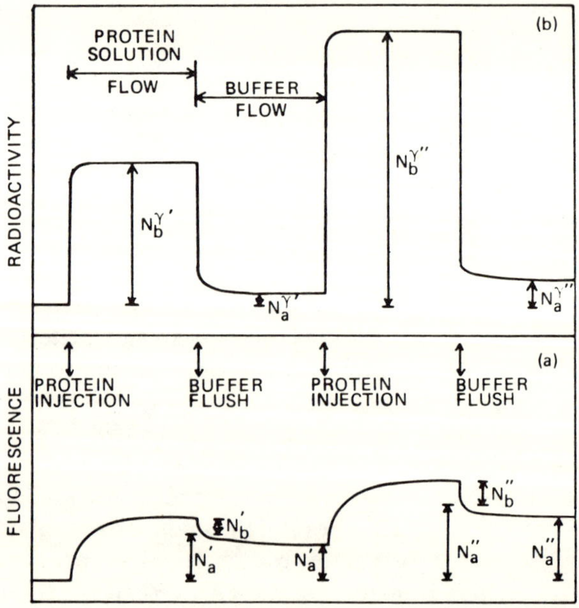

FIGURE 4. An illustration of typical TIRIF data output (a) combined with the output of the parallel radiolabeled protein adsorption experiment (b). See text for explanation of symbols used.

fluorescence. When the ^{125}I-labeled protein solution is introduced into the TIRIF flow cell, the entire depth of the cell is sensed by the scintillator PMT because of high γ-ray penetration in water. The quartz sample plate causes about 20% loss of γ-rays. Depending on the thickness of the TIRIF flow cell, the γ-ray signal from the adsorbed protein layer usually comprises about 1/100–1/30 of the total signal. Therefore, it is possible to distinguish between the bulk and surface signals only after the nonadsorbed protein has been flushed out of the flow cell by appropriate buffer or with unlabeled protein of the same concentration. In the case of low protein adsorption, either a higher specific activity or longer γ-ray signal integration time has to be used.

A typical protein adsorption experiment with the fluorescence signal and γ-ray signal both measured as a function of adsorption and desorption time is shown in Figure 4. Introduction of protein solution into the TIRIF flow cell causes an immediate rise of both signals. Due to its relative sensitivity to the evanescent volume adjacent to the interface the fluorescence intensity continues to increase with time indicating that protein adsorption is taking place. At the same time the γ-ray signal intensity is rather constant since the entire TIRIF flow cell depth is monitored. After a given time period of protein interfacial residence time, a buffer or electrolyte flush is used in order to remove the nonadsorbed protein from the cell. As the result of the flush, the γ-ray signal drops significantly while the fluorescence signal decrease is usually only a small portion of the total unless there is a relatively high proportion of scattered excitation light propagating through the flow cell. This is likely to happen due to insufficiently clean optical components. In order to obtain the fluorescence intensity proportional to the amount of protein on the surface, the bulk signal established by the buffer flush has to be subtracted from the total fluorescence signal. Desorption of protein is then followed as a function of time and shear rate.

The fluorescence emission spectra can be taken at any time during the protein adsorption experiment. In order to shorten time-consuming fluorescence emission spectra recordings and long exposure times of the interfacial protein layer to the ultraviolet light an alternative experimental TIRIF setup has been developed. Instead of a fluorescence PMT (*Y*, Figure 3) and photon-counting equipment (*Z*, Figure 3) an optical multichannel analyzer (OMA) is used. The OMA consists of an intensified photodiode array which is periodically analyzed or read by the computer. The appropriate OMA software enables virtually instant recording and easy manipulation (subtraction, adding, storing, smoothing) of fluorescence spectra.

4.2. METHODS OF TIRIF QUANTITATION

The actual observed TIRIF intensity is proportional to the product of the absorption and emission probabilities. Since absorption of the exciting evanescent light is proportional to the interfacial concentration, an opportunity exists to quantitate the TIRIF signal coming from the adsorbed protein at a given interface. Generally, the fluorescence emission intensity of wavelength λ from one molecule is governed by the product of three factors: $\phi f(\lambda)a$ where ϕ is the probability of emission (quantum yield or overall quantum yield in the case of multi-fluor molecules), $f(\lambda)$ is the fraction of the total radiation emitted at wavelength λ, and a is the fraction of the total radiation at λ that is collected by the detector. In the case of a homogeneously distributed fluorescent substance inside the TIRIF cell the fluorescence intensity N is given by:

$$N = 2.303 \cdot \phi \cdot f(\lambda) \cdot a \cdot \varepsilon \cdot c \cdot b \cdot \int_0^\infty (E^e)^2 \, dz \tag{10}$$

where c is the concentration, ε is the extinction coefficient at the exciting wavelength, and a and b are constants. The term $b \int_0^\infty (E^e)^2 \, dz$ is equivalent to the product of I_0 (incident light intensity) times cell thickness d in conventional spectrofluorometry. In the case of perpendicularly polarized incident light

$$\int_0^\infty (E_\perp^e)^2 \, dz = \int_0^\infty (E_{y0}^t \, e^{-z/d_p})^2 \, dz \tag{11}$$

For the case of a substance which adsorbs at the interface the general equation for TIRIF intensity changes to:

$$N = 2.303 a \cdot b \int_0^\infty [f(\lambda)](z)\phi(z)\varepsilon(z)c(z) \cdot (E_{y0}^t \, e^{-z/d_p})^2 \, dz \tag{12}$$

where ϕ, ε, c, and $f(\lambda)$ vary with distance z from the interface. Since some of these parameters change [ϕ, ε, $f(\lambda)$] with distance z (i.e., with the adsorption) and may not be accessible to direct experimental verification, the approach to the quantitation of the TIRIF signal intensity has followed three different routes. The first approach is graphical and relies on both conventional and TIR fluorescence of nonadsorbing macromolecules as an external standard along with the adsorbing protein TIRIF signal. The second approach utilizes the TIRIF signal emanating from the nonadsorbed bulk protein in the TIRIF cell as an internal standard together with a somewhat modified form of Eq. (12). The third approach, while still in the development phase, uses the second approach together with simultaneous detection of the γ-radiation coming from ^{125}I-labeled protein. All three approaches to TIRIF quantitation will be explained in detail.

4.2.1. Graphical Quantitation

The graphical quantitation method uses both a calibration molecule which does not adsorb on the surface but models the properties of the protein in the absence of macromolecular adsorption and extrapolation of linear quantitation plots of fluorescence versus bulk concentration into the higher concentration domain for the purpose of graphical modeling.[22] The extrapolation assumption allows graphical modeling of the fluorescent properties of the calibration solution and protein solution as though nonlinear effects do not occur, which is reasonable in light of the extremely short "path length" of the TIRIF experiment.

The graphical quantitation procedure is illustrated in Figure 5. The bulk protein concentrations which are used in the TIRIF adsorption experiment are found to produce a bulk fluorescence intensity (Figure 5a). Next, the bulk fluorescence of calibrating molecules (such as FITC-labeled dextran or *N*-acetyltryptophanyldextrans) which model bulk protein fluorescence in the TIRIF cell is measured (Figure 5b). A TIRIF calibration plot is generated next by plotting the TIRIF signal against the concentration of calibrating labeled dextran solutions. By using protein bulk fluorescence and labeled dextran bulk fluorescence plots, a concentration of calibrating labeled dextran solution is found and its appropriate fluorescence intensity is read from the TIRIF calibration plot (Figure 5c). This fluorescence intensity value represents the fluorescence which would be produced in the TIRIF cell by nonadsorbed protein since labeled dextran used as a calibration molecule was shown not to adsorb at the solution concentrations used. The protein adsorption experiment is then conducted and the previously determined nonadsorbed component is subtracted from the TIRIF output signal. Any fluorescence intensity increase above the background level minus the nonadsorbed protein component should represent the fluorescence intensity due to adsorbed protein only.

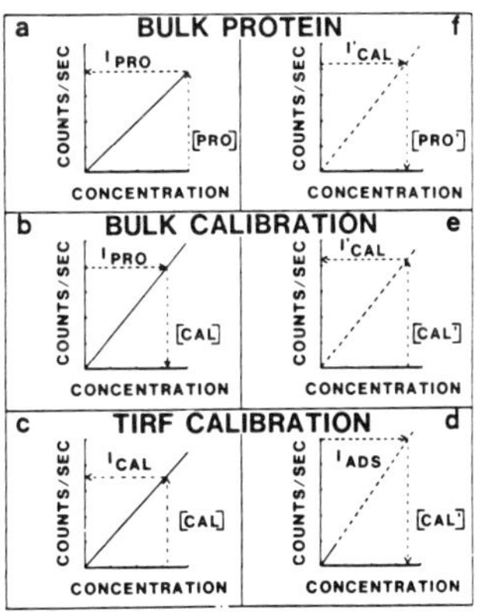

FIGURE 5. Six quantitation plots illustrating the graphical adsorption quantitation procedure; (a) bulk solution protein fluorescence, (b) bulk fluorescence of the calibrating solution, (c) TIRIF fluorescence of the calibrating nonadsorbing solution, (d)–(f) extrapolated versions of (c), (b) and (a), respectively. [Reprinted from Reference (22) by permission. Copyright by Elsevier Sequoia S.A.]

The plots are then used in reverse order to generate a surface volume concentration from the adsorption component of the TIRIF output signal. The TIRIF fluorescence intensity signal associated with adsorbed protein must now be modeled by a calibration solution concentration of equal fluorescence intensity in the TIRIF instrument. From the extended TIRIF calibration plot this calibration solution model concentration can be found (Figure 5d). The resulting calibration solution model concentration is then utilized to obtain a bulk fluorescence intensity from the extended bulk calibration plot (Figure 5e). Finally, from the extended protein bulk fluorescence plot, the protein surface volume concentration can be found from the bulk fluorescence model value (Figure 5f). Thus, a series of graphical modeling steps utilizing three calibration plots makes it possible to derive the surface volume concentration of an adsorbed protein.

The graphical quantitation scheme is primarily based upon the assumption that the calibration molecule does not adsorb on the surface. When *N*-acetyltryptophanyldextran or FITC–dextran with diffusion coefficients matching that of the protein are used in the TIRIF system, the fluorescence intensity rapidly returns to the background level after washing out the dextran solution, demonstrating the lack of irreversible adsorption of labeled dextrans. However, it is possible that a rapidly reversible adsorption process could be present. If such a mechanism were operating, it would be expected that the kinetics of the buffer wash-out phase of the TIRIF signal would change with dextran concentration due to the difference between diffusion and desorption kinetics. However, TIRIF calibration data using labeled dextran show no concentration-dependent kinetic variations; therefore, either no reversible adsorption mechanism exists or, if it does exist, the adsorption kinetics are similar to or faster than the diffusional kinetics and have no adverse effect upon the assumption made here.

4.2.2. Numerical Quantitation

The numerical approach to TIRIF quantitation uses the TIRIF signal N_b coming from excited, nonadsorbed protein molecules within the evanescent volume together with the TIRIF signal N_a emanating from the adsorbed protein layers after the nonadsorbed protein has been flushed out and replaced by buffer (Figure 4a). In the case of the adsorbed protein layer of thickness $\bar{\Delta}$, the ratio N_a/N_b is given by:

$$\frac{N_a}{N_b} = \frac{2.303ab \int_0^{\bar{\Delta}} [f(\lambda)](z)\phi(z)\varepsilon(z)c(z)(E^t_{y0}\,\mathrm{e}^{-z/d_p})^2\,\mathrm{d}z}{2.303ab \int_{\bar{\Delta}}^{\infty} [f(\lambda)](z)\phi(z)\varepsilon(z)c(z)(E^t_{y0}\,\mathrm{e}^{-z/d_p})^2\,\mathrm{d}z} \qquad (13)$$

Equation (13) can be simplified by assuming a step in the distribution of protein molecules, the concentration of adsorbed protein layer being equal to Γ_{vol} in the range $0 < z < \bar{\Delta}$ and c_p in the range $\bar{\Delta} < z < (d - \bar{\Delta})$. Parameter d represents the thickness of the TIRIF cell. It is, furthermore, assumed that c is not depleted by adsorption (which is easily achieved by flowing the protein solution of concentration c_p through the TIRIF cell throughout the course of the experiment), and that ε and $f(\lambda)$ are the same for both adsorbed and bulk protein molecules. Then,

$$\frac{N_a}{N_b} = \frac{\phi_a}{\phi_b} \cdot \frac{\Gamma_{\text{vol}}}{c_p} \cdot \frac{\int_0^{\bar{\Delta}} (E_{y0}^t \, e^{-z/d_p})^2 \, dz}{\int_{\bar{\Delta}}^{\infty} (E_{y0}^t \, e^{-z/d_p})^2 \, dz} \tag{14}$$

It is possible that protein emission probability changes due to adsorption, thus making the ϕ_a/ϕ_b ratio an unknown. As a first approximation it can be assumed that $\phi_a/\phi_b \sim 1$, i.e., that the quantum yield of adsorbed protein molecules is not altered by the adsorption. Such an assumption is certainly not justified in the case where a protein molecule has only a few intrinsic fluors because it may be expected that the presence of the interface changes both the microenvironment of the fluor(s) as well as the energy transfer from tyrosine to tryptophan residues in adsorbed proteins. Also, the degree of surface coverage (packing density of adsorbed protein molecules) may change the emission probability. As the number of intrinsic fluors in the protein molecule increases, the effect of protein adsorption on the emission ratio is expected to become smaller making $\phi_a/\phi_b \rightarrow 1$. In this case the experimental parameters N_b and N_a will enable the calculation of Γ_{vol} which in turn can provide surface area concentration of adsorbed protein Γ_p. Calculation of Γ_p from Γ_{vol} implies a knowledge of $\bar{\Delta}$. However, it can be demonstrated that Γ_p is not critically dependent on the choice of $\bar{\Delta}$ as long as $\bar{\Delta}$ is much smaller than λ_0, $\bar{\Delta} \ll \lambda_0$. It is the change in the ratio between electric field amplitude integrals in Eq. (14) that compensates for the change in $\bar{\Delta}$ since the integration is carried from/to different $\bar{\Delta}$ values.

The biggest problem encountered in the numerical quantitation procedure is scattered excitation light, which causes increased fluorescence from nonadsorbed protein molecules that adds to the evanescently excited signal N_b.* Generally, the scattering arises from the imperfections in optical components, coupling fluid, and an imperfectly collimated excitation beam as well as from the polymeric coating at the interface. Any scattering from the proteins in the flow cell was found to be negligible while the scattering

* There is some advantage in having a high proportion of the bulk fluorescence signal originating from the scattered excitation waves. These waves propagate through the TIRIF cell volume and can be used in the study of protein flow hydrodynamics; in this case no nonadsorbing model substances like labeled dextran are necessary.

from the other sources can be minimized only by scrupulous cleaning procedures applied to all optical components.[23] As a check of the extent of excitation scattering, the fluorescence emission spectra of nonadsorbing standard substance such as tryptophan or its derivatives are usually recorded with the excitation wavelength included in the spectra.

The fact that the scattered excitation light has a propagating nature can be used in order to correct experimental N_b for its scattering-originated part.[23] Namely, by using increasing concentrations of nonadsorbing standards for a given thickness of TIRIF flow cell the condition can be reached where the incident intensity of the scattered excitation light is almost completely attenuated in the TIRIF cell volume, i.e., where $\varepsilon cd > 2$. In this case 99% of the scattered light is absorbed and reemitted as fluorescence with given quantum yield. Since in the TIRIF adsorption experiments the fluorescence is collected at the front face of the cell, any increase of fluorescence with the further increase of the standard concentration beyond $\varepsilon cd > 2$ is due to the evanescent excitation only. By choosing an appropriate thickness of the TIRIF flow cell, the intensity of propagating scattered excitation light can be completely attenuated in the concentration range where evanescently excited fluorescence is a linear function of the concentration, i.e., where $k_2 < 0.001$. After plotting the background-subtracted fluorescence intensity of nonadsorbing tryptophan standards as a function of εc, the linear part found at higher εc is extrapolated to $\varepsilon c = 0$ and translated along the y-axis until it passes through the origin. This straight line then represents the evanescent portion of the total TIRIF signal as a function of εc. By the same token, any protein solution of the same εc would absorb the same relative proportions of evanescent and scattered excitation light (neglecting any scattering of both by the protein molecules in the TIRIF volume) which is then emitted as bulk protein fluorescence. It is very likely that nonadsorbed protein will reemit both kinds of adsorbed light with the same quantum yield. In this case, the ratio between any evanescent N_b and total N_b can be obtained from the calibration plot at given εc. Multiplying experimental N_b obtained after protein flush-out (Figure 4a) by this ratio will result in evanescent N_b which then can be used as an internal standard in numerical quantitation. Note that experimental N_b is obtained in the presence of the adsorbed layer of given thickness $\bar{\Delta}$, while the ratio between evanescent N_b and total N_b is derived by using nonadsorbing standards in the absence of this layer. Due to this difference, evanescent N_b has to be corrected by a factor:

$$\frac{\int_0^\infty (E_{y0}^t \, e^{-z/d_p})^2 \, dz}{\int_{\bar{\Delta}}^\infty (E_{y0}^t \, e^{-z/d_p})^2 \, dz}$$

or Eq. (14) has to be used in the form of:

$$\frac{N_a}{N_b} = \frac{\phi_a}{\phi_b} \cdot \frac{\Gamma_{\text{vol}}}{c_p} \cdot \frac{\int_0^{\bar{\Delta}} (E_{y0}^t \, e^{-z/d_p})^2 \, dz}{\int_0^{\infty} (E_{y0}^t \, e^{-z/d_p})^2 \, dz} \tag{15}$$

4.2.3. In Situ Isotope Quantitation

This independent quantitation procedure is based on the fact that the γ-ray signal from ^{125}I-labeled protein is directly proportional to the amount of labeled protein in the sensing TIRIF volume. This volume is determined by the thickness of the TIRIF flow cell, d, and by a lead aperture of area, A, which is placed in the front of a NaI(Tl) scintillator. Figure 4b shows two distinct signals arising in the course of a protein adsorption experiment: N_b^γ coming from nonadsorbed labeled protein after the protein is introduced into the TIRIF cell and N_a^γ arising from adsorbed protein after the nonadsorbed protein has been flushed out. In the quantitation procedure it is assumed that the radioisotope-labeled protein molecules are not preferentially adsorbed so that their distribution across the TIRIF cell represents the distribution of nonlabeled protein. This assumption is easily checked by using protein solutions having different degrees of radioisotope labeling. Several studies have shown no preferential adsorption in the case of ^{125}I-BSA and ^{125}I-collagen adsorption on polymer-coated and clean glass surfaces.[24,25] However, different adsorption isotherms for the same proteins were also reported when either radioactivity measurements or UV spectroscopy were used as a detection technique.[26]

In the case of nonpreferential adsorption, the signal N_b^γ can be used as an internal standard in calculating the amount of adsorbed protein. It can be written:

$$N_b^\gamma = k \cdot m_b; \qquad m_b = c_p \cdot V = c_p \cdot r^2 \pi \cdot d \tag{16a}$$

$$N_a^\gamma = k \cdot m_a; \qquad m_a = \Gamma_p \cdot 2 r^2 \pi \tag{16b}$$

where m_b and m_a are the amounts of protein in the sampling volume V in the bulk and at the interface of area $2\pi r^2$, respectively. Since the proportionality constant k is not changed in the course of an experiment, Eq. (16b) becomes:

$$\Gamma_p = \frac{N_a^\gamma}{N_b^\gamma} \cdot \frac{c_p \cdot d}{2} \tag{17}$$

An alternative route to the quantitation of adsorbed protein is to use ^{125}NaI of known activity as an internal standard. In this case, the degree of ^{125}I protein labeling should be known as accurately as possible. The resolution of an intermediate γ-ray signal in the course of a protein adsorption experiment depends on the degree of protein labeling. Given 1 mg/ml of ^{125}I-labeled protein solution in the sensing volume of the TIRIF cell of priming volume 0.1 ml, the overall count rate would be roughly $\sim 10^8$–10^9 cpm assuming a 25% detection efficiency. Given an adsorbed protein concentration of $\sim 0.1\ \mu g/cm^2$ the signal due to adsorbed protein would be $\sim 10^5$–10^6 cpm. Although the signal of the adsorbed protein is only a small portion of the bulk signal and cannot be significantly resolved from it as in the fluorescence case, it is sufficiently above the background signal ($\sim 10^3$ cpm) to be detected upon flushing the nonadsorbed ^{125}I-labeled protein solution out of the cell. Protein exchange may be determined by flushing the cell with a cold labeled protein solution of the same bulk concentration. In this way, the protein desorption or exchange process can be followed. By obtaining N_a, N_b, N_a^γ, and N_b^γ simultaneously as a function of time, the γ-ray signal can be used to calibrate the TIRIF signal and this in turn makes the calculation of the ϕ_a/ϕ_b ratio possible. It is important to stress that the radioisotope approach can be used as a calibration approach for a particular protein-surface combination and only at a given set of conditions (pH, type of buffer/electrolyte, temperature, etc.).

4.3. FLOW CELL HYDRODYNAMICS

An attractive feature of the TIRIF approach is the ability to study the real-time adsorption and desorption behavior of proteins. To do this it is necessary to understand the hydrodynamics in the flow field of the TIRIF cell, in order to determine whether transport to the surface or actual adsorption is the rate-determining step in the process. This has been done by studying the fluorescence of high molecular weight, FITC-labeled, nonadsorbing dextrans having diffusion coefficients similar to the proteins of interest. Figure 6 illustrates a typical flow cell characterization experiment where extrinsic fluorescence intensity is given as a function of $(\text{time})^{1/2}$ for a FITC–dextran with an average molecular weight of 64,000. The diffusion coefficient is approximately identical to that of IgG ($D \sim 4 \times 10^{-7}\ \text{cm}^2\ \text{s}^{-1}$). The data are presented at high time resolution. The bump observed during the injection of dextran at about four seconds is due to bulk solution FITC–dextran fluorescence excited by scattered light generated at the interface. The total internal reflection process assumes an optically perfect interface, but as previously discussed, some imperfections are always present; i.e., the quartz prism, sample plate, and optical coupling fluid all scatter

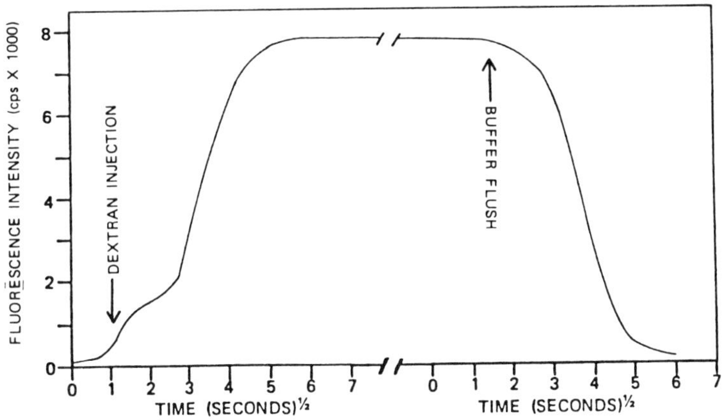

FIGURE 6. Results of a TIRIF flow cell characterization experiment using an extrinsic fluor-labeled dextran to determine the time course of convective diffusion transport to the interface.

light and any polymer film on the surface will result in additional light scatter. Scattered light is a controllable component which can be used advantageously in rigorous quantitation. After approximately the first seven seconds, one observes the FITC–dextran encountering the evanescent wave region adjacent to the flow cell wall. The remaining 29 s required to reach a plateau concentration is attributed to diffusion of the FITC–dextran into the 300-nm thick evanescent wave region where fluorescence can be excited by the exponentially decaying evanescent wave. Under these conditions of flow where the shear stress, γ, is $210\ s^{-1}$ the interface is fully equilibrated with protein within 36 s of injection. The desorption sequence is then initiated via buffer flush at $210\ s^{-1}$. Although the kinetics are slightly different due to the different boundary conditions, roughly 35 s after the bulk FITC–dextran has been flushed from the cell the interfacial molecules have completely diffused from the boundary layer into the bulk and have been carried away by the flush. Consequently, during the first 35 s of both the injection and flush process, signal would not be due to the kinetics of IgG adsorption or desorption but rather to the fluid mechanics and diffusion process involved. The fact that the data are relatively linear between roughly 7 and 16 s suggests a diffusion-controlled $[(\text{time})^{1/2}]$ process. A high injection and flush shear rate ($210\ s^{-1}$) are utilized in order to minimize the time required for diffusion to get protein to and from the surface so adsorption could be clearly discerned. The lower shear rate domain ($16\ s^{-1}$) is used to prevent bulk protein depletion and to carry away desorbed protein during PBS flush.

5. RESULTS AND DISCUSSION

5.1. ADSORPTION ISOTHERMS

The TIRIF technique lends itself well to the determination of protein adsorption isotherms on both silica and polymer thin films. Isotherms may be generated in two ways. A conventional isotherm can be derived by utilizing separate sample plates for each bulk protein concentration to be investigated. Consequently, at each new concentration protein interacts with a clean, uniform, and hopefully identical substrate. The disadvantages are that the approach is time-consuming and the different substrate samples have to be assumed to be or shown to be uniform. In contrast, the step isotherm approach uses a single substrate with the isotherm developed by introducing the lowest bulk protein concentration first, allowing that to adsorb to equilibrium, flushing out bulk protein, noting the signal, determining the amount bound, and then introducing the next higher bulk protein concentration. This approach suffers from the problem that with the exception of the first protein sample, subsequent higher bulk protein concentrations encounter a surface which has been altered by prior protein adsorption and possible conformational changes.

Figure 7 shows step isotherms for IgG adsorption onto both hydrophilic silica and a 50 nm thick, hydrophobic polystyrene film deposited on silica. Each adsorption kinetic experiment was allowed to run until an obvious plateau had been reached at a given bulk concentration, c_{p}.. At low concentrations this required more than one hour. Background subtracted data are plotted as post-flush adsorbed protein fluorescence intensity vs. bulk protein concentration. Bulk protein concentrations as low as 10 μg/ml are readily detected. It is calculated that at a bulk protein concentration of 2 μg/ml, it would be possible to detect a minimum of 0.002 $\mu g/cm^2$ of adsorbed protein on the polystyrene surface. The results are somewhat better for the adsorption of IgG onto clean quartz due to the lower background scatter signal.

The role of polymer film thickness is of considerable interest. Different polystyrene film thicknesses were produced by dipping quartz plates in four different polystyrene solutions: 1, 2, 3, and 5 wt. % polystyrene in toluene. The 1%, 2%, and 3% solutions gave film thicknesses of roughly 50, 100, and 150 nm, respectively, as determined gravimetrically. Adsorption experiments were conducted at two bulk protein concentrations, 0.05 and 1.0 mg/ml. It was possible to detect IgG adsorption onto polystyrene thin films at bulk concentrations as low as 0.05 mg/ml. The higher polymer concentrations yielded films which were considerably thicker with resultant higher background scatter signals and a loss of protein detectability. Also, as polystyrene film thickness increased, the exciton fluorescence emission

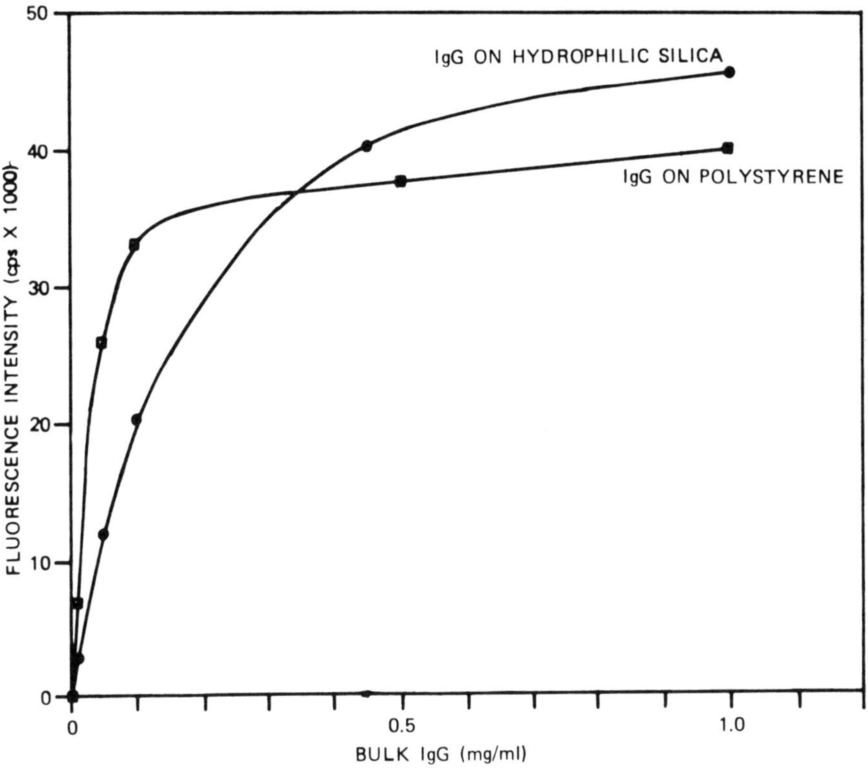

FIGURE 7. Step adsorption isotherms for IgG hydrophilic amorphous silica and hydrophobic polystyrene, PBS, pH = 7.4.

peak at about 323 nm also increased significantly, contributing an even higher background to the signal. This would probably not be the case with other polymer films which do not have this fluorescence mechanism. Optimum results were obtained for the 50 nm thick film. The fluorescence due to adsorbed protein was the predominant signal with such films. With thicker films the film fluorescence was larger than the adsorbed protein fluorescence.

5.2. DESORPTION RATE STUDIES

A study of protein desorption rates as a function of surface residence time provides information about protein accommodations to surfaces. Figure 8 illustrates four discrete stages in the first 5.4 min of IgG bulk cell flush and surface desorption from polystyrene. At time zero a PBS buffer flush at 210 s^{-1} is initiated. The flush continues for about 60 s at which time

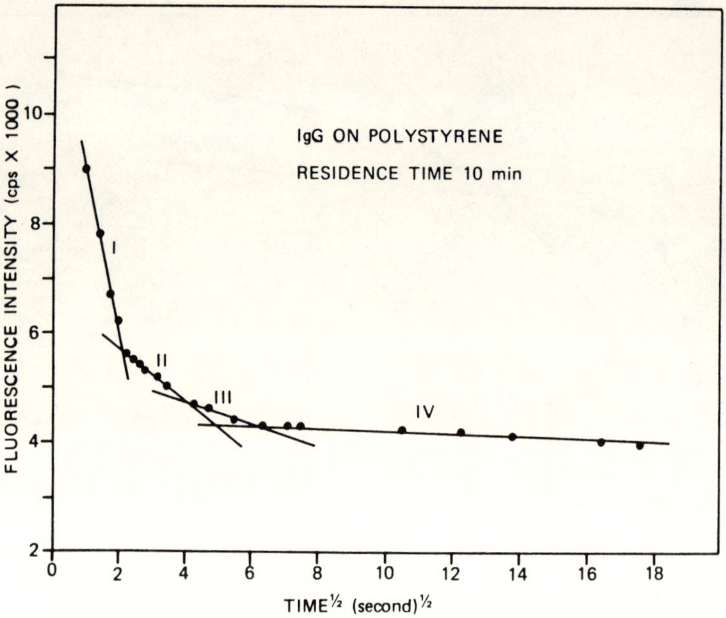

FIGURE 8. Four stages in the bulk IgG removal and desorption of IgG from a polystyrene surface.

the PBS rate is reduced to give a fluid shear of $16\ s^{-1}$. During the high shear flush there appear to be three distinct rates I, II, and III, which are attributed to removal of the protein from the TIRIF flow cell. Domain I corresponds to bulk flow cell protein removal. Domain II corresponds to diffusion of protein out of the evanescent wave region and into bulk and domain III corresponds to the subsequent removal of this protein from the bulk flow cell. The time required to remove all of the unbound IgG at a fluid shear of $210\ s^{-1}$ is about 35–40 s, which corresponds very well to the labeled dextran flow studies summarized in Figure 6.

Domain IV of Figure 8 corresponds to true IgG desorption from the polystyrene surface. Figure 9 illustrates the same desorption rate information except now time zero is defined as beginning after all the unbound protein is flushed from the flow cell and the abcissa scale is given in $min^{1/2}$ rather than in $s^{1/2}$. It is obvious that IgG having a 10-min residence time on polystyrene exhibits two well-defined desorption rates over a time course of two hours. The first domain of protein adsorption (0–7 min) has a much greater rate (slope $= -278 \pm 39$) than the second domain (slope $= -13 \pm 8$). This would seem to suggest that there are either at least two distinct populations of surface-bound IgG on the polystyrene having different binding affinities to a uniform surface or there is a single kind of protein

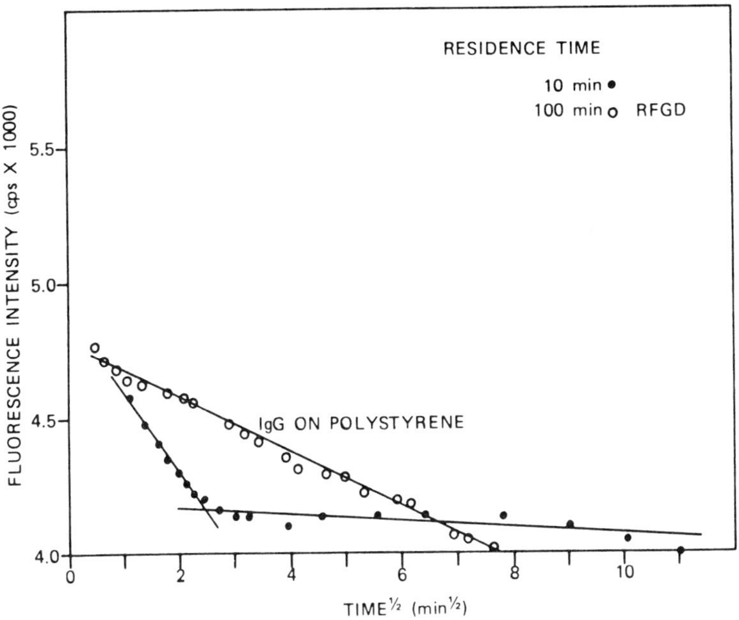

FIGURE 9. Desorption of IgG from a hydrophobic polystyrene surface and a hydrophilic radio frequency glow discharged polystyrene surface.

desorbing from a heterogeneous surface having at least two different binding sites (affinities) for IgG. Note also in Figure 9 that when polystyrene was radio frequency glow discharged in an oxygen plasma, the surface was rendered completely hydrophilic for both advancing and receding contact angles and there was a single IgG desorption rate (slope $= -100 \pm 4$).

Figure 10 illustrates the effect of IgG surface residence time on desorption from hydrophilic amorphous silica. For protein residence times of 1.5, 10, and 100 min the corresponding slopes are: -117 ± 5, -133 ± 18, and -102 ± 16, respectively. There is no significant difference between the slopes indicating that residence times of 1–100 min do not significantly alter the desorption behavior of IgG. However, when IgG resides on the surface of hydrophilic silica for 1000 min, there are two distinctly different slopes (-186 ± 54 versus -15 ± 8) and both are significantly different from the data representing residence times of 1–100 min. Note also that the time resolution for the 100-min residence time data of Figure 10 are not good in the early desorption period of 0–9 min. In fact, the data point at 0.8 min suggests the presence of another second slope in the early desorption phase. It may be concluded that IgG desorbs from hydrophilic silica at a constant rate, irrespective of whether or not residence time is 1.5 or 10 min. At longer surface residence times of perhaps 100 and certainly 1000 min there are two

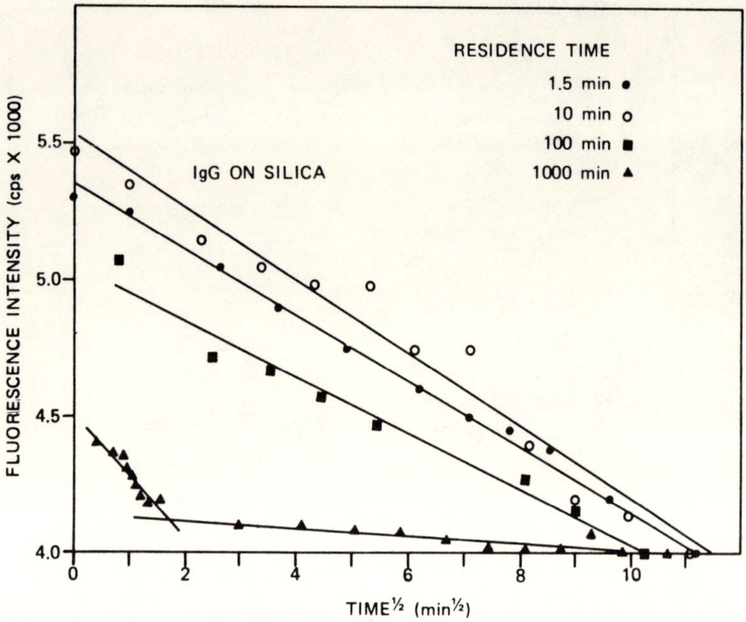

FIGURE 10. Variation of IgG desorption rate as a function of residence time on hydrophilic amorphous silica.

distinct phases in the desorption process with a transition time at about 3 min. This would suggest two different types of bonding of IgG to hydrophilic amorphous silica and because residence time appears to be a factor this would in turn suggest that at times exceeding 100 min some of the IgG molecules are adapting to the surface via conformational alterations. Note also that the latter desorption rate, corresponding to a slope of -15 ± 8, is identical to the second rate of IgG desorption from hydrophobic polystyrene (-13 ± 8).

Figure 11 summarizes the results of the IgG residence time study on hydrophobic *n*-pentyl silanized silica. Note that the first desorption phase slopes increase with increasing residence time, i.e., -164 ± 47, -292 ± 77, and -343 ± 124 for 1, 10, and 100 min residence times, respectively. This tends to drive the breakpoint between the two phases back from 9 min to 3 min. The second desorption phase is basically identical in terms of rate whether the residence time is 1.0, 10.0, or 100 min, i.e., -37 ± 16, -95 ± 19 and -45 ± 8 for 1, 10, or 100 minutes, respectively. There are several possible interpretations of such data: (1) there are two distinct protein fractions desorbing from a homogeneous *n*-pentylsilane surface, (2) there are two or more distinct binding sites on the *n*-pentylsilane surface, or (3) a combination of (1) and (2). The fact that increasing residence time results in a more

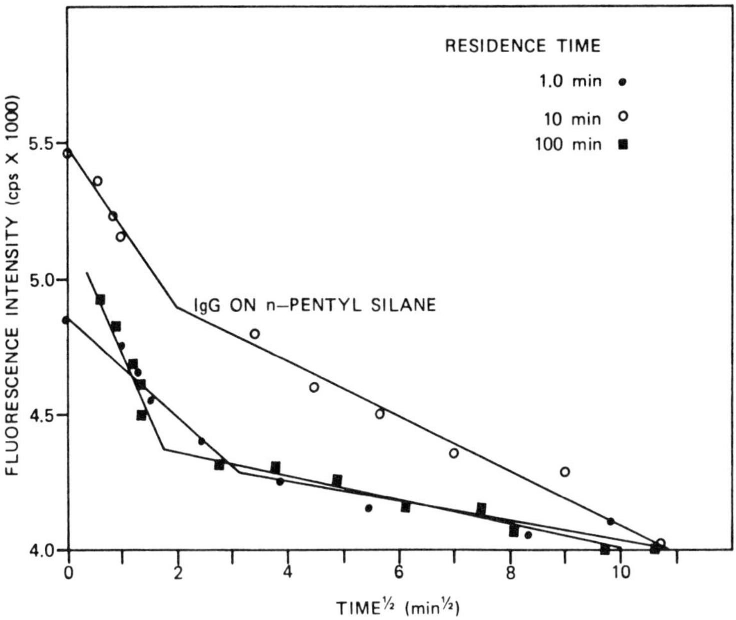

FIGURE 11. Variation of IgG desorption rate as a function of residence time on hydrophobic *n*-pentyl silanized silica.

rapidly desorbing component in the early phase of desorption would seem to suggest that IgG does indeed undergo a surface-induced conformation alteration which is enhanced with increasing contact times.

5.3. MINIMIZING PROTEIN ADSORPTION

TIRIF has been utilized to study the minimization of protein adsorption on polymer thin films. The object of the study was to minimize the nonspecific adsorption of proteins which would compromise the effectiveness of homogeneous fluorimmunoassays based upon protein fluorescence.[27] Amorphous silica served both as a substrate for polymeric thin films as well as a reference material for protein adsorption.

The general experimental protocol is depicted in Figure 12. The primed TIRIF flow cell exhibits a stable nonzero signal due to scattering (**1**). Tryptophan (0.1 mg/ml in PBS) is introduced into the flow cell as a calibration standard (**2**) and then removed with PBS (**3**). Human IgG (1 mg/ml in PBS) is introduced (**4**) and allowed to adsorb for 30 minutes. It is then rapidly flushed out (216 s^{-1}) with PBS and allowed to desorb for 30 min in PBS (16 s^{-1}) (**5**). Human albumin (0.4 mg/ml, PBS) (**6**) and human whole serum (**8**) are sequentially introduced and removed under the same condi-

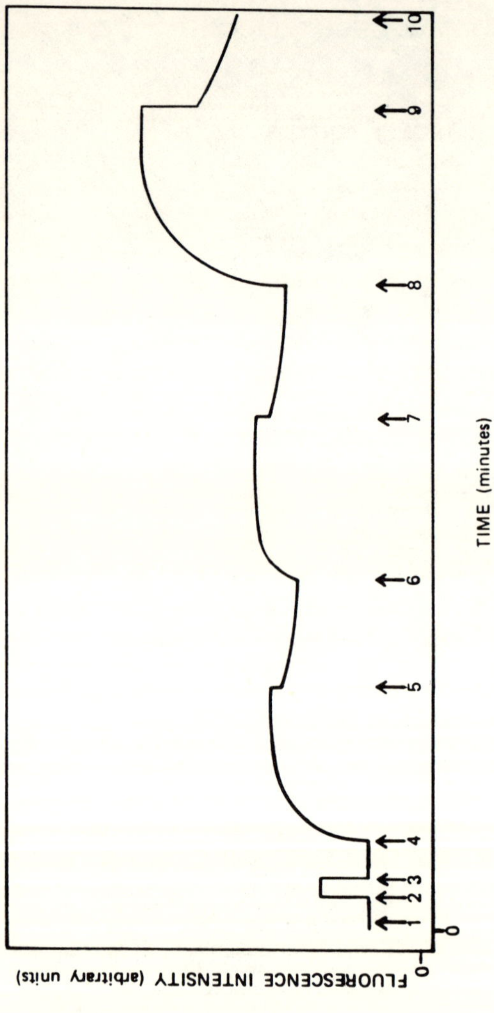

FIGURE 12. Schematic experimental sequence for the TIRIF examination of surfaces designed to minimize protein adsorption. See text for details.

tions (**7**, **9**, and **10**). The raw data at each of the ten data points noted in Figure 12 are then background-subtracted and normalized using the following equations, so that different experiments having variable degrees of scatter can be compared:

$$\text{Corrected Signal} = [\text{Raw Fluorescence–Baseline Scatter}] \times \frac{10{,}000 \text{ cps}}{\text{Trp Signal} - \text{Baseline Scatter}}$$

The value of 10,000 counts per second (cps) is an arbitrary signal level ascribed to the tryptophan standard for normalization purposes. The normalized data for IgG, albumin, and serum adsorption on various thin films and reference materials are summarized in Table 2.

Hydrophilic amorphous silica served as a hydrophilic reference material while the silanes: *n*-pentylsilane (nPS), γ-aminopropyltriethoxysilane (APS) and dichlorodimethylsilane (DDS) served as hydrophobic reference surfaces. Agarose thin films were spuncast onto silica sample plates from DMSO solvent which was subsequently evaporated. APS served as an intermediate link between the ≡Si—OH groups on the silica surface and various molecular weight polyethylene glycols (PEG). The terminal hydroxyl groups of PEG were reacted with phosgene to form *bis*-α,ω-chloroformate PEG derivatives. These were then coupled to the APS amino-functionalized silica in methylene chloride via a urethane linkage. The

TABLE 2
Average Normalized Data for TIRIF Studies on Various Hydrophilic and Hydrophobic Thin Films

Substrate	Post IgG desorption	Post albumin desorption	Post serum desorption
Hydrophilic quartz	22,423 ± 5,922 (4)	—	36,955 ± 3,603 (3)
DDS	56,947 (1)	50,131 (1)	84,905 (1)
APS	31,837 (1)	30,092 (1)	128,108 (1)
APS-glutaraldehyde	16,013 (1)	14,379 (1)	35,947 (1)
nPS	36,263 (1)	30,098 (1)	95,450 (1)
0.1% agarose	24,091 (1)	17,994 (1)	50,584 (1)
1.0% agarose	5,515 ± 1,347 (3)	4,802 ± 1,118 (3)	9,238 ± 3,116 (2)
APS-PEG 750	5,390 (1)	5,431 (1)	12,506 (1)
APS-PEG 1900	579 ± 815 (3)	407 ± 705 (1)	9,728 ± 1,210 (3)
APS-PEG 3,400	587 (1)	789 (1)	3,628 (1)
APS-PEG 5,600	441 ± 504 (4)	362 ± 551 (4)	2,046 ± 264 (4)
APS-PEG 14,000	387 (1)	387 (1)	1,610 (1)

terminal chloroformate was then hydrolyzed in the presence of water to PEG—OH, CO_2, and HCl.

The hydrophobic surfaces of nPS, γ-APS, and DDS seemed to adsorb more IgG and serum components than hydrophilic quartz. However, it should be stressed that the fluorescence intensity signal cannot be directly ascribed to the amount, and in the case of serum, the kind of protein adsorbing. One general trend throughout these studies was the finding that the 30-min albumin desorption signal was generally lower than the preceding IgG desorption signal. The albumin may possibly facilitate the desorption of IgG, or it may competitively adsorb on the surface and replace IgG, or it conceivably may bind to IgG and quench IgG fluorescence.

An agarose thin film seems to minimize protein adsorption relative to hydrophobic amorphous silica; however, a comparison of the data for 0.1% and 1.0% suggests that complete surface coverage is necessary since the 0.1% concentration probably did not completely cover the silica surface.

The data of Table 2 suggests that PEG seems highly effective in minimizing irreversible protein adsorption, with higher molecular weights being most effective. Note that the fluorescence signals arising from IgG and serum protein adsorption are reduced by factors of 60 and 23, respectively, on PEG (average MW = 14,000) versus amorphous silica. PEG probably minimizes protein adsorption by a combination of steric chain repulsion to effectively keep the protein away from the surface and a lack of secondary bonding mechanisms necessary to bind protein.

5.4. EMISSION SPECTRA

It is well established that proteins containing both Tyr and Trp amino acid residues display fluorescence emission spectra typified by a single broad emission band with a maximum varying between 310–350 nm. Based on the analogy with the solvent polarity influence on the fluorescence emission maxima of free Trp in solution, the protein emission spectrum maximum is often used as an indication of the microenvironment polarity of protein Trp residue(s). This notion can be used in the case of adsorbed protein as well, by comparing the TIRIF emission spectra of protein in solution and at the interface. Unfortunately, the changes in the emission maximum positions are in no simple way connected with the changes in the degree of exposure of Trp residues to the solvent. Therefore, no detailed conformational changes can be assessed and data can be interpreted in a qualitative way only.

Two subclasses of Class B proteins are to be distinguished: (1) those that have few tryptophan residues, like BSA, RNaseT, HSA, glucagon, etc., and (2) those having a larger number of tryptophans like IgG, fibrinogen, fibronectin, etc.

It is expected that any changes in the Trp residue microenvironment of the former class would result in a more pronounced change of emission maxima positions. In the case of the latter, large shifts are not expected since the possibility that every fluor changes its microenvironment polarity is hardly foreseeable, unless adsorption brings about extensive protein denaturation. It is also to be expected that the polarity of the interface will influence the emission maxima positions of the tryptophan residues. By measuring fluorescence emission shifts of an adsorbed protein, a combined effect of the exposures of the tryptophan residues to the solvent and/or interface is obtained. While the solvent is usually polar (protein adsorption is more often measured from aqueous solutions than from nonpolar solvents), the polarity of the adsorbing surface can be changed by choosing different polymer coatings.

Some examples of fluorescence emission maxima shifts in the course of protein adsorption are given below:

Case 1: Bovine serum albumin which has two Trp residues/molecule shows a solution fluorescence maximum at 342 nm (at pH = 6.2, 0.1 *M* NaCl). The BSA TIRIF emission spectra in the bulk prior to the 1-h adsorption (B) and after the nonadsorbed protein has been flushed out (A) are shown in Figure 13. The spectra both normalized to unity were obtained using 295-nm excitation wavelength, the adsorbing surface was hydrophilic amorphous silica, and the adsorption of BSA was carried out from 2 mg/ml

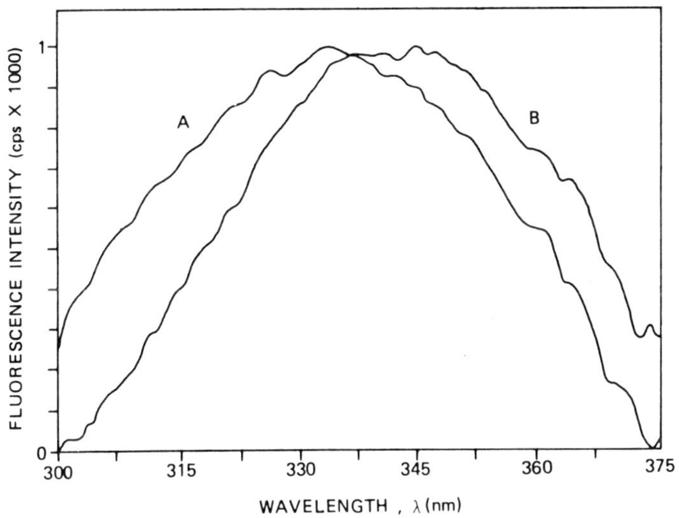

FIGURE 13. Emission spectra of BSA in the TIRIF cell; (B) at the onset of the adsorption on hydrophilic silica, (A) after the nonadsorbed protein has been flushed out of the cell. Both spectra are normalized to unity, $\lambda_{ex} = 295$ nm, 0.1 *M* NaCl, pH = 6.2, $c_{BSA} = 2$ mg/ml.

BSA solution of pH = 6.2 in 0.1 *M* NaCl. It is evident that the maximum of emission is shifted by adsorption toward shorter wavelengths. The 9-nm blue shift between B and A indicates a change in the exposure of two Trps residues from a solvent-exposed polar environment to a more apolar one. This blue shift gives no evidence for denaturation of BSA by adsorption similar to that observed in the case of BSA adsorption on copolypetide and silicon surfaces.[28] However, no TIRIF data concerning BSA adsorption onto silanized surfaces are available to date.

Case 2: Fibronectin with 25 Trp residues/molecule shows a solution fluorescence maximum in PBS buffer (pH = 7.4) at 321 nm, while denaturation by 3 *M* guanidine hydrochloride shifts the maximum to 350 nm.[29] The TIRIF emission spectra of adsorbed fibronectin on hydrophilic and nPS-silanized amorphous silica surfaces show maxima at 321 nm and 326 nm, respectively. Both spectra are shown in Figure 14. The 5-nm red shift in the fluorescence maximum found in the case of hydrophobic, silanized silica suggests some conformational change in which some portion of the Trp residues are exposed to a more polar environment than in bulk solution or when adsorbed on the hydrophilic silica surface. The influence of surface hydrophobicity is also demonstrated by the difference in antibody binding affinity of the adsorbed fibronectin as well as by a decrease in cell

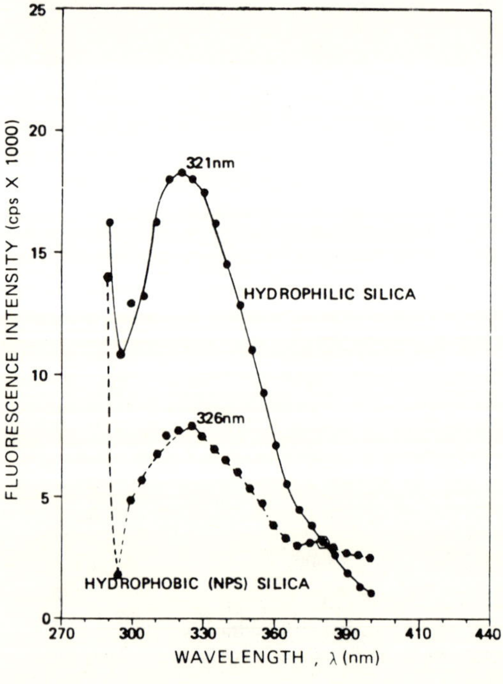

FIGURE 14. TIRIF emission spectra of fibronectin adsorbed onto hydrophilic silica (λ_{em}^{max} = 321 nm) and onto hydrophobic silica (λ_{em}^{max} = 326 nm). λ_{ex} = 280 nm in both cases. [Reprinted from Reference (29) by permission. Copyright by Academic Press Inc.]

interactions with fibronectin adsorbed at hydrophobic vs. hydrophilic surfaces.[30,31]

5.5. LIMITATIONS OF TIRIF

Although TIRIF offers a number of advantages over the conventional techniques for studying protein adsorption, there are some limitations to TIRIF applicability and widespread use. Proteins fluoresce in the ultraviolet range (300–400 nm) when excited by UV radiation (250–300 nm). Therefore, appropriate experimental equipment is necessary, i.e., a stable UV radiation source, excitation and emission monochromators with large UV radiation throughputs, high-gain UV-sensitive photomultipliers, and quartz optical components. Compared with extrinsic TIRF, the cost of such equipment may outweigh its advantages.

Any method based on total internal reflection is limited by the rather narrow choice of surfaces which can be used. TIRIF requires that all optical components of the TIRIF cell be UV transparent, low in impurity, and nonfluorescent in the UV range. Another important parameter is the index of refraction of the material to be used. In order to get more surface-sensitive resolution between adsorbed and nonadsorbed protein, the index of refraction of the IRE optical element should be as high as possible. Only a few materials satisfy these criteria: fused synthetic quartz (SiO_2), sapphire (Al_2O_3), and diamond (C). Similar requirements also prevail for the polymeric coatings that are to be tested for protein adsorption. Polymer materials used as the adsorbing surface should be free of any fluorescent impurities and particulate matter which could cause significant scattering of the excitation UV light. The preparation of clean polymer surfaces is not a trivial problem.[32] An additional problem for some types of polymers is exciton fluorescence.[33] The index of refraction of most polymers is similar to that of fused quartz. In the case when the surface of higher refraction index material (like sapphire) is coated with polymer, the advantage of better spacial resolution is often lost if the reflection is occurring at the polymer/water interface. On the other hand, if the reflecting interface is sapphire/polymer, the sensitivity of TIRIF will decrease, since the most sensitive part of the evanescent volume will be occupied by polymer film and not by adsorbing proteins.

Another important parameter is the index of refraction of the adsorbed protein film. In the course of deriving the TIRIF equations of Section 3, it was tacitly assumed that the index of refraction of the protein film is equal to that of the protein solution. If this is not the case, a more complex theory developed for evanescent wave spectroscopy at multi-layered interfaces should be applied.[15]

No matter how attractive the notion of intrinsic protein fluorescence may appear for TIRIF use, such an approach also has certain limitations among which is the occurrence of protein interfacial photochemistry during the adsorption. Since the protein fluorescence is almost wholly due to the Trp residues, the photolysis of Trp in aqueous solution has been studied extensively.[34–38] The photolysis of Tyr and Phe has also been studied.[39–42] The overall reaction of photolysis appears to be photooxidation with formation of products such as *N*-formylkynurenine.[43] Tryptophan lysis can also proceed via a competitive free radical mechanism under anaerobic conditions. An additional complicating factor in tryptophan photodecomposition is that at least one lysis product—*N*-formylkynurenine—can itself act as a photosensitizer.[44] This feature is particularly important in experiments where conditions allow a significant buildup of *N*-formylkynurenine. Photochemistry of di-, tri-, and polypeptides containing Phe, Try, and Trp has not received much attention until recently.[34,39,40,42,45] The exact mechanisms of Trp, Phe, and Tyr residue destruction in small peptides are not fully understood. This also applies to the photolysis of the same amino acid residues in proteins where such studies are scarce and the interpretations are hindered due to the complexity of the protein molecule. In the past it has been mainly changes in enzyme activity that were measured as a function of UV irradiation. Studies conducted on lysozyme and L-glutamate dehydrogenase showed a concomitant loss of enzyme activity and a decrease in fluorescence emission to be correlated with irradiation time.[45,47] Low-intensity UV irradiation of human gamma-globulin and the effect of various compounds on the UV radiation-induced changes in this protein were also studied. It was concluded that irradiation induced formation of molecular species and complexes whose fluorescence characteristics were different from those of native gamma-globulin.[48]

The susceptibility of adsorbed proteins to photodecomposition by the untraviolet evanescent wave in the course of TIRIF experiments is particularly enhanced due to several reasons: (1) relatively high intensity of the UV evanescent wave in the close proximity of the interface, (2) relatively immobilized structure of the adsorbed layer with possible buildup of *N*-formylkynurenine-like products and (3) comparatively slow replenishment of adsorbed proteins by bulk molecules.

The results of a typical TIRIF bovine serum albumin adsorption experiment, in which the photodecomposition of tryptophan residues was the dominating feature, is shown in Figure 15. Instead of the expected increase in fluorescence signal intensity due to BSA adsorption at the quartz/electrolyte interface, an immediate decrease of fluorescence was observed. It should be mentioned that the dye laser used in this study was pumped by a pulsed N_2 laser and the UV beam was blocked when the fluorescence spectra were not being recorded by the optical multichannel analyzer. The energy level of a 6-ns pulse at 295 nm was of the order of 20 μJ.

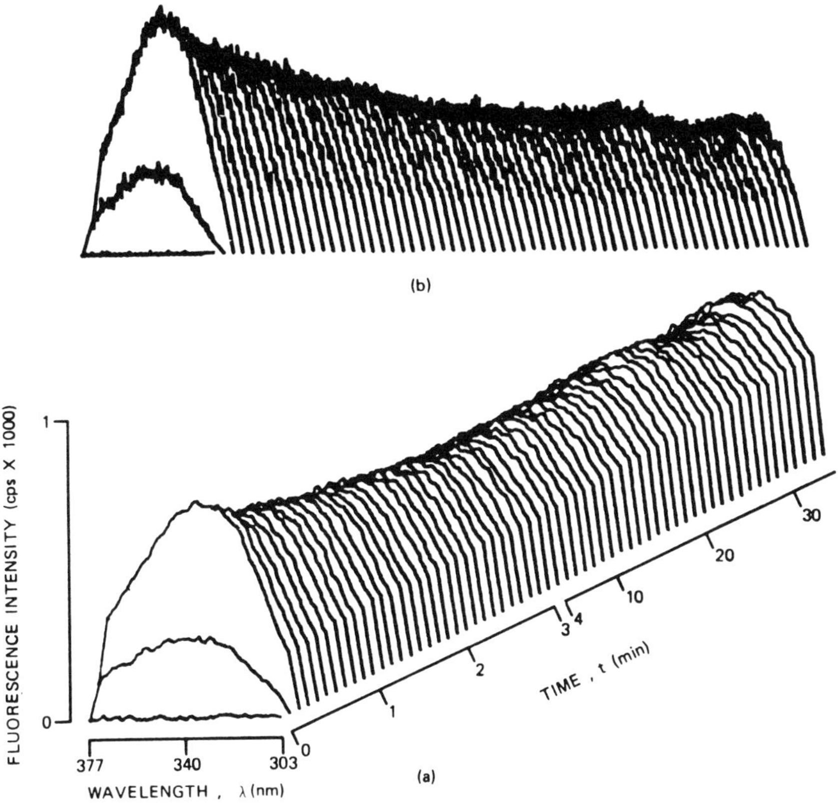

FIGURE 15. TIRIF emission spectra taken during BSA adsorption onto hydrophilic silica; $\lambda_{ex} = 295$ nm, 0.01 *M* NaCl, pH = 4.5, $c_{BSA} = 0.85$ mg/ml; (a) smoothed data; note the two parts of the time axis, (b) same data but unsmoothed and at different viewing angle; note the decrease in emission maxima with time.

Photodestruction of intrinsic protein fluors will eventually result in both plateauing of the TIRIF signal with time (indicating, thus, false equilibrium) and a continuous decrease in signal during the desorption phase (indicating erroneous desorption kinetics). In both cases, there would be no direct correlation between the TIRIF signal intensity and the amount of adsorbed protein at the interface; hence, protein photochemistry must be carefully avoided by all means.

The following approaches were found to be useful in preventing UV-induced protein photochemistry at the interface:

(1) use a continuous light excitation source with a chopper, and choose the highest possible excitation wavelength,

(2) use low excitation light intensity together with the collection of TIRIF signal from as large an area as possible in order to obtain favorable signal-to-noise ratio,
(3) deareate all solutions prior to use.

There is one application in which protein extrinsic fluorescence photodecomposition has been successfully employed. The fluorescence photobleaching recovery (FPR) method uses an increase in extrinsic fluorescence TIRF signal as a measure of lateral surface diffusion of adsorbed protein molecules following laser flash photobleaching of adsorbed protein.[4]

6. CONCLUSIONS

Total internal reflection fluorescence (TIRF) promises to be a powerful method for studying proteins at interfaces. Both the intrinsic and extrinsic fluorescence approaches have advantages and disadvantages; however, for the immediate future the extrinsic approach employing a covalently bound label will be the most widely accepted. Extrinsic TIRF (Chapter 3) is appealing because (1) selective labeling provides for the possibility of competitive adsorption studies, (2) the visible light used can lower cost and complexity of the equipment while yielding higher signed levels, and (3) there is less likelihood of photochemical bleaching in the visible range of the spectrum. The most important need in extrinsic TIRF is to show that the presence of the fluor does not alter protein physiochemical properties which may influence adsorption.

Intrinsic TIRF has most of the features of the extrinsic approach and it also has the potential of providing information concerning the conformational state of adsorbed proteins. Due to the complexity of such protein fluorescence from a heterogeneous population of intrinsic fluors and the greater potential for UV photodegradation, it will be some time yet before intrinsic TIRF finds wide application.

Both approaches to TIRF would benefit greatly if a method of *in situ* quantitation were developed which could give a reliable indication of the amount of protein bound irrespective of fluorescence quenching and photobleaching. Such work is now in progress.[23]

ACKNOWLEDGMENTS

This work was supported by NIH Grant HL-18519, a University of Utah Faculty Research Grant, a grant from Becton Dickinson, Inc., and a contract with Biomaterials International, Inc. The authors acknowledge with appreciation the awarding of the Fulbright Program Grant to Vladimir Hlady. The help of Don Gregonis, William M. Reichert, and Jeff Geisler was greatly appreciated.

REFERENCES

1. A. G. Tweet, G. L. Gaines, Jr., and W. D. Bellamy, Fluorescence of chlorophyll *a* in monolayers, *J. Chem. Phys.* **40**, 2596-2600 (1964).
2. N. J. Harrick and G. I. Loeb, Multiple internal reflection fluorescence spectroscopy, *Anal. Chem.* **45**, 687-691 (1973).
3. B. J. Zdasiuk, *Total Internal Reflection Fluorescence Spectroscopy Studies of Protein Adsorption*, M. Sc. Thesis, University of Utah (1980).
4. T. P. Burghardt and D. Axelrod, Total internal reflection/fluorescence photobleaching recovery study of serum albumin adsorption dynamics, *Biophys. J.* **33**, 455-467 (1981).
5. R. L. Beissinger and E. F. Leonard, Immunoglobulin sorption and desorption rates on quartz: Evidence for multiple sorbed states, *Am. Soc. Art. Int. Org. J.* **3**, 160-175 (1980).
6. R. W. Watkins and C. R. Robertson, A total internal reflection technique for the examination of protein adsorption, *J. Biomed. Materials Res.* **11**, 915-938 (1977).
7. G. K. Iwamoto, R. A. Van Wagenen, and J. D. Andrade, Insulin adsorption: Intrinsic tyrosine interfacial fluorescence, *J. Colloid Interface Sci.* **86**, 581-585 (1984).
8. T. P. Burghardt and D. Axelrod, Total internal reflection fluorescence study of energy transfer in surface-adsorbed and dissolved bovine serum albumin, *Biophys. J.* **22**, 979-985 (1983).
9. C. Allain, D. Aussere, and F. Rondelez, Direct optical observation of interfacial depletion layer in polymer solutions, *Phys. Rev. Lett.* **49**, 1694-1697 (1982).
10. Y.-L. Cheng, B. K. Lok, and C. R. Robertson, Interactions of macromolecules with surfaces in shear fields using visible wavelength total internal reflection fluorescence, in: this volume, Chapter 3.
11. F. W. J. Teale and G. Weber, Ultraviolet fluorescence of the aromatic amino acid, *Biochem. J.* **65**, 476-482 (1957).
12. F. W. J. Teale, The ultraviolet fluorescence of proteins in neutral solution, *Biochem. J.* **76**, 381-388 (1960).
13. J. W. Longworth, Luminescence of polypeptides and proteins in: *Excited States of Proteins and Nucleic Acids* (R. F. Steiner and I. Wienryb, eds.), pp. 319-484, Plenum Press, New York (1971).
14. K. Knutson and W. M. Reichert, *Total Internal Reflection Spectroscopy, Part I. Basic Concepts of Total Internal Reflection*, Plenum Press, New York, to be published (1985).
15. K. Knutson and W. M. Reichert, *Total Internal Reflection Spectroscopy, Part II. Principles of Total Internal Reflection Spectroscopy*, Plenum Press, New York, to be published (1985).
16. D. Axelrod, T. P. Burghardt, and N. T. Thompson, Total internal reflection fluorescence, *Ann. Rev. Biophys. Bioeng.* **13**, 247-268 (1984).
17. M. A. Born and E. Wolf, *Principles of Optics*, p. 273, Pergamon Press, New York (1980).
18. N. J. Harrick, *Internal Reflection Spectroscopy*, Interscience, New York (1967).
19. R. A. Van Wagenen, S. Rockhold, and J. D. Andrade, Probing protein adsorption: total internal reflection intrinsic fluorescence, in: *Biomaterials: Interfacial Phenomena and Applications* (S. L. Cooper and N. A. Peppas, eds.), *Adv. Chem. Ser.* **199**, 351-370 (1981).
20. W. M. Hunter and F. C. Greenwood, Preparation of iodine-131 labeled human growth hormone of high specific activity, *Nature* **194**, 495-496 (1962).

21. D. R. Phillips and M. Morrison, Exposed protein on the intact human erythrocyte, *Biochemistry* **10**, 1766–1771 (1971).
22. S. A. Rockhold, R. D. Quinn, R. A. Van Wagenen, J. D. Andrade, and M. Reichert, Total internal reflection fluorescence (TIRF) as a quantitative probe of protein adsorption, *J. Electroanal. Chem.* **150**, 261–275 (1983).
23. V. Hlady, unpublished results.
24. P. Van Dulm and W. Norde, The adsorption of human plasma albumin on solid surfaces, with special attention to the kinetic aspects, *J. Colloid Interface Sci.* **91**, 248–255 (1983).
25. G. Penners, Z. Priel, and A. Silberberg, Irreversible adsorption of triple-helical soluble collagen monomers from solution to glass and other surface, *J. Colloid Interface Sci.* **80**, 437–444 (1981).
26. H. G. W. Lensen, D. Bargeman, P. Bergveld, C. S. Smolders, and J. Feijen, High-performance liquid chromatography as a technique to measure the competitive adsorption of plasma proteins onto latices, *J. Colloid Interface Sci.* **99**, 1–8 (1984).
27. D. E. Gregonis, D. E. Buerger, R. A. Van Wagenen, S. K. Hunter, and J. D. Andrade, Poly(ethylene glycol) surfaces to minimize protein adsorption, *Transactions of the Second World Congress on Biomaterials*, Washington, D.C., April 27–May 1 1984, p. 266.
28. M. E. Soderquist and A. G. Walton, Structural changes in proteins adsorbed on polymer surfaces, *J. Colloid Interface Sci.* **75**, 386–397 (1980).
29. G. I. Iwamoto, L. Winterton, R. S. Stoker, R. Van Wagenen, J. D. Andrade, and D. Mosher, Fibronectin adsorption detected by interfacial fluorescence, *J. Colloid Interface Sci.*, in press.
30. F. Grinnell and M. K. Feld, Fibronectin adsorption on hydrophilic and hydrophobic surfaces detected by antibody binding and analysed during cell adhesion in serum-containing medium, *J. Biol. Chem.* **257**, 4888–4893 (1982).
31. U. Jonsson, B. Ivarsson, I. Lundstrom, and L. Berghem, Adsorption behaviour of fibronectin on well characterized silica surfaces, *J. Colloid Interface Sci.* **90**, 148–163 (1982).
32. J. A. Koutsky, Cleaning of polymer surfaces in: *Surface Contamination, Genesis, Detection and Control* (K. L. Mittal, ed.), Vol. 1. pp. 351–359, Plenum Press, New York (1979).
33. J. M. Torkelson, S. Lipsky, M. Tirrell, and D. A. Tirrell, Fluorescence and absorbance of polystyrene in dilute and semidilute solutions, *Macromolecules* **16**, 326–330 (1983).
34. Y. Lion, M. Kuwabara, and P. Riesz, Spin-trapping and ESR studies of the direct photolysis of aromatic amino acids, dipeptides, tripeptides and polypeptides in aqueous solutions—III. Tryptophan and related compounds, *Photochem. Photobiol.* **35**, 53–62 (1982).
35. E. Amouyal, A. Bernas, and D. Grand, On the photoionization energy threshold of tryptophan in aqueous solution, *Photochem. Photobiol.* **29**, 1071–1077 (1979).
36. D. Grand, A. Bernas, and E. Amouyal, Photoionization of aqueous indole; conduction band edge and energy gap in liquid water, *Chem. Phys.* **44**, 73–79 (1979).
37. R. F. Borkman, Ultraviolet action spectrum for tryptophan destruction in aqueous solution, *Photochem. Photobiol.* **26**, 163–166 (1977).
38. H. Templer and P. J. Thistlethwaite, Flash photolysis of aqueous tryptophan, alanyl tryptophan and tryptophanyl alanine, *Photochem. Photobiol.* **23**, 79–85 (1976).
39. Y. Lion, M. Kuwabara, and P. Riesz, Spin-trapping and ESR studies of the direct photolysis of aromatic amino acids, dipeptides, tripeptides and polypeptides in aqueous solution—II. Tyrosine and related compounds, *Photochem. Photobiol.* **35**, 43–52 (1982).
40. Y. Lion, M. Kuwabara, and P. Riesz, Spin-trapping and ESR studies of the direct photolysis of aromatic amino acids, dipeptides, tripeptides and polypeptides in aqueous solution—I. Phenylalanine and related compounds, *Photochem. Photobiol.* **34**, 297–307 (1981).
41. C. Hasselmann and G. Laustriat, Photochimie des acides amino aromatiques en solutions—I. D,L-phénylalanine, D,L-tyrosine et L-Dopa, *Photochem. Photobiol.* **17**, 275–284 (1973).
42. J. F. Baugher and L. I. Grossweiner, Photolysis mechanism of aqueous tyrosine and tyrosyl peptides, *Photochem. Photobiol.* **28**, 175–184 (1978).

43. A. Pirie, Formation of N'-formylkynurenine in proteins from lens and other sources by exposure to sunlight, *Biochem. J.* **125**, 203-208 (1971).
44. P. Walrant, R. Santus, and L. Grossweiner, Photosensitizing properties of N-formylkynurenine, *Photochem. Photobiol.* **22**, 63-65 (1975).
45. J. D. Tassin and R. F. Borkmann, The photolysis rates of some di- and tripeptides of tryptophan, *Photochem. Photobiol.* **32**, 577-585 (1980).
46. J. E. Churchich, Fluorescence changes in ultraviolet irradiated muramidase, *Biochim. Biophys. Acta* **126**, 606-608 (1966).
47. R. F. Chen, Photoinactivation of L-glutamate dehydrogenase in a spectrophotofluorimeter, *Biochem. Biophys. Res. Commun.* **17**, 141-145 (1964).
48. D. G. Wickens, A. G. Norden, J. Lunec, and T. L. Dormandy, Fluorescence changes in human gamma-globulin induced by free-radical activity, *Biochim. Biophys. Acta* **742**, 607-616 (1983).

3

Interactions of Macromolecules with Surfaces in Shear Fields Using Visible Wavelength Total Internal Reflection Fluorescence

Yu-Ling Cheng, Brent K. Lok, and Channing R. Robertson

1. INTRODUCTION

Synthetic and biological polymers, or more generally, macromolecules, adsorb onto solid surfaces from solution. This occurs in a number of applications. Proteins adsorbing onto membranes in ultrafiltration processes cause fouling of the membranes and reduce membrane performance.[1-3] Polymer solutions are used to enhance oil recovery, but expensive polymer is often lost due to adsorption onto underground geological formations.[4] In biomedical applications, adsorption of proteins onto synthetic biomaterials is believed to play an important role in determining the ultimate thrombogenicity of the material.[5-10]

The adsorption process depends on the physicochemical properties of the solid surface and those of the macromolecule. These properties and the conditions under which adsorption occurs determine the amounts of macromolecules that adsorb and rates at which they adsorb. Surface properties that have been implicated include hydrophobicity, critical surface tension, surface charge, and surface roughness.[11-14] The commonly encountered solution conditions that are important parameters in adsorption are pH, temperature, ionic strength, concentration of macromolecule in solution, and the nature of any external shear field that the macromolecules

---

Yu-Ling Cheng, Brent K. Lok, and Channing R. Robertson • Department of Chemical Engineering, Stanford University, Stanford, California 94305. Brent K. Lok's present address is Chevron Research Company, Richmond, California 94802. Yu-Ling Cheng's present address is ALZA Corp., 950 Page Mill Road, Palo Alto, California 94303.

are subjected to while adsorption occurs.[15-17] Shear fields have a particularly complex effect on macromolecules. Macromolecules in solution subjected to an external shear field may become oriented, or may experience changes in conformation such as elongation in extensional flow. When macromolecules absorb onto solid surfaces in the presence of a shear field, all the physicochemical factors that characterize the surface and the macromolecule, molecular orientation, and both chemical and shear-induced conformational effects combine to present a seemingly intractable problem. This chapter attempts to untangle the complexities in a systematic manner, to set forth the relevant issues and remaining unanswered questions, and to demonstrate how total internal reflection fluorescence can be used to address some of these questions.

Macromolecular adsorption, like all processes, is governed by thermodynamics. For adsorption to occur spontaneously, the net Gibbs free energy change for the process, ΔG, must be negative. This includes the free energy change for the macromolecule, the surface, solvent molecules, and any co-solutes present in the solution. Gas adsorption onto solid surfaces is a much simpler analogue of macromolecular adsorption. It is instructive, therefore, to first consider several models used to describe gas adsorption on solid surfaces before analyzing macromolecular adsorption. In addition to the obvious advantage of clarity, this also illustrates the differences between gas and macromolecular adsorption, and thus points out the major limitations in using gas adsorption isotherm models to describe macromolecular adsorption.

1.1. GAS ADSORPTION ON SOLID SURFACES

Gas molecules adsorb onto solid surfaces to reduce the surface tension existing at the solid–gas interface. The process continues until the free surface energy of the system, due to the imbalance of surface forces, reaches a minimum value. Isotherms—relationships between the pressure of the adsorbate gas and the amount of gas adsorbed at constant temperature—are generally used to describe gas adsorption. Three phenomena may be involved in physical adsorption of gases: monomolecular adsorption, multimolecular adsorption, and condensation in capillaries or pores.[18,19] There are three major types of isotherms:

(1) Monomolecular adsorption isotherm—Langmuir isotherm (Figure 1A). Adsorption in gas/solid systems described by this model exhibits saturation at monolayer coverage. This isotherm was theoretically derived by Langmuir[20] with the assumptions that (i) adsorption is reversible, and dynamic equilibrium between adsorption and desorption steps can be established such that rates of adsorption and desorption are equal, (ii) adsorbed species are noninteracting and all empty sites are equivalent such that the

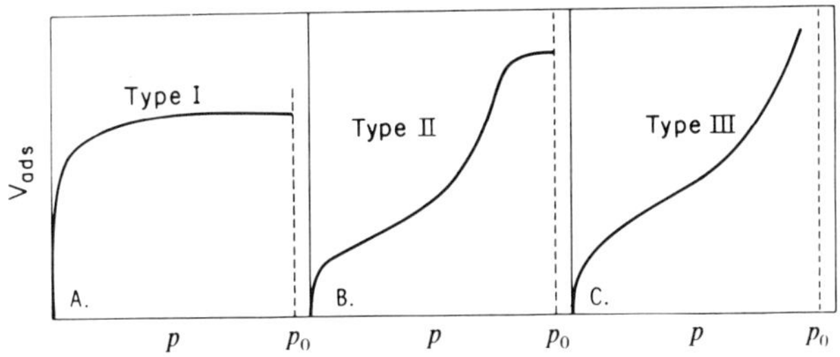

FIGURE 1. The three major types of isotherms: (A) Langmuir isotherm; (B) Multilayer adsorption isotherm; (C) Sigmoidal isotherm. Monolayer adsorption at low pressures, multilayer adsorption at medium pressures, and condensation at high pressures.

heat of adsorpton is independent of surface coverage, and (iii) a gas molecule encountering an occupied surface site does not become adsorbed—this is the monolayer assumption. Using these assumptions, together with second-order adsorption and first-order desorption rate expressions, the classic Langmuir isotherm can be derived:

$$C_s = Kp/(1 + Kp) \tag{1}$$

where C_s is the surface concentration of adsorbed gas molecules, K the equilibrium constant for the adsorption/desorption processes, and p the pressure. The Gibbs free energy change can be obtained from the equilibrium constant.

(2) Multilayer adsorption isotherm (Figure 1B). This type of isotherm is characteristic of gas–solid systems in which more than one layer of gas molecules can adsorb on the solid surface. The isotherms include plateaus that are indicative of saturation for each layer.

(3) Sigmoidal isotherms (Figure 1C). Systems characterized by these isotherms exhibit monomolecular adsorption at low pressures, multilayer adsorption at medium pressures, and condensation at high pressures.

The type of adsorption that occurs for a particular system depends on the nature of the interactions among adsorbate molecules. If adsorbate interactions are repulsive, then the surface saturates at monomolecular coverage corresponding to a close-packed monolayer or less. If the adsorbate interactions are attractive, then multilayer coverage and/or condensation may occur.

1.2. ADSORPTION OF MACROMOLECULES ON SOLID SURFACES

Adsorption of macromolecules is sufficiently complex that the simple isotherms used to describe gas adsorption cannot be readily adopted, and

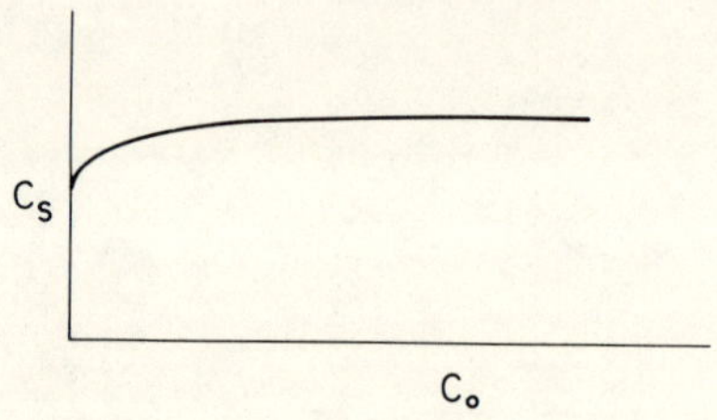

FIGURE 2. Idealized high affinity isotherm.

interpretation of experimentally observed isotherms is, at the present, ambiguous. The two features of macromolecular adsorption that contribute to this complexity are (i) the size of the macromolecules and (ii) their conformational adaptability.

Due to the large size of macromolecules, many different segments of a macromolecule may become attached to the surface during adsorption.[17,21] Thus, even if ΔG for the attachment of one segment is only slightly negative, the overall ΔG for all the segments to attach may be large, resulting in a significant driving force for macromolecular adsoption. This is manifested by "high affinity" isotherms, often seen in macromolecular adsorption. High affinity isotherms have the feature that there is a very sharp upward slope in the low bulk concentration region of the isotherm, equivalent in effect to an instantaneous step at zero concentration (Figure 2).

The conformational adaptability of macromolecules gives rise to the consequences that (i) adsorption characteristics may be sensitive to the conformational state of the macromolecule in solution, which in turn is dependent on the solution conditions—temperature, ionic strength, pH, and solution concentration, (ii) the conformation of a macromolecule may change from its solution state to one that is more energetically favorable as adsorption occurs, and (iii) after a macromolecule has adsorbed, conformational changes, reversible or irreversible, may continue to take place to maximize favorable macromolecule/surface interactions.

1.2.1. Experimental Variables

The manner in which different experimental parameters—pH, ionic strength, and temperature—influence adsorption is evident in the adsorption isotherms. With traditional techniques, the amount of protein adsorbed under various conditions can be conveniently determined.

1.2.1.1. pH. The interactions between protein molecules and the surface can be influenced by pH. At pH values greater than the isoelectric point, the net charge on a protein molecule is negative, and at pH values lower than the isoelectric point, the net charge is positive. Hlady and Furedi-

Milhofer[22] found that the adsorption of human serum albumin on negatively charged, precipitated hydroxyapatite decreased with increasing pH for pH values greater than the isoelectric point, indicating that, in this case, charge repulsion is a dominant factor. The effect of pH is not always so straightforward. Some studies show that the amounts of protein adsorbed onto negatively charged surfaces do not increase with decreasing pH, but rather exhibit maxima at the isoelectric point and decrease as pH either increases or decreases.[16,23,24] Soderquist and Walton[23] found for the adsorption of the plasma proteins albumin, gamma-globulin, and fibrinogen on copolymers of polypeptides and silicone surfaces that at the isoelectric point of these proteins more protein is adsorbed than at other pH levels. Since at the isoelectric point there is minimal intramolecular electrostatic repulsion, molecules in solution are more compact. Charge repulsion between neighboring molecules adsorbed on the surface is also reduced, so adsorbed molecules can assemble themselves more closely to one another. Both of these effects contribute to the observation that more protein can be accommodated on the surface at the isoelectric point. Bagchi and Birnbaum[16] examined the effect of pH on the adsorption of goat and rabbit IgG on anionic poly(vinyltoluene) model latex particles and observed the same trend: more IgG is adsorbed at the isoelectric point. In addition, they correlated the surface area occupied per molecule at the different pH levels using dimensions of the protein in solution determined from intrinsic viscosity measurements. Norde and Lyklema[24-28] studied the adsorption of albumin and ribonuclease on polystyrene latices. Albumin adsorption was maximal at its isoelectric point in agreement with other observations. Ribonuclease adsorption, however, showed very little dependence on pH. This is consistent with the fact that ribonuclease is conformationally stable in solution. Thus, its solution conformation changes little with shifts in pH, and consequently its adsorption behavior remains unaffected.

Forces other than electrostatic interactions have also been shown to influence adsorption. It has been observed that a negatively charged protein molecule may adsorb onto a negatively charged adsorbent surface with the evolution of heat.[24] This finding was interpreted to mean that hydrophobic rather than charge interactions are the dominant factors in protein adsorption. Mizutani[29] found that adsorption of proteins onto silicone-coated glass surfaces occurs even in the presence of 7 *M* urea, and thus concluded that hydrophobic interactions rather than hydrogen bonding (which is disrupted by the presence of urea) are the dominant factor in adsorption.

1.2.1.2. Ionic Strength. Ionic strength influences adsorption primarily by exerting shielding effects on charges present on the protein molecules and on the surface. Thus it can affect the charge interactions between neighboring protein molecules on the surface and between the surface and protein molecules, as well as change the conformation of protein molecules by influencing intramolecular electrostatic interactions.[22-24] In addition, it has been suggested that ions may exert a more direct influence by co-adsorbing onto the surface with proteins.[30]

1.2.1.3. Temperature. The conformation of macromolecules in solution depends on temperature, and since adsorption depends on solution conformation, temperature can be an important determinant in adsorption. In addition, since the Gibbs free energy of adsorption, ΔG, is related to the enthalpy, ΔH, and entropy, ΔS, of adsorption by $\Delta G = \Delta H - T\Delta S$, changing the temperature has the effect of increasing or decreasing the relative importance of the entropic contribution to adsorption. The process of adsorption involves transferring a protein molecule from solution in which it is freely rotating and translating to a two-dimensional surface on which its motions become hindered. Thus the protein molecule loses some degrees of freedom following adsorption. To optimize favorable interactions with the surface, however, the protein molecule will undergo some conformational changes, and in so doing, loses some of the rigid, ordered structure that defines its native conformation in solution, and thus gains some entropy in the process. A complete thermodynamic analysis requires consideration of several processes, including (i) the dehydration of the protein molecule, (ii) the removal of water from the surface at the site where a protein molecule eventually adsorbs, (iii) desorption of small ions or co-solutes from the adsorption site on the surface, (iv) removal of small ions and co-solutes from the hydration shell of the protein molecule, (v) the actual adsorption of the protein molecule from the solution to the surface, and (vi) as described above, the changes in configuration that the protein molecule experiences to facilitate adsorption. Norde and Lyklema[31] attempted a "simplified" thermodynamic analysis of protein adsorption and concluded that adsorption is entropically driven, with the gain in entropy arising in large part from the dehydration of the adsorbent surface, as well as from the internal structural arrangements within the protein molecule. Some experimental studies also support the notion that adsorption is entropically driven.[23] Attempts have been made to assess the importance of the enthalpic and entropic contributions to adsorption by measuring the heats of adsorption by microcalorimetry.[28,32,33] Although no conclusive experimental evidence exists, it is generally believed that the net change in entropy for the entire system following adsorption is positive, and adsorption is, to a large extent, entropically driven.

With these factors in mind, protein adsorption isotherms that have been reported in the literature will be examined. The relevant issues are reversibility of adsorption, surface mobility of adsorbed molecules, and conformational changes experienced by adsorbed molecules and how they affect reversibility, surface mobility, and types of isotherms observed.

Macromolecular adsorption isotherms have been described as "Langmuirian" in many instances,[34,35] and "kinked" or "bimodal" in other cases

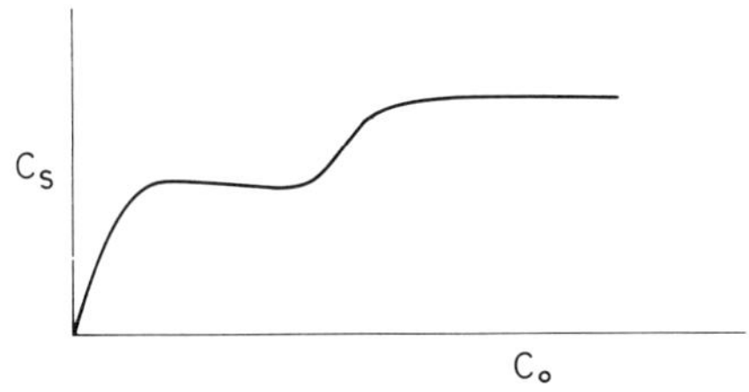

FIGURE 3. Idealized bimodal isotherm.

(Figure 3).[22,23,36] Both high affinity isotherms characteristic of polymer adsorption, and low affinity isotherms, with small slopes at the low bulk concentration region of the isotherm, characteristic of the adsorption of small molecules, have been reported.

Isotherms having the Langmuir form are often interpreted in the same manner as the gas adsorption analogues. It is being recognized with increasing frequency, however, that the premises on which the Langmuir isotherm is based are often not applicable to protein adsorption systems, and it is only coincidental that protein adsorption isotherms are sometimes described by the Langmuir model. The main difficulty in applying the Langmuir model to macromolecular adsorption lies with the reversibility assumption. This will be discussed in detail in the next section.

"Kinks" or "steps" in the so-called bimodal isotherms are universally attributed to surface transitions.[22,23,26] Below the lower plateau, protein molecules are randomly adsorbed, but at a critical surface concentration, corresponding to the lower plateau, a surface transition phenomenon occurs. This transition has been described as surface nucleation or crystallization, and is thought to lead to a more densely packed surface layer. The molecular basis of this surface transition is still uncertain. Although it would be natural to extend the interpretation of multilayer adsorption isotherms often observed in gas/solid systems (Figure 1B) to explain these bimodal isotherms (Figure 3), macromolecular surface layer thickness measurements reported in these studies showed no evidence of multilayer adsorption.

1.2.2. Reversibility

Protein adsorption has been described as totally irreversible, partially reversible, or completely reversible. In instances where reversibility has been reported, the rates of desorption or exchange (sometimes called turnover) from hydrophobic surfaces are reported to be slow,[34,37] and vary from rapid to slow from hydrophilic surfaces.[38-42] Adsorption is often found to be irreversible, or only slowly reversible, when the adsorbed layer comes in contact with a more dilute solution, but presenting a high concentration protein solution to a surface that has been pre-equilibrated with a low concentration protein solution causes more protein to be adsorbed. Conflict exists as to whether isotherms obtained by successively exposing the same surface to protein solutions of increasing concentration, so-called "successive addition" isotherms, are the same as isotherms obtained by conducting several individual experiments in which a clean surface is exposed to only one protein solution—"single shot" experiments.[23,36]

Several points ought to be considered regarding these observations. If adsorption is irreversible, why are "isotherms" observed? For irreversible adsorption, adsorption should proceed until all surface sites are filled, or until saturation occurs. As long as there are protein molecules remaining in solution, adsorption should continue, and the final surface concentration corresponding to low and high protein concentration solutions should ultimately be the same. On the other hand, if adsorption is reversible and "isotherms" are indeed manifestations of a dynamic equilibrium between adsorbed proteins and proteins in solution, then one should be able to move freely along an isotherm simply by increasing or decreasing the protein concentration of the solution in contact with the surface, and "successive addition" experiments would give the same isotherms as "single shot" experiments. One difficulty that may invalidate some of the reported results is that the rates of adsorption at very low solution concentrations, as well as the rates of desorption and exchange, are very slow. Typical first-order desorption rate constants are of the order 10^{-6} s^{-1}.[34,37] If experiments are terminated before either saturation or equilibrium is attained, then interpretation of the experimental results will be ambiguous, and the reversibility question cannot be settled. All these arguments are made with the assumption that denaturation does not occur on the surface. If adsorbed protein molecules undergo conformational changes, then the arguments require modification. This will be described later.

The reversibility of adsorption with respect to solution concentration has been discussed. One can also speak of adsorption reversibility with respect to changes in other experimental parameters, for example, pH. As mentioned earlier, the amount of protein adsorbed depends on the pH of the protein solution, with maximum adsorption occurring at the isoelectric point. The reversibility question now becomes: if a protein layer adsorbed at a certain pH is exposed to a protein solution at a different pH, would the resulting layer be the same as if adsorption had occurred from the second solution? One study reported in the literature[16] indicates that adsorption is "semi-reversible" or "quasi-reversible" to changes in pH; that is, if adsorption occurs at the isoelectric point, then shifting to a different pH causes little or no protein to be desorbed. On the other hand, if adsorption occurs at a pH level other than the isoelectric point, and the pH is changed to the isoelectric point after saturation or equilibrium has been reached at the first pH, then the more protein becomes adsorbed. The final layer, however, contains less protein than if adsorption had occurred onto a clean surface from a protein solution at the isoelectric point. The same questions raised earlier regarding changes in solution concentration are also applicable here. If adsorption is completely reversible, then the characteristics of the surface layer should only be a function of the final pH, independent of the pH history that the surface has experienced. The response of the adsorbed layer to changes in pH may be so slow, however, that equilibrium is never observed. Conformational changes of the adsorbed proteins would again complicate the issue.

It is possible, as previously discussed, that the observed lack of complete reversibility is merely a result of experimental difficulties in detecting changes in surface concentration due to the slow rates of desorption or exchange. A systematic study of the rates of desorption and exchange would give insight into the mechanism of exchange. It has been reported that the longer an adsorbed layer has been adsorbed on the surface, the slower it desorbs.(23) Also, both the fraction of reversibly adsorbed protein and the rate of exchange increase with shear rate and the concentration of the exchanging protein solution.(34)

1.2.3. Surface Mobility

The ability to attain a new surface layer by challenging a pre-adsorbed surface to different solution concentrations or different pH values can be explained by the existence of a dynamic equilibrium between adsorption and desorption processes, or, alternatively, by assigning a certain mobility to molecules adsorbed on the surface. If molecules are mobile rather than "frozen" on the surface, then the state of the adsorbed layer is not entirely dependent on the conditions prevailing when the layer was adsorbed.

By assuming that surface diffusion did not occur, Bagchi and Birnbaum(16) were able to model their results of the effect of changing pH on the adsorption/desorption of IgG pre-adsorbed on PVT latex surfaces. They argued that this assumption is reasonable since one would expect many points of surface attachment for protein molecules. In contrast, surface mobility has been invoked to explain the surface transition manifested as the inflection in bimodal isotherms.(23) TIRF is the only technique to date used to measure surface diffusion of adsorbed protein molecules.(38,39) These studies will be described later.

1.2.4. Conformation Changes/Denaturation

It is almost universally agreed that some degree of denaturation occurs following adsorption, and it has been reported that the degree of denaturation increases with surface residence time.(23) Measurements of adsorbed layer thickness, one of the few quantitative indications of "conformation," show, however, that the thickness of adsorbed protein layers corresponds closely to one of the dimensions of the native protein in solution.[36,43,44] This implies that the degree of conformational change is not extensive enough to make the dimensions of the adsorbed molecule significantly different from those in solution.

The possibility of gradual denaturation of surface-adsorbed proteins can have an effect on adsorption isotherms. For example, assume that adsorbed molecules denature to optimize interactions with the surface in a way that molecules become increasingly spread, and that this process is defined by a rate—R(denaturation). At high solution concentrations, the rate of adsorption is so fast that all empty sites are filled with protein molecules in their native conformation before the molecules that are adsorbed begin to denature and spread. At very low solution concentrations, however, where the rate of adsorption is slow, the rate of denaturation may be sufficiently rapid relative to the rate of adsorption that adsorbed molecules denature and spread to occupy some of the empty surface sites before additional molecules can adsorb from solution, thereby resulting in a lower surface concentration at saturation. This may explain the paradoxical observations reported regarding "reversibility." No conclusive evidence exists to prove or disprove this model as only scattered observations have been reported. Fair and Jamieson[36] did not observe any difference in layer thicknesses at different surface coverages, a finding that is not consistent with this picture. Modeling using these assumptions, however, has been used to explain some kinetic and isotherm results obtained with TIRF.[40-42] These studies will be described in detail in a later section.

1.2.5. Kinetics of Adsorption

In addition to adsorption isotherm determination, measurements of rates of adsorption, desorption, and exchange can be made. Some desorption and exchange-rate studies were discussed earlier. Only a handful of adsorption-rate studies have been reported in the literature, mainly because most techniques used to study protein adsorption do not permit continuous *in situ* measurement of adsorption. Adsorption occurs much faster than desorption or exchange, so that discrete methods cannot be applied to measuring adsorption rates. This is one of the significant advantages of TIRF. The kinetic results that have been obtained with TIRF will be discussed in detail later.

1.2.6. Competitive Adsorption

In all practical applications of macromolecular adsorption, a heterogeneous mixture of macromolecules is present in solution. This can be a polydisperse sample of a synthetic polymer, or a mixture of numerous proteins such as found in blood plasma. It is of interest to study how mixtures of such proteins adsorb onto surfaces—which species adsorb first; do the initially adsorbed species block others from adsorbing, or does replacement occur; and if replacement occurs, how fast and to what extent.

It has been theoretically predicted that in a polydisperse mixture of a synthetic polymer, at equilibrium, the highest molecular weight species will be preferentially adsorbed.[45] The lowest molecular weight molecules, however, are the fastest diffusing and would therefore adsorb onto the surface first. The rate at and the extent to which replacement of the low molecular weight species by the high molecular weight species occurs would determine the nature of the adsorbed polymer film, and how it changes with time.[26,46,47]

It becomes more complicated when macromolecules of different molecular weight as well as different chemical composition are involved. This is the case for adsorption from multiple protein solutions. A small number of multiple protein solution studies have been reported.[48,49] TIRF can be conveniently used to address the questions posed above. Some studies that have demonstrated the application of TIRF to multiple protein studies will be described in a later section.

1.3. SHEAR EFFECTS ON MACROMOLECULES IN SOLUTION

As discussed previously, adsorption can be influenced by the conformation of the protein molecules in solution. Protein conformation in solution can be affected by factors such as pH and ionic strength; it can also be affected by shear. Shear fields have two possible effects on macromolecules in solution—orientation and denaturation.

If a macromolecule is asymmetric, then it is possible to orieint it in solution by shear. If the orienting shear field is sufficiently strong to overcome the randomizing thermal motion of the macromolecule, characterized by its rotational diffusion coefficient, θ, then the molecule will be oriented. The dimensionless Péclet number is a measure of this balance ($Pe = \gamma/\theta$ where γ is the shear rate). High Péclet number indicates that shear orientation dominates, and low Péclet number indicates that random thermal motion is dominant. Shear orientation is usually measured by flow birefringence.[50]

In addition to orienting macromolecules, shear can change the structure and conformation of macromolecules in solution. For example, polymers subjected to extensional flow, such as during extrusion processes, may experience elongation due to the flow field.[51] In some extreme cases of very high shear rates and/or fragile, ultrahigh molecular weight macromolecules, degradation may occur. This is frequently seen in experimental procedures involving DNA,[52] and is one of the problems with

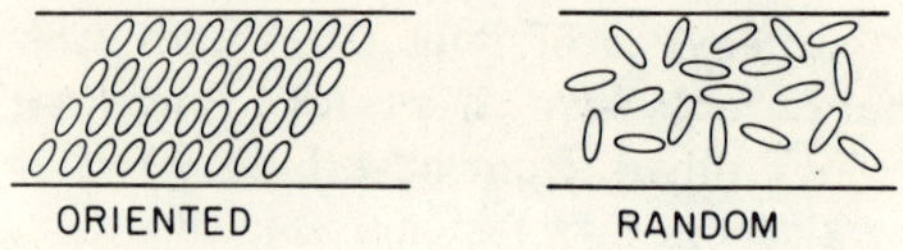

FIGURE 4. Oriented surface vs. randomly packed surface.

chromatographic analysis of ultrahigh molecular weight synthetic polymers.[53]

1.4. SHEAR EFFECTS ON THE ADSORPTION PROCESS AND ON ADSORBED MACROMOLECULES

Hydrodynamic shear can influence adsorption in several ways. As described above, macromolecules in solution can be oriented, denatured, or degraded by shear. Shear also affects the rate at which macromolecules are transported to the solid surface prior to adsorption, and thus may be an important determinant of the rate at which adsorption occurs. Once near the surface, the initial contact that the macromolecule makes with the surface depends on any orientation or conformation changes that it has experienced in solution; this in turn may determine how the molecule is adsorbed. Once adsorption occurs and the macromolecule becomes surface-bound, orientation becomes much more likely than in solution. The rotational diffusivity of a molecule on the surface will be much lower than the rotational diffusivity of a molecule in solution, reflecting the loss of rotational freedom due to adsorption. Orientation is then determined by the balance between the shear rate and a "hindered rotational diffusion coefficient"—a surface Péclet number. Surface orientation has an influence on isotherms since orientation can affect how macromolecules are concentrated on the surface. A more oriented and, therefore, closer-packed surface would correspond to a higher surface concentration than an unoriented or more randomly packed surface (Figure 4). Finally, surface-adsorbed molecules, in addition to denaturing due to chemical factors, may also be denatured by shear. Some preliminary evidence exists to indicate that this indeed does occur.[54,55]

2. VISIBLE TIRF METHODOLOGY

TIRF has been used to address some of the issues discussed above. A discussion of TIRF methodology will be presented first. A summary of results obtained using TIRF will be used to demonstrate the application of TIRF to protein adsorption problems.

2.1. THEORY

TIRF theory has been discussed in considerable detail in Chapter 2 and elsewhere[56,57] and will, therefore, only be briefly outlined here. When a beam of light is incident on an interface between two transparent media (with indices of refraction n_1 and n_2) at an angle α, such that $\alpha > \sin^{-1}(n_2/n_1)$, the beam totally internally reflects at the interface between the two media. At the point of total internal reflection, an evanescent wave penetrates into the medium of lower refractive index (Figure 5). Two important properties of the evanescent wave are that (i) it is a stationary electromagnetic wave initiated at the interface and (ii) the intensity of this stationary wave in the rarified medium decays exponentially with distance normal to the interface. The depth of penetration, d_p, the distance at which the intensity of the evanescent field has decayed to e^{-1} of its interfacial value, depends on the incident wavelength, the angle of incidence, and the refractive index ratio, n_{21} $(= n_2/n_1)$, and is given by:

$$d_p = \frac{\lambda_1}{2\pi(\sin^2\alpha - n_{21}^2)^{1/2}} \tag{2}$$

Making allowance for the presence of an adsorbed layer characterized by thickness d and refractive index n_3 between media 1 and 2, the electric field intensities in the rarer medium at the interface are

$$E_{0\perp} = \frac{2\cos\alpha}{(1 - n_{21}^2)^{1/2}} \tag{3}$$

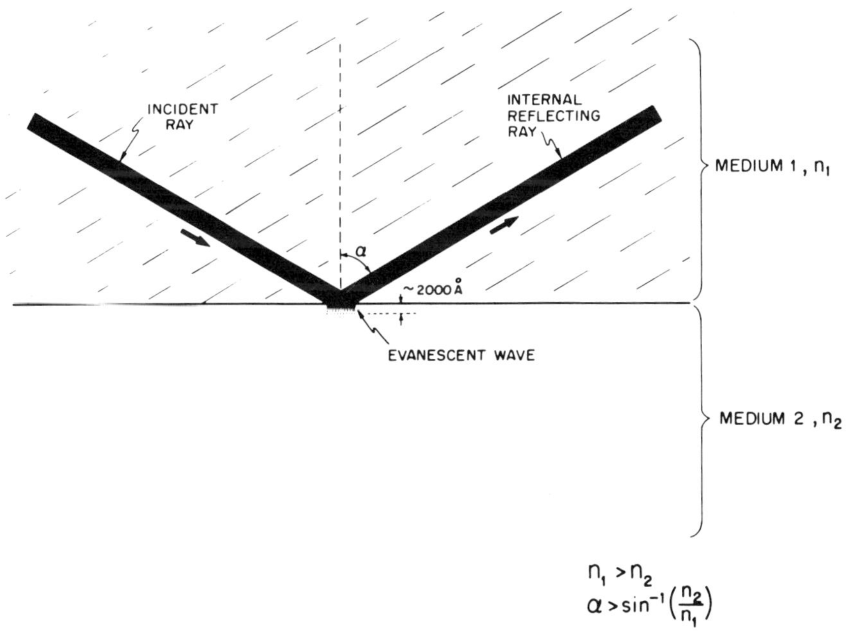

FIGURE 5. Geometry of total internal reflection.[64] Reprinted with permission.

for perpendicularly polarized incident light, and

$$E_{0\parallel} = \frac{2\cos\alpha[(1+n_{23}^4)\sin^2\alpha - n_{21}^2]^{1/2}}{(1-n_{21}^2)^{1/2}[(1+n_{21}^2)\sin^2\alpha - n_{21}^2]^{1/2}} \quad (4)$$

for parallel polarized incident light.

2.2. APPLICATION OF TIRF TO PROTEIN ADSORPTION

Following introduction of a macromolecule solution into medium 2 (Figure 5) macromolecules will diffuse to, and then adsorb on, the interface between media 1 and 2. If these macromolecules contain fluorescent moieties, either intrinsic or extrinsic, that absorb energy at the wavelength of the incident light, then the adsorbed macromolecules along with any macromolecules in solution within the evanescent field will fluorescence. Using appropriate calibration techniques to be discussed later, fluorescence intensities due to surface-adsorbed macromolecules can be related to absolute surface macromolecule concentrations.

By choosing the angle of incidence and the incident wavelength properly, an experimental apparatus can be designed such that the evanescent field encompasses the entire thickness of a surface-adsorbed protein or polymer layer with minimal penetration into the bulk solution, thus making TIRF a surface-sensitive technique.

Two of the commonly encountered amino acids, tyrosine and tryptophan, are fluorescent, and can act as the fluorescent moieties in the TIRF experiments. The advantages of using intrinsic fluorescence are that (i) there is no question of added chromophores interfering with adsorption experiments and (ii) fluorescence spectra can be obtained that reflect the environment of the protein, and not of the extrinsic dye. Intrinsic TIRF is covered in Chapter 2.

Extrinsic chromophores offer some advantages over intrinsic chromopores. By choosing efficient chromophores or using high labeling densities, greater sensitivity can be achieved than is possible with intrinsic chromophores. Multiple protein studies can be conducted by selective labeling of the various protein species. In addition, selective labeling can be used to distinguish different "layers" or proteins adsorbed on different surface sites.

The basic experimental considerations are similar for extrinsic and intrinsic chromophores. Intrinsic fluorescence necessitates using UV wavelength incident light owing to the excitation wavelengths of tryptophan and tyrosine.[58-60]. Extrinsic dyes permit more flexibility. All work with extrinsic fluorescence thus far has been in the visible region.[38-42,56,61-65]

2.3. TIRF APPARATUS—OPTICS

The basic optical requirements of TIRF experiments are (i) a light source, usually an arc lamp, with or without a monochromator, or a laser, (ii) a means of directing the incident beam to achieve a desired incident angle so that total internal reflection occurs at the solid surface/protein solution interface, and (iii) a detector for measuring the intensity of fluorescence emitted by adsorbed proteins. Photodiodes, photomultiplier tubes, and fluorescence microscopes have been used for this purpose.

In addition to these basic requirements, polarizers can be included for fluorescence polarization studies, grids for photobleaching recovery experiments, and monochromators for collecting fluorescence spectra. Optical filters for eliminating stray light, focusing lenses, apertures, collimators, and beam choppers have all been used to optimize the optical performance of TIRF apparatuses. Schematic figures of some TIRF optical designs are shown in Figures 6A-8A and Figure 9.

2.4. PROTEIN ADSORPTION APPARATUS

The design of a protein adsorption apparatus is constrained by the physical process that one wants to examine, and the necessity of coupling the physical and optical portions of the experimental equipment. This usually involves using a transparent slide (glass or quartz) that can be optically coupled to a prism. The surface of the slide facing away from the prism can be coated with a thin, transparent film of the polymer of interest, so that the TIRF incident beam can be transmitted through the prism, the slide, and the polymer coating to be totally internally reflected at the polymer/protein solution interface. The design of the cell that contains the protein solution to be brought in contact with the polymer coating varies depending on the type of experiment. Schematic figures of some cells that have been used are shown in Figures 6B to 8B.

2.5. TIRF PROTEIN ADSORPTION—EXPERIMENTAL CONSIDERATIONS

Since labels may affect the adsorption behavior of macromolecules,[(65,66)] the use of extrinsic labels often raises doubt regarding the validity of the results. This objection can be answered by conducting adsorption experiments at different labeling ratios, and demonstrating that fluorescence intensities obtained from TIRF adsorption experiments vary linearly with labeling ratio.[(56,62-65)] Physical characterizations, including dynamic light scattering and electrophoretic mobility,[(63-65)] of the protein molecules before and after labeling should also be conducted to show that no changes occurred with labeling.

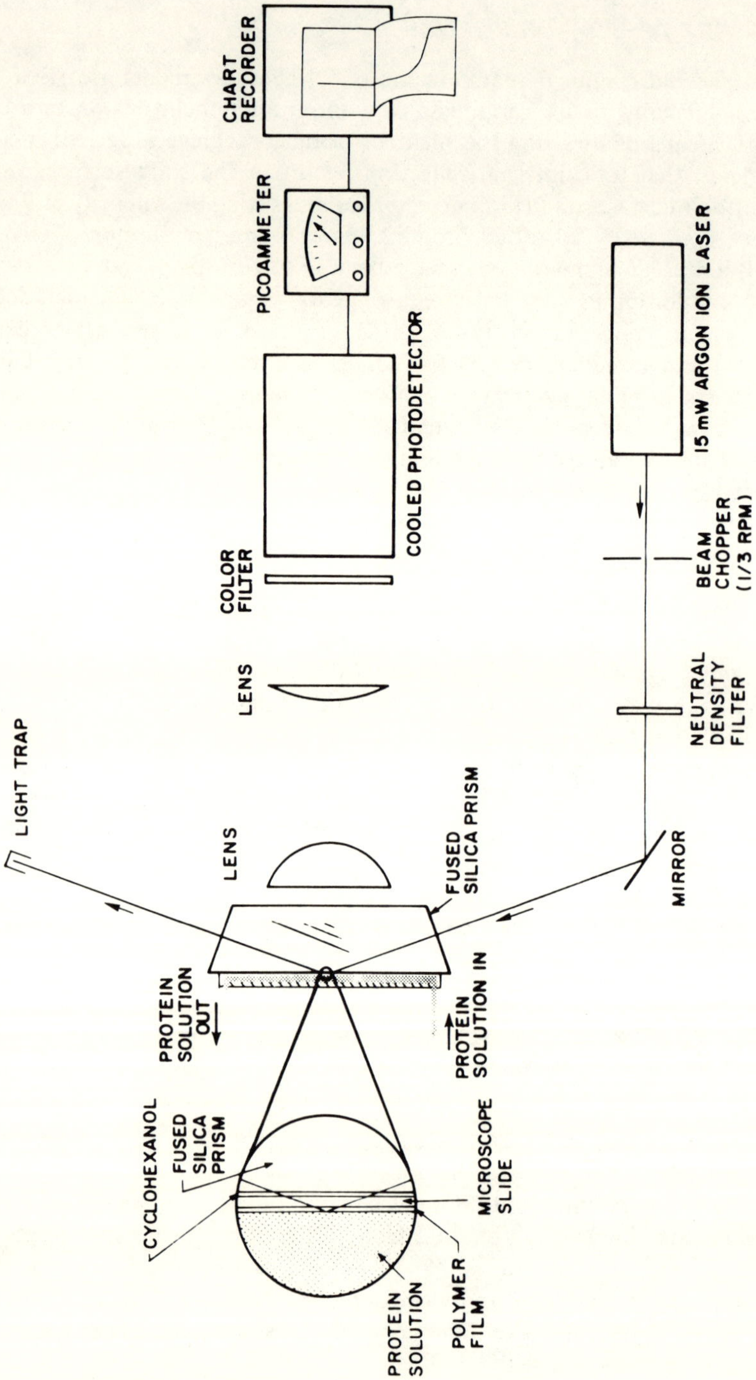
CHART RECORDER
PICOAMMETER
COOLED PHOTODETECTOR
COLOR FILTER
LENS
15 mW ARGON ION LASER
BEAM CHOPPER (1/3 RPM)
NEUTRAL DENSITY FILTER
LIGHT TRAP
LENS
FUSED SILICA PRISM
MIRROR
PROTEIN SOLUTION OUT
PROTEIN SOLUTION IN
CYCLOHEXANOL
FUSED SILICA PRISM
MICROSCOPE SLIDE
PROTEIN SOLUTION
POLYMER FILM

FIGURE 6(A)

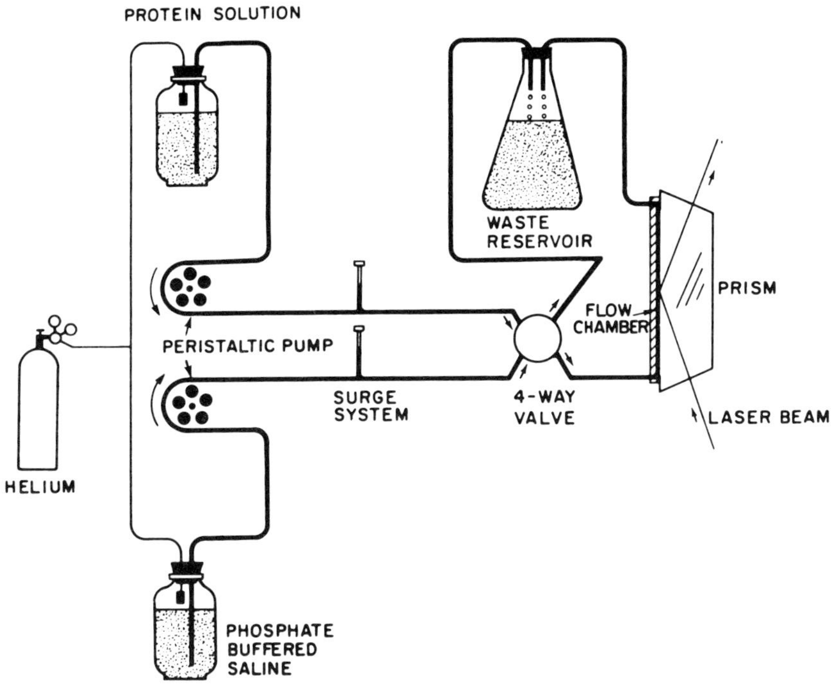

FIGURE 6(B)

FIGURE 6. (A) TIRF optical apparatus used by Lok *et al.*[63-65] The beam from a 15-mW argon ion laser is directed into a 70° dove-type prism (2.9 cm × 3 cm × 9 cm). A neutral density filter is used to attenuate the laser beam. A beam chopper is used to minimize the possibility of chromophore photobleaching during experiments. The beam enters and exists normal to the two small beveled faces of the prism. The beam transmits through the prism, a thin layer of cyclohexanol used as index matching fluid, a glass slide, a thin polymer film coated on the glass slide, and totally internally reflects at the polymer–protein solution interface. That portion of fluorescence emitted by the chromophore-protein conjugate that transmits through the prism is collimated by an aspheric lens and in turn is condensed on the photocathode face of a cooled photomultiplier tube by a second lens. A color filter is placed in front of the PMT to block scattered incident radiation and selectively transmit the fluorescence emission band. A picoammeter voltage output is displayed on a chart recorder. Reprinted with permission. (B) Flow system used by Lok *et al.*[63-65] incorporates two parallel flow systems, one containing pure solvent and the other containing protein solution, and a peristaltic pump that moves fluid through surge reservoirs to damp out pulsations, and then into a 4-way slide valve. The setting of this valve determines whether protein solution or blank solvent enters the TIRF flow chamber. The flow chamber consists of a polymer-coated glass slide as one interior surface, and a machined Teflon well provides the remaining interior surfaces. The Teflon flow cell and the prism are each attached to aluminum blocks and, in turn, are pressed together to provide an air- and watertight seal for the flow chamber. The flow chamber–prism unit is mounted vertically so that fluid enters from the bottom and exits at the top of the chamber. Reprinted with permission.

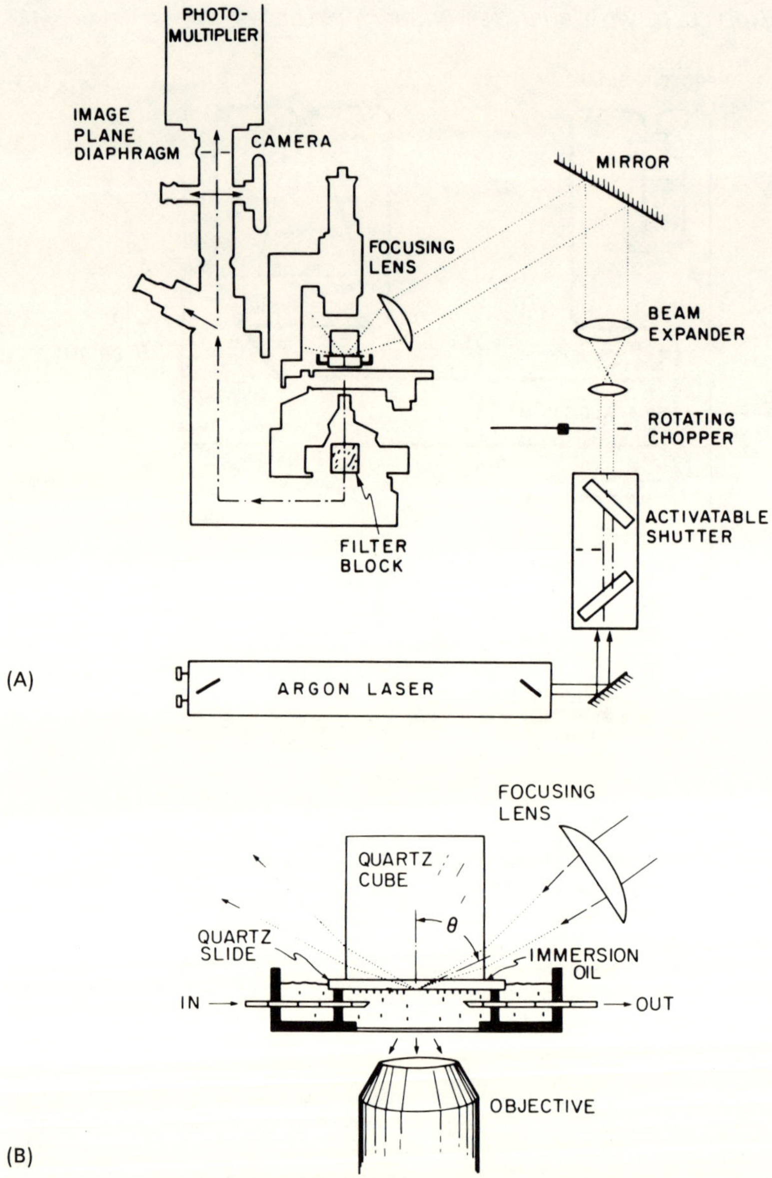

FIGURE 7. (A) Optical apparatus for total internal reflection/fluorescence photobleaching recovery used by Burghardt and Axelrod.[39] The slowly rotating chopper wheel is used to avoid photobleaching during extended observation. The filter block, composed of a dichroic mirror and barrier filter, is identical to one used for fluorescence microscopy by epi-illumination. Reproduced from the *Biophysical Journal* **33**, 455–468 (1981), by copyright permission of the Biophysical Society. (B) Details of the adsorption dish used by Burghardt and Axelrod.[39] Protein molecules adsorbed to a quartz slide are schematically indicated by *x*. The fused silica Suprasil quartz cube is 1.5 cm on each side. The hole in the bottom of the dish is covered with a glass cover slip and sealed in with encapsulating resin. Reproduced from the *Biophysical Journal* **33**, 455–468 (1981), by copyright permission of the Biophysical Society.

It is imperative that the fluorescence signals obtained from TIRF experiments be indicative of surface-adsorbed proteins only, and not be contaminated by signals from protein molecules in the bulk solution.[63–65] Bulk fluorescence originates from two sources; (i) bulk proteins within the evanescent field and (ii) bulk proteins elsewhere in the flow cell that are excited by scattered light (which consists of propagating waves). Scattered light can be minimized by the careful preparation of surface coatings to minimize scattering centers on the surface. This involves meticulous slide-cleaning as well as sophisticated surface-coating techniques. In addition, it is desirable to terminate all TIRF experiments by flowing a solution of unlabeled protein at the same concentration as the original experiment into the flow cell. This removes all bulk contribution from the final fluorescence signal.

(A)

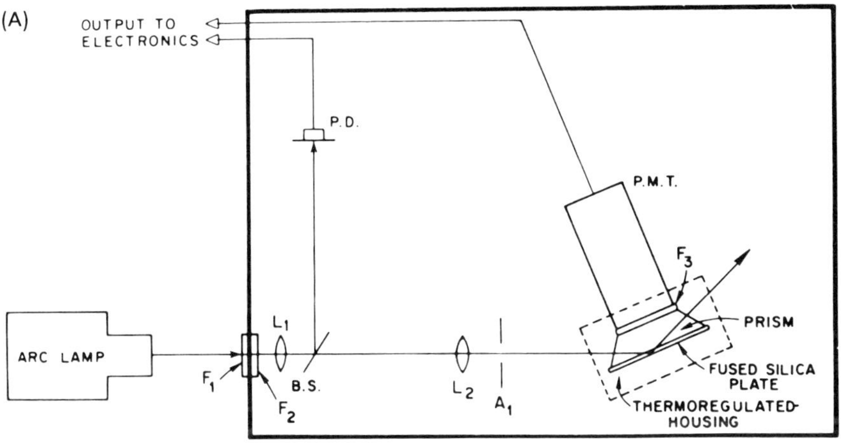

FIGURE 8. (A) Block diagram of TIRF apparatus used by Watkins and Robertson.[56] A beam of light from a high-intensity mercury arc lamp is directed into the otherwise light-tight housing containing the remainder of the apparatus through a polarizing filter (F1) which selects the component of light perpendicular to the point of reflection, and a blue filter (F2). Once inside, the beam passes through a 75-mm focal length lens (L1) and is separated into two portions by a beam splitter (BS). One portion is directed onto the photodiode (PD) surface to monitor the arc intensity while the remainder passes through another 75-mm focal length lens (L2), a 0.07-cm diameter aperture (A1), and onto the internal reflection prism. The emitted fluorescence passes through a filter (F3) transmits the emitted fluorescence, but prohibits incident light from reaching the PMT. Reprinted with permission. (B) Schematic diagram of the flow channel assembly used by Watkins and Robertson.[56] The prism is held in place by two machined aluminum blocks containing channels for circulating constant temperature fluid. Protein solutions may be circulated through the rectangular channel in the Teflon base shown here assembled within its thermostated aluminum housing. A thin layer of cyclohexanol provides optical coupling between the prism and the removable fused silica plate when assembled. The fused silica plates can be coated with a wide range of polymeric coatings. Reprinted with permission.

(B)

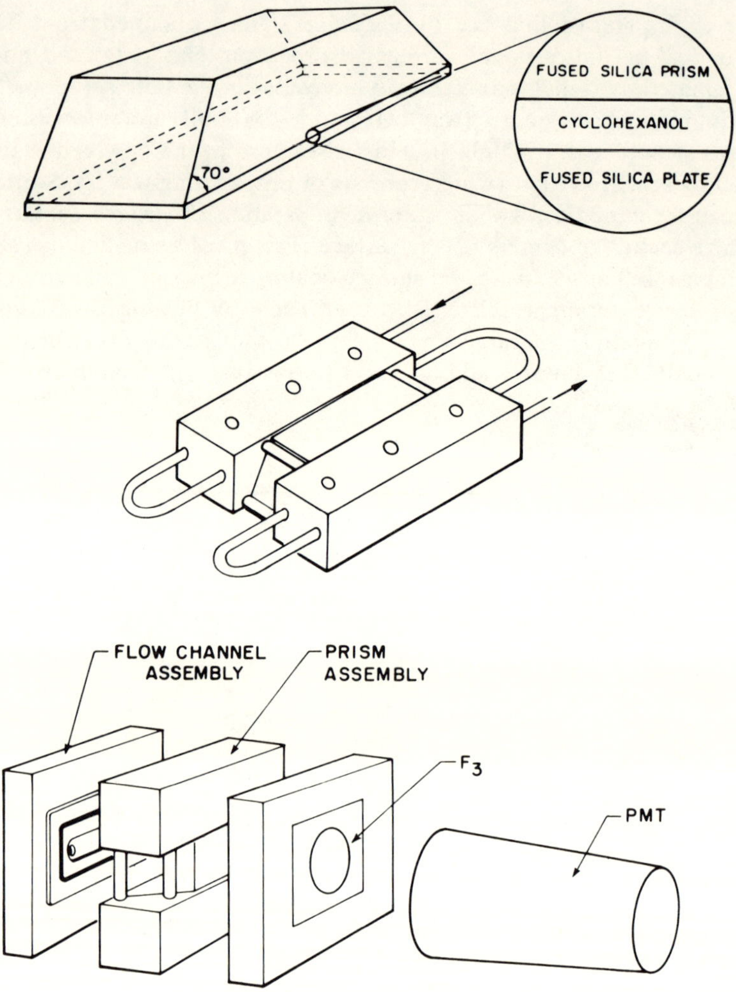

FIGURE 8. (cont.)

2.6. CALIBRATION

Using TIRF instruments and the techniques described above, the adsorption of proteins on solid surfaces can be determined in real time. To relate TIRF fluorescence signals to protein surface concentrations, calibration techniques are required. Four different techniques have been developed and reported in the literature, and will be described later. Several factors must be considered in developing calibration techniques.

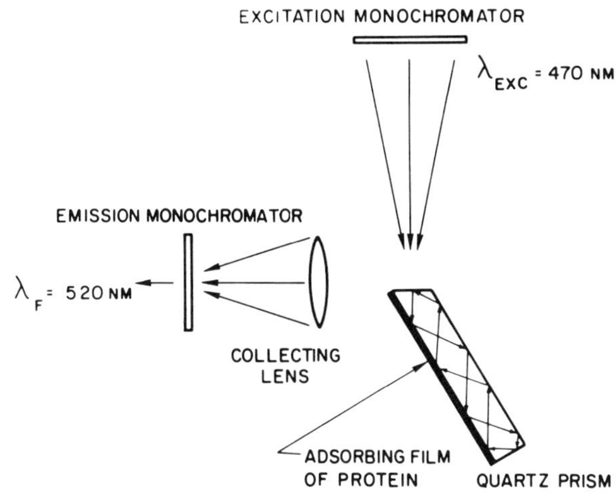

FIGURE 9. Schematic of TIRF apparatus used by Beissinger and Leonard.[41] Visible light (460 nm, 10-nm band pass) was used to excite the fluorescence (measured at 520 nm, 4-nm band pass) of protein conjugate adsorbing to one side of the multiple internal reflection prism. Reprinted with permission.

2.6.1. Linearity

It is essential that the fluorescence signal obtained from TIRF experiments be linearly related to the surface concentration.[56,62–65] Otherwise, unambiguous quantitative interpretation of TIRF results is not possible. Nonlinearity may occur at high labeling ratios if there are chromophore/chromophore interactions or shielding effects within the adsorbed layer.[63,64] Nonlinearity may also occur if the concentration of adsorbing fluorophores is sufficiently high to perturb the evanescent field.[39] It is important that low enough values of the fluorophore to protein ratio, F/P, and total fluorophore concentration be used to insure that calibration results remain linear.

2.6.2. Evanescent Field Intensity: Polymer Film Thickness

Fluorescence from adsorbed proteins is proportional to both the amount of adsorbed protein and the intensity of the evanescent field at the interface. The evanescent field intensity is in turn related to the incident beam intensity. The incident beam intensity can be measured and controlled. As the incident beam transmits through the polymer film, however, it is attenuated due to scattering by imperfections in the polymer film. Thicker and less uniform polymer coatings scatter more light and, therefore, result in a less intense evanescent field. This gives rise to reduced fluorescence signals for the equivalent amount of protein adsorbed.[37] Accurate TIRF calibration necessitates that uniform polymer coatings of the same thickness be used in each experiment. This, again, involves meticulous slide-cleaning and sophisticated surface-coating techniques.[63–65]

2.6.3. Quantum Yield

Quantum yield, the number of photons emitted by fluorescence for each photon adsorbed by a fluorophore, is very sensitive to the environment of the fluorophore.[68,69] Certain environmental conditions offer relaxation pathways other than fluorescence for the excited fluorophore. In general, the quantum yield for a fluorescent protein molecule changes with the conformational state of the molecule. Thus, it is to be expected that protein molecules in solution will not have the same quantum yield as protein molecules adsorbed on surfaces. In addition, the quantum yield of adsorbed proteins is expected and has been found to change if the adsorbed layer is challenged with variations in temperature, pH, and ionic strength of the bathing solution,[37] or even the presence or lack of protein in the bathing solution.[63–65] In attempting to relate TIRF fluorescence signals to protein surface concentrations, it is important that the quantum yield remains unchanged within each set of calibrations. Conversely, if it is desired to use TIRF to study adsorption as a function of variables such as pH or ionic strength that are known to cause quantum yield changes, calibrations must be conducted at each pH and ionic strength.

2.6.4. Calibration Techniques

Keeping these considerations in mind, calibration techniques that have been used will be described.

Watkins and Robertson[56,62] were the first to attempt a calibration of TIRF results. Surfaces having known protein surface concentrations were prepared by evaporating fluorescein isothiocyanate (FITC)-labeled gamma-globulin dissolved in salt-free isopropanol (to facilitate even spreading). The fluorescence emitted from these slides was measured in a TIRF apparatus and corrected for the difference in refractive index of air and PBS.

Burghardt and Axelrod[39] used a similar idea to calibrate fluorescence emission from tetramethylrhodamine-labeled BSA (R–BSA). R–BSA was adsorbed on 13 quartz slides in a watertight chamber with two of the slides serving as end windows of the chamber such that a total of 24 surfaces were exposed to R–BSA. The slides were then washed with buffer to remove all but irreversibly adsorbed R–BSA. The absorbance at 550 nm (rhodamine adsorption maximum) of the 24 protein-coated surfaces was measured. The total protein surface concentration for the 24 surfaces is directly proportional to absorbance. One of the chamber window slides (coated only on one side) was removed and placed in a TIRF apparatus, and the fluorescence emission from the protein layer adsorbed on this slide was measured to obtain the proportionality between TIRF fluorescence and surface concentration.

In both of these calibration techniques, the adsorbed protein layer was exposed to air during the procedure. This undoubtedly caused conformational and thus quantum yield changes. Therefore, one should be cautious in applying these calibration results directly to fluorescence signals obtained under actual experimental conditions.

Beissinger and Leonard[41,42] and Lok *et al.*[63–65] overcame the problem of change in quantum yield using two different approaches with the same basic idea. Instead of measuring the fluorescence emission from slides with known protein surface concentrations as had been done previously, TIRF experiments were conducted first to obtain fluorescence signals directly from adsorption experiments, then the slide was removed from the TIRF apparatus and the amount of protein adsorbed on the surface, corresponding to the TIRF fluorescence signal, determined.

Beissinger and Leonard[41,42] measured the fluorescence after 20 min of adsorption from a 1.6 mg/ml solution of FITC–albumin on quartz. The adsorbed protein was chemically removed and oxidized (in 70% perchloric acid for 24 h), and the amount of nitrogen recovered was determined chemically. The protein concentration was then estimated from the known nitrogen content of proteins. This technique is still in the developmental stage. The problems are lack of sensitivity and possible incomplete removal of adsorbed protein.

Lok *et al.*[63–65] measured the fluorescence emitted from BSA labeled with both FITC and ^3H adsorbed onto silicone rubber surfaces. The polymer-coated slides were removed from the TIRF apparatus and placed in a sample oxidizer. The adsorbed protein was oxidized to form CO_2 and tritiated water. The tritiated water was collected and placed in a scintillation counter. Knowing the tritium labeling density of the proteins, the protein surface concentration corresponding to the measured fluorescence was then determined. It was shown that this procedure resulted in nearly 100% recovery of the adsorbed protein. It was also demonstrated that fluorescence was linearly related to the surface concentration. This technique appears to have overcome all of the reservations associated with obtaining accurate calibrations for use in TIRF.

3. TIRF RESULTS

3.1. ISOTHERMS

TIRF, properly calibrated, can be used easily to obtain isotherm data. All studies with TIRF have reported both reversibly and irreversibly adsorbed proteins. Watkins and Robertson[56,62] reported the presence of a tightly

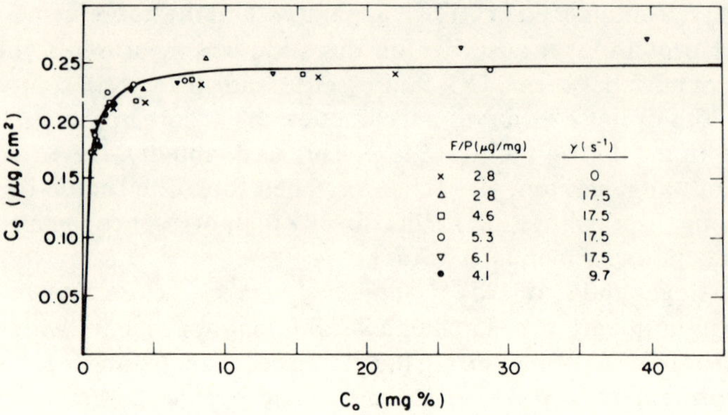

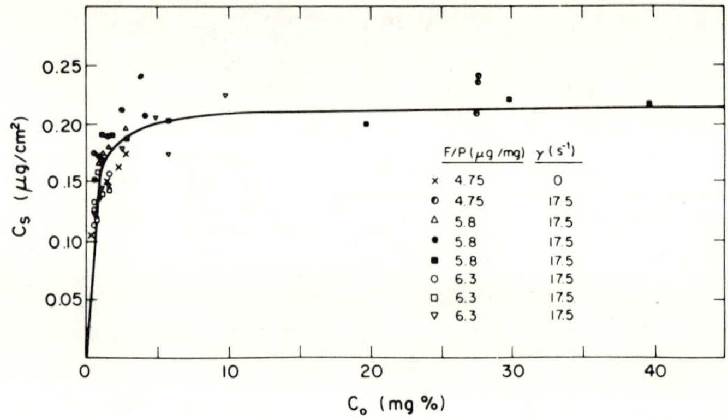

FIGURE 10. Surface concentration of gamma-globulin adsorbed on silicone rubber as a function of bulk concentration. (a) 25°C; (b) 37°C. The solid line is given by Langmuir's isotherm model: $C_s = C_s^* KC_0/(1 + KC_0)$, where C_s = surface concentration, C_s^* = maximum surface concentration, K = the adsorption/desorption equilibrium, and C_0 the bulk concentration. Reprinted with permission from Watkins and Robertson.[56]

bound and a loosely bound layer of bovine gamma-globulin adsorbed on silicone rubber. When a saturated surface was brought in contact with a PBS solution devoid of protein, the fluorescence signal decreased slowly indicating that protein molecules were desorbing from the surface. The layer remaining after PBS rinses, or the tightly bound layer, was found to obey Langmuir's isotherm model (Figure 10), and the amount of loosely bound protein was found to be proportional to the bulk protein concentration (Figure 11). The decrease in fluorescence during PBS rinses was later found to be partly due to changes in quantum yield when the bathing solution contacting the protein layer was changed from a protein-containing solution to protein-free PBS.[63–65]

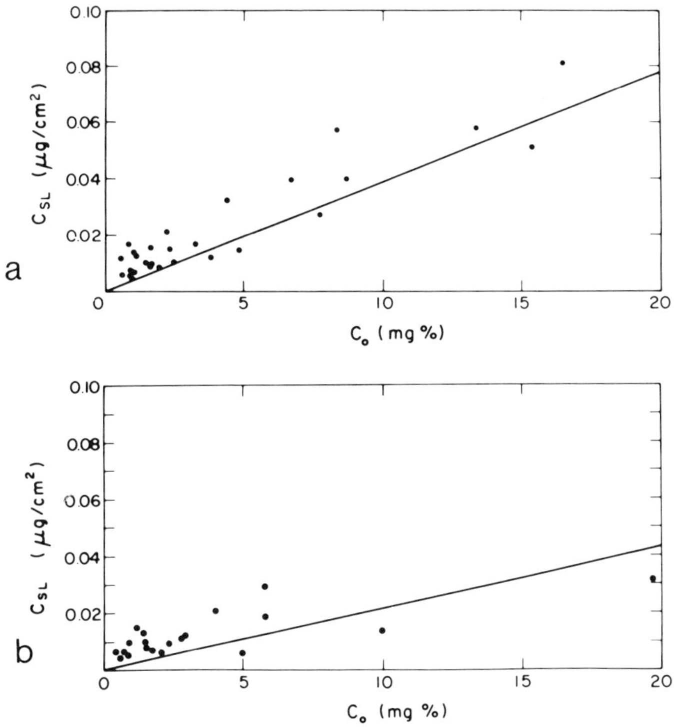

FIGURE 11. Net amount of loosely held gamma-globulin, C_{sL}, on silicone rubber as a function of bulk concentration: (a) 25°C; (b) 37°C. Reprinted with permission from Watkins and Roberston.[56]

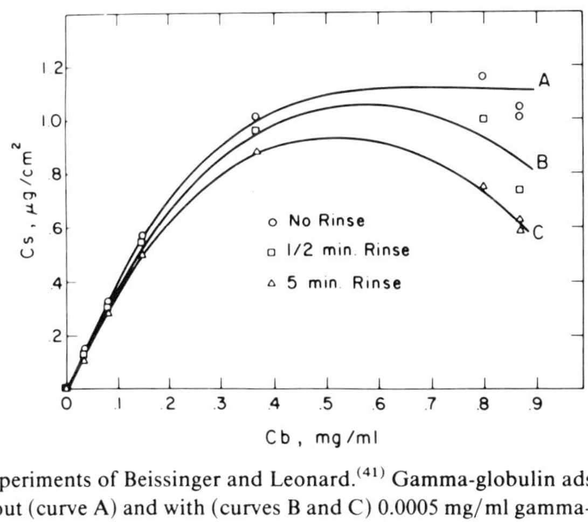

FIGURE 12. Experiments of Beissinger and Leonard.[41] Gamma-globulin adsorption isotherm on quartz without (curve A) and with (curves B and C) 0.0005 mg/ml gamma-globulin solution rinsing of the flow chamber following adsorption. Reprinted with permission.

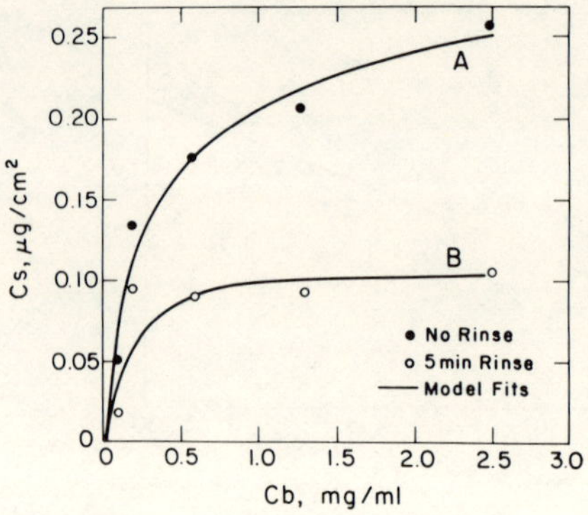

FIGURE 13. Experiments of Beissinger and Leonard.[42] Unrinsed adsorption isotherm of albumin on quartz derived from 20-min kinetic adsorption data (Curve A). Rinsed adsorption isotherm of A made up of the same kinetic sorption data as above followed by 5-min rinsing of the albumin-adsorbed surfaces with 0.005 mg/ml albumin solution (Curve B). Model predictions of the unrinsed and rinsed isotherms are also represented. Reprinted with permission.

Beissinger and Leonard[40–42] also observed both a tightly bound and a loosely bound layer of gamma-globulin and albumin adsorbed on quartz (Figures 12 and 13). In agreement with the results of Watkins and Robertson,[56,62] they found the loosely bound layer of gamma-globulin to be linearly related to the bulk concentration (Figure 14). The loosely bound layer of albumin on quartz, however, was not linearly related to bulk concentration (Figure 15). Also, the rates of desorption were considerably faster than those reported by Watkins and Robertson.[56,62] An additional aspect of their results is the presence of a maximum in the "tightly bound" isotherm at an intermediate bulk concentration of gamma-globulin. This maximum was fitted by a seven-parameter model that will be discussed in the next section.

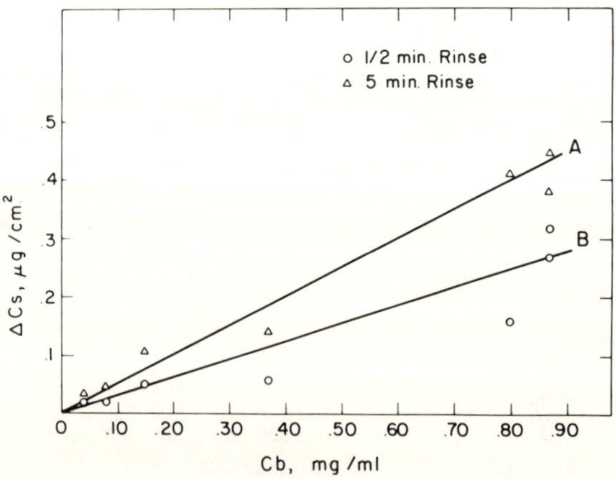

FIGURE 14. Experiments of Beissinger and Leonard.[41] Isotherm of gamma-globulin loosely associated with quartz. These isotherms were approximately linear with the solution concentration used during adsorption. Reprinted with permission.

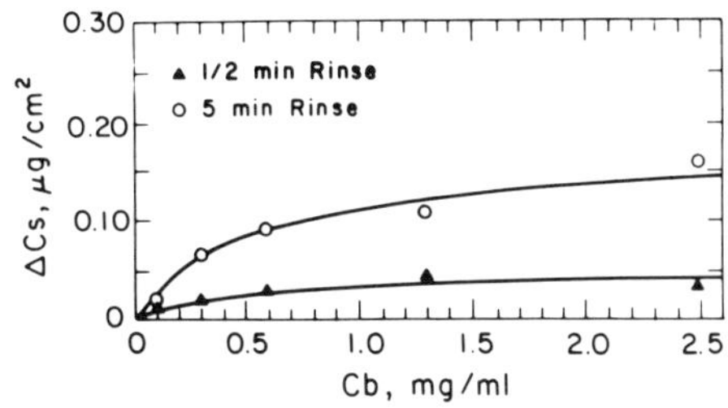

FIGURE 15. Experiments of Beissinger and Leonard.[42] Isotherms of loosely adsorbed albumin on quartz. Amounts of albumin loosely adsorbed were quantified by measuring the reduction in adsorbed surface concentration after either 0.5- or 5-min rinsing of the albumin-adsorbed surfaces with 0.005 mg/ml albumin solution. Reprinted with permission.

Using total internal reflection coupled with fluorescence photobleaching recovery, Burghardt and Axelrod[39] detected three components of BSA adsorbed on quartz: irreversible, slowly reversible, and rapidly reversible components (Figure 16).

Lok *et al.*[63-65] found that isotherms describing BSA and fibrinogen adsorption on silicone rubber were best described by a model incorporating a Langmuir reversible component and an irreversible component (Figures 17 and 18). Assuming 26% of the saturated BSA layer to be irreversibly adsorbed gave the best fit of the observed results (Figure 17). Long exchange experiments were conducted to determine the exchangeable fraction of adsorbed protein. It was found that at a bulk concentration of 10 mg%, a wall shear rate of 58 s^{-1}, and a temperature of 37°C, approximately 40% of the adsorbed layer was replaced after 100 h of exchange (Figure 19).

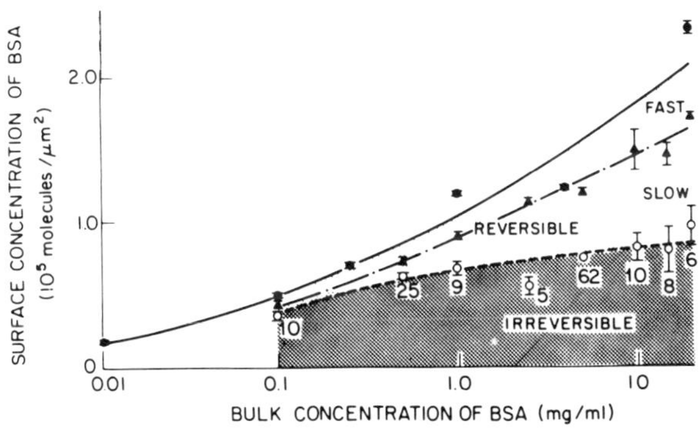

FIGURE 16. Experiment of Burghardt and Axelrod.[39] The irreversibly adsorbed BSA concentration (open circles), the sum of the irreversible and slowly reversible portions (closed triangles), and the total adsorbed BSA concentration (closed circles), as a function of bulk BSA concentration. The irreversibly adsorbed portion reaches a quasi-plateau at higher bulk concentrations whereas the total adsorption (reversible + irreversible) continues to rise. The points representing total adsorption are averages of 5–11 fluorescence readings taken at different locations on a single quartz slide, with standard error bars indicated. Reproduced from the *Biophysical Journal* **33**, 455–468 (1981), by copyright permission of the Biophysical Society.

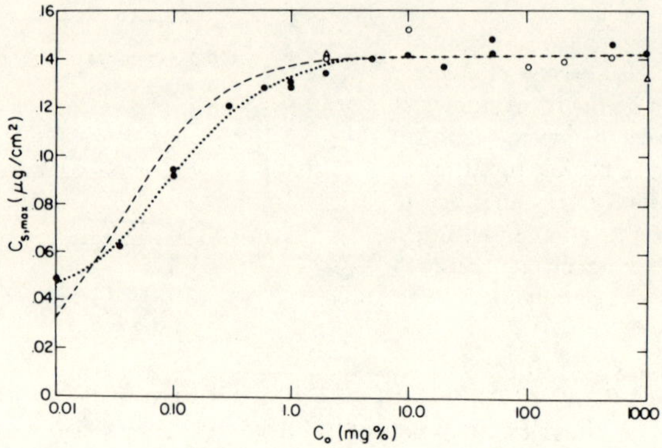

FIGURE 17. Results of Lok *et al.*[65] for BSA/silicone rubber isotherm. Equilibrium surface concentration vs. solution concentration; closed circle, 100% labeled protein; open circle, 50% labeled protein; triangle, 25% labeled protein; dotted line given by $C_s = C_s^* \rho_I + [(1 - \rho_I)KC_0/(1 + KC_0)]$ with $\rho_I = 0.26$; dashed line, $\rho_I = 0$; where C_s^* is the maximum surface coverage, ρ_I is the irreversibly adsorbed fraction of the total adsorbed layer, K the Langmuir constant for the reversibly adsorbed fraction of the adsorbed layer, and C_0 the bulk protein concentration. Reprinted with permission.

Discrepancies exist with regard to the surface concentrations observed by different research groups, even when the same proteins and surfaces are used. This is probably due to deficiencies in the calibration techniques used.

3.2. KINETICS MODELING

One of the primary advantages in using TIRF to examine protein adsorption is that continuous measurements can be made *in situ* and in real time so that adsorption rates can be determined. Rates of protein adsorption are governed by transport of protein to the surface followed by

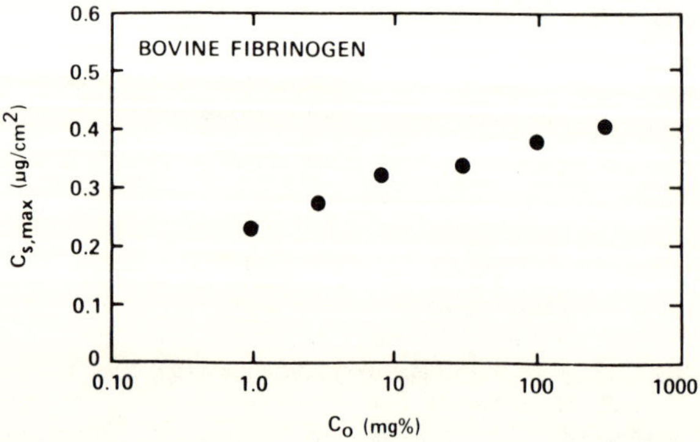

FIGURE 18. Results of Lok *et al.*[65] Fibrinogen on silicone rubber isotherm; equilibrium surface concentration vs. solution concentration. Reprinted with permission.

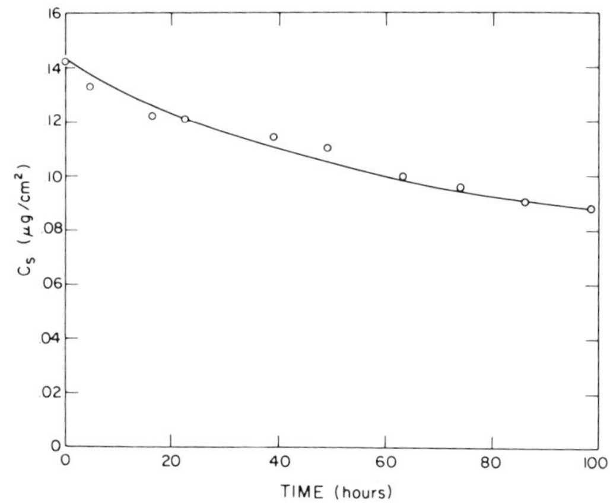

FIGURE 19. Results of Lok *et al.*[65] Long BSA exchange, adsorbed FITC-BSA concentration vs. time; $\gamma = 58\ s^{-1}$. Reprinted with permission.

adsorption. The adsorption process can be described by using an intrinsic adsorption rate constant characteristic of the protein/surface pair under the conditions of the experiment. Watkins and Robertson[56,62] applied classic transport analysis to their system and determined that in the case of either total diffusion or kinetic limitations, analytical solutions describing the rates of adsorption could be obtained. In the diffusion limit, the rate of diffusion of protein molecules to the surface is much slower than the intrinsic kinetic rate of adsorption, and the observed adsorption rate is a measure of the rate of diffusion of protein to the surface. Intrinsic kinetics cannot be observed in this case. In the kinetic limit, macromolecular diffusion to the surface occurs much faster than intrinsic adsorption, and the solution concentration of protein near the surface is essentially identical to that in the bulk. The observed adsorption rate in this case is the intrinsic adsorption rate; no mass transfer limitations are present. If both diffusion and intrinsic adsorption kinetics are important, that is, if diffusion rates and intrinsic adsorption rates are about the same order of magnitude, then a transport model applicable to the flow conditions of the system with boundary conditions describing surface adsorption kinetics must be solved numerically to obtain the intrinsic rate constant for adsorption. Watkins and Robertson[56,62] assumed a second-order irreversible adsorption rate expression and numerically solved the model governing their system to obtain the intrinsic rate constant for gamma-globulin adsorbing onto silicone rubber: $k_1 = (1.3 \pm 0.45) \times 10^{-4}$ cm/s. Their experiments were conducted at shear rates of 0, 9.7, and 17.5 s^{-1}, bulk protein concentrations ranging from 0.5 to 2.5 mg%, and temperatures of 25 and 37°C. No significant difference was observed between results at the two temperatures.

Lok *et al.*[63-65] examined the adsorption of BSA and fibrinogen onto silicone rubber surfaces at 37°C with a wall shear rate of $58\ s^{-1}$. It was found that for both of these proteins, adsorption under these conditions was diffusion-limited, and at low bulk concentrations, adsorption rates agreed with the predictions of the Lévêque solution to the time-independent transport model with diffusion limit boundary conditions.[70] At higher bulk concentrations (>0.1 mg/ml), however, the time-independent assumption inherent in the Lévêque solution is no longer satisfied, and the complete time-dependent convection/diffusion model with diffusion-limit boundary conditions must be solved numerically. The numerical solution agreed well with the experimental results at high concentrations, indicating that in this region adsorption remains diffusion-limited.

Beissinger and Leonard[40-42] used multiple internal reflection fluorescence to study the adsorption of albumin and gamma-globulin on quartz surfaces. Experiments were conducted at $1000\,s^{-1}$. By comparing the observed adsorption rates with the diffusion-limit Lévêque solution, it was determined that, under the experimental conditions used, adsorption was kinetically limited, and the observed adsorption rates reflected the intrinsic kinetics of adsorption without mass transfer limitations. Beissinger and Leonard then attempted to model the surface kinetics assuming that adsorbed proteins exist in two different conformational states on the surface. Molecules were assumed to adsorb from solution in the form of state 1, which can then desorb or denature to become adsorbed state 2 which in turn can also desorb (Figure 20). State 2 is accessible only from state 1. States 1 and 2 represent distinct conformational states that occupy different surface areas per molecule. The occupied areas per molecule for both states were also parameters in this model. A parameter was also included to account for the presence of protein dimers and oligomers in solution that would give rise to an adsorption rate expression other than first order in protein concentration. Using the best least-squares fit of this model to the experimental adsorption rates, the seven model parameters were obtained (Table 1). With these seven parameters, isotherms could be predicted by

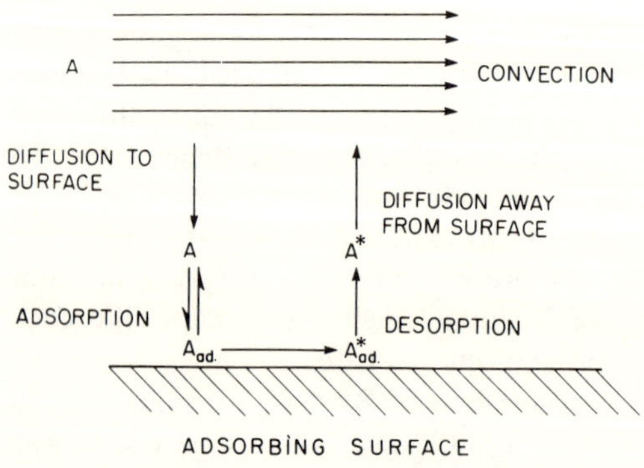

FIGURE 20. Model of Beissinger and Leonard.[41] Convection of species A through the flow cell with diffusion of A near surface, followed by state 1 adsorption on the surface. At this point A can either desorb or react to enter state 2 adsorption, followed by desorption. Reprinted with permission.

TABLE 1
Best Parameter Values[a,b]

Parameter	Value	95% confidence interval
B_1	4.28	±0.63
B_2	0.35	±0.13
B_3	1.16	±0.27
B_4	0.02	±0.01
B_5	0.87	±0.11
B_6	0.45	±0.17
B_7	1.71	±0.17

[a] Best parameter values for a model used to describe gamma-globulin adsorption on quartz:

$$\frac{d\theta_1}{dt} = B_1 Cb^{B_7}[1 - (\theta_1 + \theta_2)] - B_2\theta_1[1 - (\theta_1 + \theta_2)] - B_3\theta_1$$

$$\frac{d\theta_2}{dt} = B_6 B_2 \theta_1[1 - (\theta_1 + \theta_2)] - B_4\theta_2$$

where θ_1 is the fraction of surface area containing molecules in state 1, θ_2 is the fraction of surface area containing molecules in state 2. The seven parameters have the following physical significance: B_1 is the rate constant for adsorption as state 1; B_2 is the rate constant for transition from adsorption state 1 to state 2; B_3 is the rate constant for desorption from state 1; B_4 is the rate constant for desorption from state 2; B_5 is the surface density of molecules adsorbed in state 1; B_6 is the ratio of the adsorbed surface area per molecule in state 2 to that in state 1; B_7 accounts for a suspected nonlinear dependence of sorption rate on solution concentration, C_b.

[b] Reprinted with permission from Reference (41).

carrying the numerical calculations to long times. They were also able to predict desorption rate constants for both state 1 and state 2, with state 2 desorbing thirty times more slowly than state 1. A puzzling result of this approach is that the surface area occupied by each molecule in state 1 (the unsaturated state) is larger than that occupied by each molecule in state 2, the denatured state. It would seem that the more tightly bound (slower desorbing) state would involve more surface attachment sites as a result of the molecule denaturing and spreading to occupy a larger surface area. This work was not supported by any independent studies of molecular conformation. One could argue that there should be continuous denaturation on the surface, and that desorption of protein molecules in all stages of denaturation would be possible. Caution should be exercised, however, so as not to attribute too much physical significance to the experimental results by excessive modeling. A similar approach was used to model the results obtained for albumin adsorption on quartz.[42]

3.3. SURFACE DIFFUSION

As discussed earlier, "reversibility" can be explained by the establishment of dynamic equilibrium between adsorption and desorption processes, or it can be explained by surface diffusion of adsorbed molecules. Using total internal reflection coupled with fluorescence photobleaching recovery, Burghardt and Axelrod[39] determined that serum albumin adsorbed on quartz can diffuse on the surface. A low-intensity beam was used to monitor adsorption until equilibrium was achieved, then a flash of a high-intensity beam was used to photobleach the fluorophores within the adsorbed layer. This resulted in decreased fluorescence (Figure 21). After this maneuver, the low-intensity beam was again used to record the fluorescence recovery. The fluorescence recovery is a measure of adsorption/desorption (exchange) and surface diffusion rates. Burghardt and Axelrod found that their recovery curves could be approximated by a model that included irreversible, slowly reversible, and rapidly reversible components within the adsorbed layer. The two reversible components were each characterized by a recovery time constant. There are two contributions to fluorescence recovery—exchange of photobleaching adsorbed proteins with fluorescent proteins in solution, and the surface diffusion of photobleached adsorbed protein out of, and fluorescent adsorbed protein into the detection region. Using two different beam widths to examine the relative importance of surface diffusion to exchange, they found that the slowly reversible component was independent of beam width, indicating that this class of adsorbed protein was not mobile on the surface. The rapidly reversible component, on the other hand,

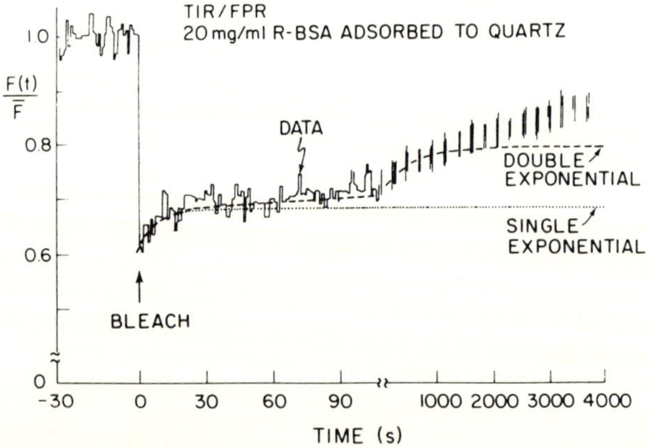

FIGURE 21. Experiment of Axelrod.[39] An actual TIR/FPR curve for R-BSA (20 mg/ml bulk BSA concentration) adsorption on quartz. Two types of computer-generated nonlinear least-squares fit are also shown. Reproduced from the *Biophysical Journal* **33**, 455–468 (1981), by copyright permission of the Biophysical Society.

displayed a faster recovery rate in the narrow-beam experiments relative to the wide-beam experiments. This indicates that this class of adsorbed proteins is mobile on the surface, and that surface diffusion of these adsorbed proteins contributed to fluorescence recovery in the narrow-beam experiments. The surface diffusion coefficient for this component was determined to be 5×10^{-9} cm^2/s.

3.4. MULTIPLE PROTEIN STUDIES

Extrinsic TIRF is ideally suited for the study of multiple protein adsorption. Lok *et al.*[63,65] and Beissinger and Leonard[41,42] have demonstrated the use of TIRF in multiple protein studies.

Lok *et al.* examined the adsorption of a fibrinogen and albumin mixture (100 mg% BSA, 8 mg% fibrinogen, $\gamma = 58\ \text{s}^{-1}$) on poly(dimethylsiloxane) and found that albumin adsorbed first with rates similar to what would be expected in the absence of fibrinogen. Fibrinogen adsorption followed, on surfaces now partially covered with BSA, at rates slower than fibrinogen adsorbing alone (Figure 22). Fibrinogen adsorption continued long after BSA adsorption had equilibrated at 0.12 μg/cm^2; after six hours, the fibrinogen surface concentration was 0.06 μg/cm^2 and still increasing. The partitioning between surface-adsorbed and solution proteins indicates that fibrinogen is preferentially adsorbed over BSA.

Beissinger and Leonard[41,42] examined the adsorption of albumin and gamma-globulin on quartz. Both sequential and simultaneous exposure experiments were conducted. Their results are shown in Figure 23. An

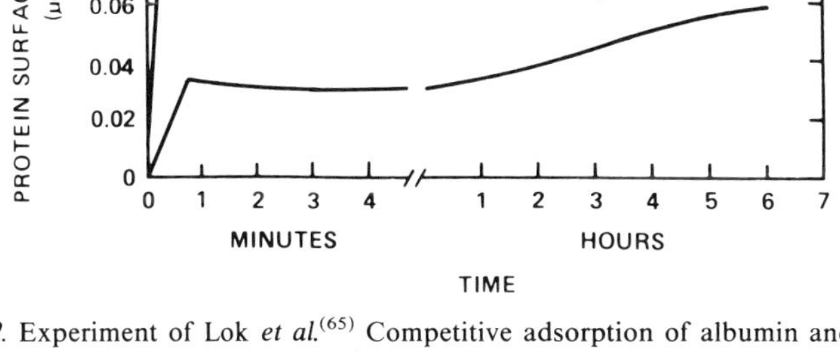

FIGURE 22. Experiment of Lok *et al.*[65] Competitive adsorption of albumin and fibrinogen on silicone rubber; $T = 37°\text{C}$, $\gamma = 58\ \text{s}^{-1}$; protein surface concentration vs. time. Reprinted with permission.

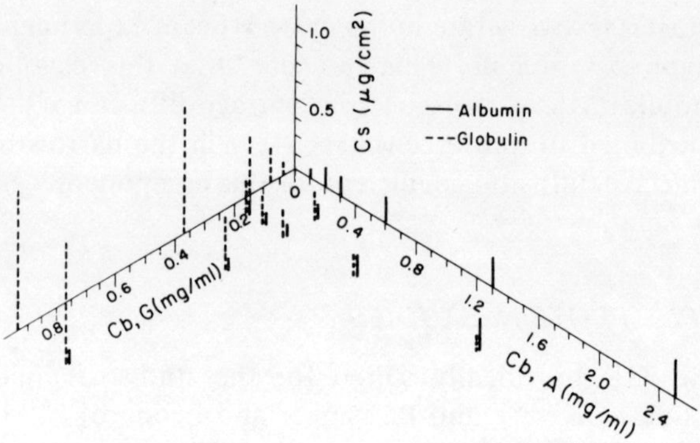

FIGURE 23. Experiments of Beissinger and Leonard.[42] Adsorption of albumin and gamma-globulin on quartz. Competitive adsorption isotherms are shown three-dimensionally and are based on 20-min binary kinetic adsorption data. Reprinted with permission.

attempt to model results from this multicomponent system using the parameters obtained from single-component experiments was only moderately successful. They, therefore, concluded that protein/protein interaction terms are required for multicomponent modeling.

4. TIRF FUTURE

TIRF has proved to be a useful tool in obtaining protein adsorption isotherm and rate results. The interpretation of these results in terms of the structure of the surface-adsorbed layer has thus far been speculative. Several possibilities exist, however, for extending the use of TIRF to retrieving structural information. These include surface fluorescence spectroscopy, fluorescence polarization, fluorescence energy transfer, layer-thickness measurements, and the use of quantum yield changes as a probe of adsorbed protein conformation.

4.1. FLUORESCENCE SPECTRA OF ADSORBED PROTEIN LAYERS (see Chapter 2)

A monochromator placed between the protein layer and the fluorescence detector can be used to obtain fluorescence spectra of the adsorbed proteins. Differences in the fluorescence spectra of adsorbed protein and native protein in solution would provide information about structural changes due to adsorption. One problem with applying this technique to visible TIRF using extrinsic labels is that the fluorescence spectra would be characteristic of the chromophore and not the macromolecule itself, and any features seen in those spectra would need to be interpreted accordingly.

4.2. FLUORESCENCE POLARIZATION

Fluorescence polarization has long been used to examine the orientation and mobility of macromolecules in solution.[68,69,71,72] It is a logical extension to apply this technique to surface studies. By examining the difference in how the adsorbed protein layer interacts with either a parallel or perpendicularly polarized incident beam, and measuring the anisotropy of the emitted fluorescence, some measure of the order and mobility of the adsorbed layer can be obtained. This may be useful, for example, in proving or disproving speculations in the literature regarding the nature of the bimodal isotherms. If the lower plateau of bimodal isotherms corresponds to a randomly adsorbled protein layer, and the upper plateau to a densely packed, "crystallized" monolayer, then the fluorescence polarization from the two surfaces should be different. This would also be useful in examining order on the surface induced by interactions of an external shear field with the adsorbed protein layer.[37]

4.3. FLUORESCENCE ENERGY TRANSFER

Fluorescence energy transfer is another technique that has long been applied to studies of macromolecule solutions.[68,69,73,74] Two fluorophores are used, a donor and an acceptor, chosen such that the emission spectrum of the donor overlaps the absorbance spectrum of the acceptor. By exciting the donor fluorescence and measuring the efficiency with which the donor fluorescence is absorbed by the acceptor which then fluoresces with its characteristic emission spectrum, the distance between the donor and acceptor fluorophores can be calculated. In extending this technique to TIRF surface studies, an estimate of intermolecular separation may be obtained. For example, by varying the ratio of donor-labeled, acceptor-labeled, and unlabeled protein in solution, adsorbed protein layers with varying distances between donor and acceptor can be obtained. The efficiency of energy transfer measured for these different adsorbed protein layers would be characteristic of the separation between donor and acceptor, which in turn would be a measure of intermolecular separation (Figure 24).[37] This would also be useful in proving or disproving the surface transition theory that is prevalent in the literature for explaining bimodal isotherms.

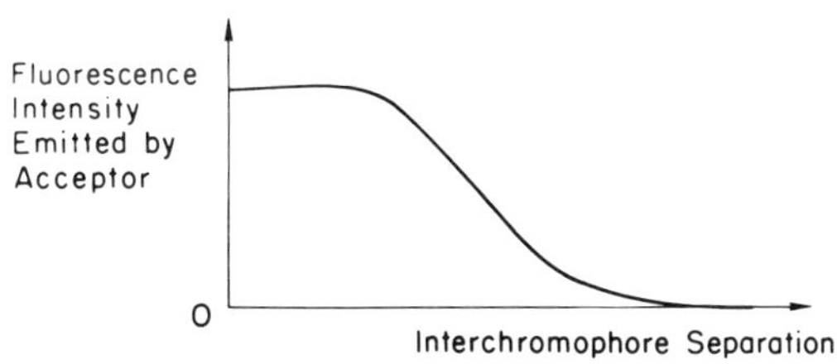

FIGURE 24. Fluorescence energy transfer efficiency vs. interchromophore separation.

4.4. QUANTUM YIELD AS A PROBE OF SURFACE STRUCTURE

The fluorescence quantum yield has been found to be highly sensitive to changes in the environment of the adsorbed protein layer. It changes with pH, ionic strength, temperature, the presence of shear fields, and surface macromolecule concentration.[37] As discussed earlier, it is important that the quantum yield remain constant for a given set of calibration experiments so that TIRF results can be quantitatively interpreted. It may be possible, however, to use these variations in quantum yield as a probe of perturbations in the structure of the surface-adsorbed protein layer. The important unsolved problem here involves trying to relate quantum yield changes to events occurring on a molecular level.

4.5. LAYER-THICKNESS MEASUREMENTS

The thicknesses of adsorbed polymer and protein layers are the most quantitative measures of "conformation" thus far reported in the literature. Thicknesses have been measured by ellipsometry,[75-77] quasi-elastic light scattering,[36,43,44] and permeability changes in capillaries or membrane pores[55,78,79] before and after adsorption. Theoretically, TIRF can also be used to measure layer thicknesses. Before adsorbing proteins onto the solid surface, a solution of nonadsorbing fluorescent molecule in the flow cell would emit a fluorescence signal proportional to the total amount of these molecules in the evanescent region. Then, after adsorbing a layer of unlabeled protein, and introducing the nonadsorbing molecule solution into the flow cell again, the fluorescence signal would now be proportional to the amount of these molecules in the evanescent field except for the region occupied by the adsorbed protein layer (Figure 25).[37] Knowing the depth of penetration of the evanescent field, the thickness of the protein layer can then be computed.

FIGURE 25. Layer-thickness measurements with TIRF. If there is no adsorbed layer on the surface, then a solution of nonadsorbing fluorescent molecules would occupy the entire evanescent region (checkered area and hatched area). The emitted fluorescence is therefore proportional to the depth of penetration. If a layer of unlabeled protein of thickness t is adsorbed on the surface, then a solution of nonadsorbing fluorescent molecules would be excluded from the region occupied by the adsorbed layer (checkered area), and the emitted fluorescence would consequently be lower than the clean surface signal. Knowing the depth of penetration of the evanescent field, the reduction in signal can be used to compute the thickness of the adsorbed layer.

5. CONCLUSIONS

The possible applications of TIRF to problems associated with protein adsorption are numerous. To date, only a limited number of studies have been reported. These studies have demonstrated the usefulness of TIRF in addressing some of the key issues in protein adsorption—reversibility, surface mobility, kinetics, and competitive adsorption. Direct investigations of adsorbed protein conformational changes using TIRF have not been reported, but several possibilities were described in the previous section. It is clear that TIRF is a useful tool with some unique advantages for protein adsorption studies, and its potential is not yet fully realized.

REFERENCES

1. H. P. Gregor and C. D. Gregor, Synthetic membrane technology, *Sci. Am.* **242** (7), 112–127 (1978).
2. A. S. Michaels, C. R. Robertson, and H. Reihanian, Evaluation Analysis and Prevention of the Fouling of Ultrafiltration Membranes by Macrosolutes, Office of Water Research and Technology Comprehensive Summary Report (June, 1980), OWRT Contract No. 14-34-001-0411.
3. A. S. Michaels and K. E. Lange, A Study of Macrosolute Transport Mechanisms in Asymmetric Ultrafiltration Membranes. I. Surface Contamination of Asymmetric Ultrafiltration Membranes During Water Filtration: Consequences and Prevention Techniques. II. Adsorption of Bovine Albumin by a Dynel Ultrafiltration Membrane, and Its Effects on Membrane Hydraulic Permeability and Solute Rejection, Office of Water Research and Technology Interim Report (February 28, 1981), OWRT Contract No. 14-34-0001-8548.
4. D. C. MacWilliams, J. H. Rogers, and J. J. West, in: *Water Soluble Polymers* (N. M. Bikales, ed.), Plenum Press, New York (1973).
5. M. A. Packham, G. Evans, M. F. Glynn, and J. F. Mustard, The effect of plasma proteins on the interaction of platelets with glass surfaces, *J. Lab. Clin. Med.* **73**, 686–697 (1969).
6. L. Vroman and A. L. Adams, Identification of rapid changes at plasma-solid interfaces, *J. Biomed. Materials. Res.* **3**, 43–67 (1969).
7. L. Vroman, A. L. Adams, and M. Klings, Interactions among human blood proteins at interfaces, *Fed. Proc.* **30**, 1484 (1971).
8. L. Vroman, A. L. Adams, M. Klings, and G. Fischer, Fibrinogen, globulins, albumin and plasma at interfaces, *Adv. Chem. Ser.* **145**, 255–289 (1975).
9. M. B. Zucker and L. Vroman, Platelet adhesion induced by fibrinogen adsorbed onto glass, *Proc. Soc. Exp. Biol. Med.* **131**, 318–320 (1969).
10. G. E. Stoner, S. Srinivasan, and E. Gileadi, Adsorption inhibition as a mechanism for the antithromobogenic activity of some drugs. I. Competitive adsorption of fibrinogen and heparin on mica, *J. Phys. Chem.* **75**, 2107–2111 (1971).
11. J. D. Andrade, Interfacial phenomena and biomaterials, *Medical Instrumentation* **7**, 110 (1973).
12. P. K. Weathersby, T. A. Horbett, and A. S. Hoffman, Fibrinogen adsorption to surfaces of varying hydrophilicity, *J. Bioeng.* **1**, 395–410 (1977).
13. S. A. Barenberg, J. S. Schultz, J. M. Anderson, and P. H. Geil, Hemocompatability: macromolecular motions and order of the polymer interface, *Trans. Am. Soc. Art. Int. Org.* **25**, 159–166 (1979).

14. S. D. Bruck, Macromolecular aspects of biocompatability—a review, *J. Biomed. Mater. Res.* **6**, 173–183 (1972).
15. R. C. Eberhart, M. E. Lynch, F. H. Bilge, and H. A. Arts, Effects of fluid shear and temperature on protein adsorption on Teflon surfaces, *Trans. Am. Soc. Art. Int. Org.* **26**, 185–193 (1980).
16. P. Bagchi and S. M. Birnbaum, Effect of pH on the adsorption of immunoglobulin G on anionic poly(vinyltoluene) model Latex particles, *J. Colloid Interface Sci.* **83**, 460–478 (1981).
17. F. E. Eirich, The conformational states of macromolecules adsorbed at solid-liquid interfaces, *J. Colloid Interface Sci.* **58**, 423–436 (1977).
18. A. W. Adamson, *Physical Chemistry of Surfaces*, Interscience Publishers, Inc., New York (1960).
19. S. Brunauer, L. S. Deming, W. E. Deming, and E. Teller, On a theory of van der Waals adsorption of gases, *J. Am. Chem. Soc.* **62**, 1723–1732 (1940).
20. I. Langmuir, *J. Am. Chem. Soc.* **60**, 309 (1938).
21. B. Vincent and S. G. Whittington, Polymers at interfaces and in disperse systems, in: *Surface and Colloid Science* (E. Matijevic, ed.), Vol. 12, pp. 1–117, Plenum Press, New York (1982).
22. V. Hlady and H. Furedi-Milhofer, Adsorption of human serum albumin on precipitated hydroxyapatite, *J. Colloid Interface Sci.* **69**, 460–468 (1979).
23. M. E. Soderquist and A. G. Walton, Structural changes in proteins adsorbed on polymer surfaces, *J. Colloid Interface Sci.* **75**, 386–397 (1980).
24. W. Norde and J. Lyklema, The adsorption of human plasma albumin and bovine pancreas ribonuclease at negatively charged polystyrene surfaces: I. Adsorption isotherms. Effects of charge, ionic strength, and temperature, *J. Colloid Interface Sci.* **66**, 257–265 (1978).
25. W. Norde and J. Lyklema, The adsorption of human plasma albumin and bovine pancreas ribonuclease at negatively charged polystyrene surfaces: II. Hydrogen ion titrations, *J. Colloid Interface Sci.* **66**, 266–276 (1978).
26. W. Norde and J. Lykema, The adsorption of human plasma albumin and bovine pancreas ribonuclease at negatively charged polystyrene surfaces: III. Electrophoresis, *J. Colloid Interface, Sci.* **66**, 277–284 (1978).
27. W. Norde and J. Lyklema, The adsorption of human plasma albumin and bovine pancreas ribonuclease at negatively charged polystyrene surfaces: IV. The charge distribution in the adsorbed state, *J. Colloid Interface Sci.* **66**, 285–294 (1978).
28. W. Norde and J. Lyklema, The adsorption of human plasma albumin and bovine pancreas ribonuclease at negatively charged polystyrene surfaces: V. Microcalorimetry, *J. Colloid Interface Sci.* **66**, 295–302 (1978).
29. T. Mizutani, Adsorption of proteins on silicone-coated glass surfaces, *J. Colloid Interface Sci.* **82**, 162–166 (1981).
30. P. Van Dulm, W. Norde, and J. Lyklema, Ion participation in protein adsorption at solid surfaces, *J. Colloid Interface Sci.* **82**, 77–82 (1981).
31. W. Norde and J. Lyklema, Thermodynamics of protein adsorption: theory with special reference to the adsorption of human plasma albumin and bovine pancreas ribonuclease at polystyrene surfaces, *J. Colloid Interface Sci.* **71**, 350–366 (1979).
32. E. Nyilas, T-H. Chiu, and G. A. Herzlinger, Thermodynamics of native protein/foreign surface interactions: I. Calorimetry of the human r-globulin/glass system, *Trans. Am. Soc. Art. Int. Org.* **20**, 480–490 (1974).
33. T-H. Chiu, E. Nyilas, and L. R. Turcotte, Microcalorimetric and electrophoretic studies of protein sorption from plasma, *Trans. Am. Soc. Art. Int. Org.* **24**, 389–402 (1978).
34. J. L. Brash and Q. M. Samak, Dynamics of interactions between human albumin and polyethylene surface, *J. Colloid Interface Sci.* **65**, 495–504 (1978).

35. H. Y. K. Chuang, W. F. King, and R. G. Mason, Interaction of plasma proteins with artificial surfaces: protein adsorption isotherms, *J. Lab. Clin. Med.* **92**, 483–496 (1978).
36. B. D. Fair and A. M. Jamieson, Studies of protein adsorption on polystyrene latex surfaces, *J. Colloid Interface Sci.* **77**, 525–534 (1980).
37. Y.-L. Cheng and C. R. Robertson, unpublished results.
38. N. L. Thompson, T. P. Burghardt, and D. Axelrod, Measuring surface dynamics of biomolecules by total internal reflection fluorescence with photobleaching recovery or correlation spectroscopy, *Biophys. J.* **33**, 435–454 (1981).
39. T. P. Burghardt and D. Axelrod, Total internal reflection/fluorescence photobleaching recovery study of serum albumin adsorption dynamics, *Biophys. J.* **33**, 455–468 (1981).
40. R. L. Beissinger and E. F. Leonard, Immunoglobulin sorption and desorption rates on quartz: evidence for multiple sorbed states, *Am. Soc. Art. Int. Org. J.* **3**, 160–175 (1980).
41. R. L. Beissinger and E. F. Leonard, Plasma protein adsorption and desorption rates on quartz: approach to multi-component systems, *Trans. Am. Soc. Art. Int. Org.* **27**, 225–230 (1981).
42. R. L. Beissinger and E. F. Leonard, Sorption kinetics of binary protein solutions: general approach to multicomponent systems, *J. Colloid Interface Sci.* **85**, 521–533 (1982).
43. B. W. Morrissey and C. C. Han, The conformation of γ-globulin adsorbed on polystyrene latices determined by quasielastic light scattering, *J. Colloid Interface Sci.* **65**, 423–431 (1978).
44. E. E. Uzgiris and H. P. M. Fromageot, Thickness and density of protein films by optical mixing spectroscopy, *Biopolymers* **15**, 257–263 (1976).
45. M. A. Cohen Stuart, J. M. H. M. Scheutjens, and G. J. Fleer, Polydispersity effects and the interpretation of polymer adsorption isotherms, *J. Polym. Sci., Polym. Phys. Ed.* **18**, 559–573 (1980).
46. L. K. Koopal, The effect of polymer polydispersity on the adsorption isotherm, *J. Colloid Interface Sci.* **83**, 116–129 (1981).
47. K. Furusawa, K. Yamashita, and K. Konno, Adsorption of monodisperse polystyrene on porous glass. I. Preferential adsorption and displacement of high-molecular weight species, *J. Colloid Interface Sci.* **86**, 35–42 (1982).
48. R. G. Lee, C. Adamson, and S. W. Kim, Competitive adsorption of plasma proteins onto polymer surfaces, *Thrombosis Research* **4**, 485–490 (1974).
49. S. W. Kim, R. G. Lee, C. Adamson, and D. J. Lyman, Competitive adsorption of plasma proteins onto polymer surfaces in: *Polymer Science and Technology* (H. P. Gregor, ed.), Vol. 7: Biomedical application of polymers, pp. 69–74, Plenum Press, New York (1975).
50. R. F. Cerf and H. A. Scheraga, Flow birefringence in solutions of macromolecules, *Chem. Rev.* **51**, 185–261 (1952).
51. R. B. Bird, O. Hassager, R. C. Armstrong, and C. F. Curtiss, *Dynamics of Polymeric Liquids*, Vol. 1: Fluid mechanics, John Wiley & Sons, Inc., New York (1977).
52. C. Levinthal and P. F. Davison, *J. Mol. Biol.* **3**, 674–683 (1961).
53. E. L. Slagowski, L. J. Fetters, and D. McIntyre, Upper molecular weight limit for the chracterization of polystyrene in gel permeation chromatography, *Macromolecules* **7**, 394 (1974).
54. G. G. Fuller and J. J. Lee, Department of Chemical Engineering, Stanford University, personal communication, 1982.
55. D. P. Yavorsky and J. A. Quinn, Static and hydrodynamic studies of adsorbed protein conformation at the solid/liquid interface, presented at the session on Recent Developments in Colloidal Phenomena, AIChE 74th Annual Meeting, New Orleans, November 8–12, 1981.
56. R. W. Watkins and C. R. Roberston, A total internal reflection technique for the examination of protein adsorption, *J. Biomed. Materials. Res.* **11**, 915 (1977).

57. N. J. Harrick, *Internal Reflection Spectroscopy*, Interscience Publishers, Wiley & Sons, New York (1967).
58. This volume, Chapter 2.
59. G. K. Iwamoto, R. A. Van Wagenen, and J. D. Andrade, Insulin adsorption: intrinsic interfacial fluorescence, *J. Colloid Interface Sci.* **86**, 581–585 (1982).
60. R. A. Van Wagenen, S. Rockhold, and J. D. Andrade, Probing protein adsorption. Total internal reflection intrinsic fluorescence, in: *Biomaterials: Interfacial Phenomena and Applications* (S. L. Cooper and N. A. Peppas, eds.), *Adv. Chem. Ser.* **199** (1981), pp. 351–370.
61. N. J. Harrick and G. I. Loeb, Internal reflection fluorescence, *Anal. Chem.* **45**, 687 (1973).
62. R. W. Watkins, A Total Internal Reflection Fluorescence Technique for the Study of Protein Adsorption, Ph.D. Dissertation, Stanford University (1976).
63. B. K. Lok, Protein Adsorption onto Cross-linked Polydimethyl Siloxane Using Total Internal Reflection Fluorescence, Ph.D. Dissertation, Stanford University (1981).
64. B. K. Lok, Y.-L. Cheng, and C. R. Robertson, Total internal reflection fluorescence: a technique for examining interactions of macromolecules with solid surfaces, *J. Colloid Interface Sci.* **91**, 87–103 (1983).
65. B. K. Lok, Y.-L. Cheng, and C. R. Robertson, Protein adsorption on grosslinked polydimethylsiloxane using total internal reflection fluorescence, *J. Colloid Interface Sci.* **91**, 104–116 (1983).
66. E. Brynda, J. Drobnik, J. Vacik, and J. Kalal, Protein sorption on polymer surfaces measured by fluorescence labels, *J. Biomed. Materials Res.* **12**, 55–65 (1978).
67. A. van der Scheer, J. Feijen, J. Klein Elhorst, P. G. L. C. Krugers Dabneaux, and C. A. Smolders, The feasibility of radiolabeling for human serum albumin (HSA) adsorption studies, *J. Colloid Interface Sci.* **66**, 136–145 (1978).
68. D. Freifelder, *Physical Biochemistry. Applications to Biochemistry and Molecular Biology*, W. H. Freeman & Co., San Francisco (1976).
69. C. R. Cantor and P. R. Schimmel, *Biophysical Chemistry, Part II: Techniques for the Study of Biological Structure and Function*, W. H. Freeman & Company, San Francisco (1980).
70. M. Lévêque, in: *Ann. Mines* **13**, 14 (1928).
71. J. Yguerabide, H. F. Epstein, and L. Stryer, Segmental flexibility in an antibody molecule, *J. Mol. Biol.* **51**, 573–590 (1970).
72. G. Weber, Polarization of the fluorescence of macromolecules: 1. Theory and experimental methods; 2. Fluorescent conjugates of ovalbumin and bovine serum albumin, *Biochem. J.* **51**, 145–155, 155–167 (1952).
73. L. Stryer, Fluorescence spectroscopy of proteins, *Science* **162**, 526–540 (1968).
74. L. Stryer, Fluorescence energy transfer as a spectroscopic ruler, *Ann. Rev. Biochem.* **47**, 819–846 (1978).
75. B. W. Morrissey, L. E. Smith, R. R. Stromberg, and C. A. Fenstermaker, Ellipsometric investigation of the effect of potential on blood protein conformation and adsorbance, *J. Colloid Interface Sci.* **56**, 557–563 (1976).
76. P. A. Cuypers, W. T. Hermens, and H. C. Hemker, Ellipsometric study of protein film on chromium, *Ann. N.Y. Acad. Sci.* **283**, 77–85 (1979).
77. R. B. Davis, R. M. A. Azzam, and G. Holtz, Ellipsometric observations of surface adsorption and molecular interactions of native and modified fibrinogen and factor VIII, *Surface Sci.* **96**, 539–554 (1980).
78. R. Varoqui and P. Dejardin, Hydrodynamic thickness of adsorbed polymers, *J. Chem. Phys.* **66**, 4395–4399 (1977).
79. Ph. Gramain and Ph. Myard, Adsorption studies of polyacrylamides in porous media, *J. Colloid Interface Sci.* **84**, 114–126 (1981).

4

Surface Enhanced Raman Spectroscopy of Biological Macromolecules

Therese M. Cotton

1. INTRODUCTION

This review summarizes applications of the newly developed spectroscopic technique, surface enhanced Raman scattering (SERS), to the study of biomolecules. Surface enhanced Raman scattering is the giant increase in Raman scattering intensity observed when molecules are adsorbed onto certain metal surfaces. The enhancement can be as large as 10^6 and has been the most extensively characterized for surfaces of Ag, Cu, and Au. To present the current status of the field in proper perspective, a review of the discovery of this effect will be given, followed by a brief summary of the theories which have been proposed to explain the effect. The origin of the enhancement process is not yet fully understood, and an attempt will be made to indicate those areas which are still controversial. This should enable the reader to appreciate the fact that the full potential of SERS for spectroscopic analysis of molecules at surfaces cannot be realized without further experimental and theoretical elaboration. Because both the experimental and theoretical aspects of the field are currently in a state of flux, it might be considered premature to attempt to apply SERS to complex biomolecules. This does not appear to be the case, however, as much valuable information has already been obtained from these complex systems. Evidence to support this point will be given in the examples of the various types of information

Therese M. Cotton • Department of Chemistry, Illinois Institute of Technology, Chicago, Illinois 60616. The author's present address is Department of Chemistry, University of Nebraska, Lincoln, Nebraska 68588.

obtained to date. In addition, those areas which are in need of further research will be underscored.

Many of the SERS studies which have been undertaken thus far have focused on molecules adsorbed onto electrode surfaces. The relevance of such an experimental approach to the study of biomolecules might be questioned at this juncture. In response to this point, it should be noted that many biological reactions occur at membrane surfaces and these often involve electrical phenomena at the morphological level (e.g., cell subdivision, organization, and differentiation of tissues) or the physicochemical level (e.g., respiration, muscle function, photoelectric reactions in vision and photosynthesis, neural impulse transmission, membrane function in ion transport, and anabolic and catabolic processes concerned with energy storage or utilization). Consequently, scientists have long recognized the relevance of electrochemical theory and methods in providing an understanding of these processes *in vivo* and in model systems. The enormous scope of the field of bioelectrochemistry attests to the relevance of this approach. In general, bioelectrochemistry can be subdivided into five major categories of study[1]:

(1) electrochemical stimulation of living systems;
(2) electrochemical analysis of biological systems;
(3) electrochemical studies of membrane biophysics;
(4) faradaic reactions of biological macromolecules;
(5) faradaic reactions involving small molecules.

The first category has practical applications, including the use of direct (implanted electrodes) or indirect (induced electric fields produced by external electrodes) electrochemical stimulation of bone growth or tissue regeneration.[2] Examples of studies in the second category include the analytical determination of redox active molecules in whole organisms, tissues, cells, or cellular fluids, such as the determination of neurotransmitter concentration in rat brain[3] and the use of enzyme electrodes to detect low concentrations of substrates.[4] There are many examples of electrochemical studies relating to membrane biophysics, category 3. This approach has been popular because many of the important functions of membranes can be analyzed in terms of electrochemical observables such as surface charge effects, capacitance, potential gradients, and ion transport.[5] This category can also be extended to encompass the effects of surfaces in general on the behavior of biomolecules. In addition to natural membrane surfaces, polymer, metal, and inorganic surfaces are important in many biomedical applications. The interaction of proteins with these surfaces in prosthetic devices and implants determines their acceptability. Finally, categories four and five of bioelectrochemistry include studies of electron transfer reactions in redox proteins and their isolated prosthetic groups.[6,7]

Recent literature contains examples of the application of SERS to the study of several different classes of biomolecules adsorbed onto electrode surfaces. Nucleic acid bases, DNA, proteins, and prosthetic groups have all been examined. As will be seen, the main advantages associated with the use of this technique include the enormous increase in sensitivity, the experimental flexibility, rejection of fluorescence background, which is so often associated with complex molecules, and the availability of molecular information at surfaces.

2. BACKGROUND

2.1. HISTORICAL PERSPECTIVE

Approximately ten years ago, Fleischmann and coworkers[8] undertook experiments to study the electrode–electrolyte interface by Raman spectroscopy. The advantages of using Raman methods have been cited frequently[9] and one of the most important in this endeavor is the lack of interference by water, the solvent used in the electrochemical studies. In addition, the vibrational information which results may be used as a fingerprint to identify molecules and to determine information about their environment. However, a disadvantage of Raman spectroscopy, as compared to infrared spectroscopy, which also gives vibrational information and has been used to study molecules at surfaces,[10] is that it is relatively insensitive. The quantum yield for Raman scattering is extremely low and often sample concentrations as high as 1 *M* are required. Thus, the results of the initial experiments of the Fleischmann group were not unexpected with only weak Raman signals (signal-to-noise ratio of 30) observed for several molecular layers of mercury halides adsorbed onto a mercury-coated Pt electrode. Since these compounds are strong Raman scatterers, these researchers recognized that methods for increasing the signal intensity were needed in order for this approach to be generally useful. In a subsequent study, they reported Raman spectra for pyridine adsorbed on a Ag electrode and these results were quite surprising in that a very strong Raman spectrum was observed following an electrochemical anodization of the Ag.[11] The spectrum was thought to arise from a monolayer of adsorbed pyridine and the increased Raman intensity was attributed to an increase in the surface area as a result of the anodization procedure.

In retrospect, it is apparent that the choice of electrode material by Fleischmann *et al.* was extremely fortuitous. Jeanmaire and Van Duyne[12] repeated these experiments and observed that an increase in surface area could not in itself account for the strong Raman signals exhibited by pyridine on Ag, since an even stronger spectrum resulted when the extent of surface roughening was decreased. It was calculated that an enhancement in Raman scattering intensity of approximately six orders of magnitude was exhibited by the surface-bound pyridine. An increase in the polarizability of the pyridine molecules near the electrode surface as a result of the large electric field gradient was suggested as the possible origin of the enhancement effect. Albrecht and Creighton reported at about the same time an enhancement of 10^5 in Raman scattering intensity for pyridine on Ag.[13] Similar observations by other groups followed rapidly.[14,15]

2.2. SURFACE ENHANCED RAMAN SCATTERING THEORIES

Since the first observations of an enhancement effect for certain molecules on Ag, there has been a rapid growth in the number of research groups involved in the investigation of this effect. A large fraction of the research effort has been directed toward developing an understanding of the physical basis for the enhancement mechanism, and the experimentalists have concentrated on relatively simple molecules (most often pyridine). A number of theories have been proposed and these are continuously being revised as new data become available. These have been reviewed recently by Metiu[16] and categorized into two groups: (1) purely electrodynamic mechanisms (PEDM); and (2) purely molecular mechanisms (PMM). This categorization is based upon the origin of normal Raman scattering, which arises from the product of the electric field experienced by a molecule and its polarizability.

The first category, PEDM, applies to those theories which invoke local field enhancement to explain the anomalously strong Raman signals resulting from molecules such as pyridine adsorbed onto Ag. The local field results from the laser electric field, as modified by the surface, as well as the image field which is induced in the metal by the molecular dipole at some distance from the surface. The laser field can be enhanced through excitation of electromagnetic resonances in the metal, or by a high surface curvature. The experimental predictions of the variants of this theory include the following:

(1) long-range enhancement of molecules many angstroms from the surface of the metal;
(2) minimal enhancement for molecules adsorbed on flat surfaces;
(3) selection rules which show the greatest enhancement for those highly polarizable modes which are perpendicular to the surface;
(4) frequency dependence of the enhancement factor, EF, which is determined by the metal and not the adsorbate, i.e., metals with a small imaginary dielectric constant should produce the largest EF.

The theories proposed in the second category, PMM, claim that the strong Raman signals resulting from molecules adsorbed on certain metals arise from a profound modification of the molecular polarizability. Examples of these types of theories include the image field effect (IFE) and those which invoke vibrational modulation of electrons in the metal through charge transfer or nuclear vibrations. As indicated by Metiu,[16] experimental predictions are difficult to deduce from PMM theories. An exception to this is the IFE which predicts experimental observables in considerable detail. For example, this theory predicts that enhancement should depend strongly upon the distance of the molecule from the metal surface, with the greatest enhancements resulting from molecules which are less than two angstroms from the metal surface. The theory also predicts the excitation frequency dependence of the EF and the predicted relationship agrees well with experimentally observed excitation profiles.[17]

At the present time, many theorists would agree that the SER effect probably arises from a complex interplay of several of the proposed mechanisms. The controversy appears to reside primarily with allocating the relative contributions of the various mechanisms to the overall effect.

2.3. CORRELATIONS BETWEEN EXPERIMENTAL RESULTS AND THEORIES

None of the experimental data available at present supports completely any of the major theories proposed thus far to explain the SER effect. This deficiency is partly due to the limitations of the various theories, but also to the choice of experimental systems examined for the most part. With respect to the former, electrodynamic theories have utilized phenomenological Maxwell's equations at close distances to the metal surface and have employed bulk dielectric constants for small metallic particles.[16] Neither approach may be applicable to the surface properties of the metal. With respect to the choice of systems, many experimentalists have concentrated their efforts on solid surfaces, both roughened and smooth, whereas the electromagnetic theories can best treat small solid particles, such as colloids. In addition, there are many basic experimental questions remaining to be answered. Those most relevant to this review include the following. What metals exhibit the enhancement effect? Are other surfaces, such as semiconductors, capable of producing enhanced Raman scattering from adsorbed molecules under the appropriate experimental conditions (e.g., laser excitation wavelength)? Do all molecules experience an enhancement or is the effect limited to certain classes of molecules? Is a chemical interaction between the adsorbate and the surface necessary for the effect to occur?

Experimental results which bear on the above questions are often controversial. Thus far, the metals which are commonly accepted to exhibit enhancement include Ag, Cu, and Au.[18] There have been reports of enhancement on other metals and surfaces, including Hg,[19] Ni,[20,21] Pt,[21,22]

Pd,[21] Li,[23] TiO_2, NiO,[24] Al,[25] and In.[26] The results on Hg have not been reproduced, however.[27] The method of pretreatment of the metal appears to be an important factor. Most investigators agree that, in the case of an electrode, anodization both cleans the surface as well as provides an increase in surface roughness. The contribution of surface roughness to the magnitude of the enhancement is under dispute, however. Schultz *et al.*[28] attribute only an increase of 100-fold in Raman scattering intensity to surface roughness, while attributing the major source of enhancement to the increase in polarizability of the molecule at the electrode surface as a result of the image field effect.[17]

Currently, SERS studies indicate that only some molecules exhibit enhanced Raman scattering when adsorbed onto SERS-active metals. At least, enhancement factors vary considerably for different classes of molecules. Otto and coworkers[27] regard the lack of enhancement for certain molecules, such as hydrocarbons, as evidence for the requirement of an interaction between the adsorbed molecule and special active sites on the metal surface. These active sites are proposed to consist of atomic-scale roughness, or adatoms which chemically interact with the adsorbate. The specification of active sites distinguishes the adatom model from other PMM which have been proposed.

The difficulties encountered in attempting to compare the various experimental SERS results can largely be attributed to the numerous experimental parameters which influence the Raman intensity enhancement. These include, in addition to the electrode material and its pretreatment, the electrode potential, the nature and concentration of the supporting electrolyte, the bulk solution concentration of the adsorbate, and the laser excitation wavelength. In those cases where the adsorbate is photosensitive, the laser power also affects the SERS results.[29]

2.4. TYPES OF SUBSTRATES

The observation of SERS has not been limited to metal electrodes. Enhancement has also been observed for pyridine on Ag and Au sols,[30,31] citrate on Ag sols,[32] $Au(CN)_2$ on Au particles,[33] and halide ions on Ag and Au sols.[34] Sols are important in that they can be used to test electrodynamic theory which has been well-developed for spheres of varying sizes and optical constants.[35] Thus far one of the most important experimental observables in this respect is the SERS excitation profile. A number of experimental results disagree with the theoretical prediction that maximum SERS enhancement should occur at the same excitation wavelength as the absorption maximum of colloidal particles less than 5 nm. The failure to obtain experimental agreement with the theory suggests that another mechanism may be operative in addition to the electrodynamic one.[32,33]

One possibility which has been proposed[33] is that colloids contain grain boundaries or defects which result in chemical interactions with the adsorbate similar to those proposed in the adatom theory.[27] In any case, the colloidal state offers other advantages besides those relating to tests of theory. These include control over experimental parameters, such as the ability to control and vary particle size, and experimental simplicity (ease of preparation and the ability to flow the samples in the laser beam and thereby minimize possible photodamage).

Other types of metal substrates which have been shown to exhibit enhancement of Raman scattering for adsorbed molecules include matrix-isolated metal clusters,[37] powders,[38] metal-capped polymer posts,[39] holographic metal gratings,[40] vacuum-deposited metal films, including continuous[41,42] and island structures,[42b] smooth metal surfaces by means of attenuated total reflectance techniques,[43] single crystals,[44] and tunnel junction structures.[45] This list is not meant to be exhaustive, but rather to indicate that the enhancement process, whatever its origin, is not limited to electrode surfaces. It has been observed under a variety of experimental conditions. Because of this proven diversity of substrates, it should be feasible to observe enhancement of Raman scattering for molecules on other types of substrates by overlayering the surface with thin silver (or other SERS-active metal) films. Therein lies the potential for using this method to study complex biological systems such as proteins, membrane fragments, and cell surfaces. Moreover, this approach should prove useful in detecting adsorbed species on surfaces used in biomedical applications (e.g., polymeric and metallic implants). Of course, the full range of information which can be gleaned from this approach cannot be predicted until the fundamental mechanism is more fully understood. At the very least, however, it should be possible to identify species on surfaces by means of their characteristic "fingerprints." Through the enhancement process, surface coverages on the order of monolayers should be detectable, a feat which is not readily achieved by other surface-sensitive methods.

2.5. COMBINED SURFACE ENHANCED AND RESONANCE RAMAN EFFECTS

The Raman scattering cross section of a molecule can be increased substantially (as much as seven orders of magnitude) by utilizing a laser excitation frequency which is in resonance with an electronic transition in that molecule. This was predicted from Raman scattering theory and has been demonstrated experimentally for a very large number of molecules.[46] Indeed, resonance Raman (RR) scattering has proven to be a valuable spectroscopic method for probing chromophores in biological molecules, since these often possess strong absorption bands in the visible region of the electromagnetic spectrum.[47] An important advantage of RR spectroscopy over normal Raman (NR) for this purpose is the ability to excite selectively a given chromophore in a complex milieu (e.g., protein matrix or membrane) by means of a judicious choice of laser excitation frequency.

The increased sensitivity of RR scattering has been used advantageously to study molecules adsorbed on surfaces as well. Examples of applications to electrode and semiconductor studies include the detection of methylene blue on SnO_2,[48] rose bengal on ZnO,[49] and iron phthalocyanine on glassy carbon.[50] These studies illustrate that it is possible to determine the RR spectra of very small amounts (*ca.* monolayers) of colored species on surfaces. What detection level might then be expected from combined resonance Raman and surface enhanced Raman effects? This question cannot be answered quantitatively at this time, but some examples of molecules already studied will be given below with the estimated enhancement factors. As with SERS, there is controversy concerning both the theoretical predictions and the experimental observations reported thus far. Much of the uncertainty surrounding the experimental results could be eliminated or at least minimized if accurate surface coverage measurements were reported with the Raman results. Of course, this information is needed in SERS experiments as well, but it should be more readily acquired in the case of absorbing molecules by means of absorption spectroscopy.

Jeanmaire and Van Duyne[12] were the first to report a surface enhanced Raman spectrum for molecules in resonance (hereafter referred to as surface enhanced resonance Raman scattering, SERRS). Strong spectra were observed for both methyl orange and crystal violet adsorbed onto an anodized Ag electrode, with only 1 mW of laser power. No attempt was made to determine the enhancement factor in these experiments, but it was concluded that the surface enhanced and resonance effects were multiplicative.[29] A recent study challenges this conclusion.[51] In the latter investigation, mixtures of a dye which is resonance enhanced and a non-resonance enhanced molecule, either citrate or pyridine, were co-adsorbed onto Ag sols. A comparison was made of the intensity changes in the surface spectra of the two types of molecules with changes in the bulk solution concentration of the dye, methyl red. It was concluded that the net enhancement factor was the same for both methyl red and pyridine (or citrate) when co-adsorbed at approximately the same surface coverage. However, the increase in Raman scattering for pyridine resulting from the SERS effect was determined to be about 10^5. This is equal to the resonance enhancement factor observed for methyl red in solution. Thus, it appears that the methyl red experiences only the traditional resonance enhancement and no surface enhancement effect. Other dyes were found to exhibit similar behavior, including crystal violet and methyl orange. These results illustrate the need for experiments in which quantitative measurements are made of both the surface coverage and enhancement factors for dyes or other resonance enhanced molecules. Without such information, no definite conclusions can be drawn concerning the relative contributions of RR and surface enhanced Raman enhancement mechanisms to colored species.

Although many of the reported studies of resonance enhanced molecules at surfaces have not quantitated either the coverage or the EF's, it is, nevertheless, useful to examine the types of information which have been obtained for the different systems. The substrates have included primarily electrodes, colloids, and island films of Ag, Cu, or Au. Yeager and coworkers[(52)] have utilized SERRS techniques to obtain spectra of cobalt phthalocyanine adsorbed onto a Ag electrode as a function of electrode potential. Their objective was to determine information about the structural interactions of the CoPC with the electrode surface. Four models were proposed on the basis of the vibrational spectra, which were found to vary with electrode potential. At the most positive potential accessible on Ag (+0.2 V vs. SCE) the spectrum of the CoPC found at the surface resembled that of the solution species, indicating only weak interaction between the adsorbate and the electrode surface. Spectra observed at potentials near −0.2 V vs. SCE were similar to those observed for oxygenated CoPC in solution, which is believed to exist as a dimer. At very negative potentials (−0.7 V vs. SCE) the spectrum changed drastically and the authors proposed that the macrocycle may undergo ring cleavage under these conditions.

Free base phthalocyanine has been studied on Ag metal films and an enhancement factor of 10^3 was determined by comparing the spectrum observed for PC adsorbed on a glass substrate with that seen on Ag.[(53)] It was noted that surface roughness in the form of metal islands was crucial for the observation enhancement; thick (200 Å) relatively smooth Ag films gave no surface enhancement. Although spectra were not shown, it was reported that PC exhibits SERRS on indium substrates as well.

Lippitsch[(54)] has studied the bile pigments biliverdene and pyrromethenone adsorbed onto Ag colloids. These molecules are thought to play an important role in photochemical processes of biological relevance and spectroscopic methods should be useful in reaction pathway studies. Conventional RR techniques cannot be used to obtain spectra of these molecules, due to their high fluorescence which obscures the weaker RR signals. Adsorption of these species onto colloids resulted in fluorescence quenching and the SERRS spectra were observed in the presence of pyridine. An enhancement factor was not calculated, but it was concluded that it was quite large. This conclusion was based upon the fact that the bulk solution concentration of pyridine was three orders of magnitude greater than that of the bile pigments but its spectrum was not observed. This conclusion could be unwarranted, however, if the bile pigments have significantly greater surface concentrations than the pyridine.

2.6. ENHANCEMENT OF FLUORESCENCE

Molecules adsorbed on Ag, Cu, and Au have been observed to exhibit enhancement of other photophysical processes besides Raman scattering. These include absorption, fluorescence, and nonlinear processes.[55] The relationship between fluorescence and Raman enhancement is important in those cases where fluorescence quenching is desirable. Vacuum-deposited metal films have been particularly useful for comparing the relative enhancement factors for these two processes. Both fluorescence and resonance Raman scattering were found to be enhanced for the dyes fluorescein isothiocyanate and rhodamine 6G on Ag island films.[56] The maximum EF for both processes was found to depend upon film thickness. A thickness of 40 Å produced a sixfold increase in rhodamine 6G fluorescence, whereas an optimum resonance Raman enhancement as high as 10^5 was observed. It was proposed that the fluorescence intensity was not increased as much as the Raman intensity due to lifetime shortening near the metal. The lifetime shortening arises from two factors—energy transfer to the metal and an increase in the rate of radiative emission for the molecules near the metal islands. The enhancement mechanism at the island films was attributed partly to the excitation of transverse collective electron resonances and partly to the radiating electric moment induced in the metal islands by the Raman-excited molecules. Both processes enhance the effective electric field at the molecule. Optimal enhancement was found to depend upon maximum overlap between the excitation and emission frequencies of the films as well as upon the dielectric and structural properties. Film thicknesses greater than 120 Å produced smooth surfaces and yielded EF's which were smaller by three to four orders of magnitude.

Similar results were obtained for the dependence of fluorescence and resonance Raman enhancement on film thickness for the dyes rhodamine B and Nile Blue on Cu, Ag, and Au films.[57] Silver island films were found to increase the fluorescence quantum yield for the dye basic fuchsin by a factor of ten, from 0.01 to 0.1, whereas an increase of only a factor of three was found for rhodamine 6G.[58] Enhancement of RR scattering was generally observed to vary between 10^2 to 10^4 for these two dyes and was dependent upon film thickness and the excitation wavelength. It was concluded that molecules with a high quantum efficiency for fluorescence experience less enhancement than those with low quantum efficiency. The RR enhancement for molecules with high quantum efficiency can be as high as 10^5, however, and this difference in relative enhancement for the two photoprocesses can be used to advantage to overcome fluorescence background problems.

3. APPLICATION OF SERS AND SERRS TO ORGANIC AND BIOLOGICALLY IMPORTANT MOLECULES

A number of small organic molecules besides pyridine have been studied by SERS. Examples of the kind of information obtainable will be summarized from several different investigations. In some cases frequency shifts were observed in the Raman bands for molecules in the adsorbed state and these were interpreted in terms of the nature of the adsorbate–metal interaction. Adsorption of citrate ion on Ag colloids, for example, produced a 15 cm^{-1} shift to lower frequencies in the carbonyl mode as compared to its frequency in solution. As the colloid aged, changes were observed in the Raman spectrum which were attributed to the formation of citrate bridges between Ag atoms on different colloidal particles.(32) Acetate ions, on the other hand, exhibited no frequency shifts upon adsorption on Ag colloids. The adsorbed acetate, Ag acetate, and the acetate ion in solution all exhibited identical spectra. This was attributed to the fact that acetate is "hard" in the Pearson classification.(36)

In another example, the importance of the metal substrate surface in the adsorption interaction is illustrated. Ritchie and Chen(56) found that isonicotinic and benzoic acids adsorbed readily onto vacuum-deposited Ag films on exposure of the films to aqueous solutions of these acids, whereas pyridine did not adsorb. It was felt that these results indicated the formation of a silver oxide film on the metal surface. The carboxylic groups on the acids were capable of interacting with the oxide, whereas pyridine was not. On the other hand, all three of these molecules adsorbed onto gold films, apparently due to the lack of oxide formation on gold. Also, chemisorption appears to be important in these interactions since the authors found that if benzoic acid is first adsorbed onto glass and then overlayered with Ag islands, no Raman spectrum is observed.

3.1. AMINO ACIDS AND NUCLEIC ACID BASES

There have been several SERS investigations of small molecules which are components of high molecular weight biomolecules. These include amino acids and nucleic acid bases. Nabiev and coworkers(59) found that aqueous solutions of aromatic amino acids (histidine, tyrosine, tryptophan, and phenylalanine) produce Raman spectra on Ag electrodes on adsorption from bulk solution concentrations that were two orders of magnitude less than that required in traditional Raman studies. Nonaromatic amino acids, such as glutamine and leucine, did not produce SER spectra on Ag, however. Two proteins, leucine-isoleucine-valine(LIV)-binding and leucine-specific (LS), were also found to produce enhanced Raman scattering of the above named aromatic residues when adsorbed onto Ag.

Nucleic acid components were investigated by SERS in an effort to determine the nature of their interaction with electrodes. The results were compared with previous voltammetric studies on Hg electrodes, which were undertaken to model the surface interactions present at a biological interface. Adenine, adenosine, and 5′-adenosine monophosphate were found to adsorb spontaneously on an anodized Ag electrode from a 2.2×10^{-3} *M* solution.[60] The spectra were many times more intense than solution spectra which require concentrations of 0.21 *M* merely to reach the detection limit. The strongest features in the spectra were assignable to the adenine ring. These were the 740 cm^{-1} ring breathing mode and 1340 cm^{-1} skeletal vibration. In 5′-AMP a band near 240 cm^{-1} was observed only for the adsorbed species and was therefore assigned to Ag–PO_4 interaction. The 5′-AMP gave the strongest SERS signals and was found to produce the most intense spectrum at relatively positive potentials (−0.1 V vs. SCE), with a sizable decrease in signal intensity as the potential was made more negative. The decrease in signal intensity with potential probably results from coulombic repulsion between the PO_4^- groups and the negatively charged electrode, since those adenine derivatives which lack the PO_4^- group produced similar intensities at potentials from −0.1 V to the potential of zero charge (pzc) or −0.6 V. Only small frequency shifts with respect to solution spectra were observed on adsorption for all of the derivatives. Curiously, it was found that purine, the parent compound, did not give any SER spectrum if the electrode anodization was carried out in its presence. A weak surface spectrum resulted when the electrode was first anodized in the absence of purine and then placed in a purine-containing solution. This observation suggests that complex formation between the purine and Ag ion may occur during the anodization step and subsequently interfere with the re-reduction of the Ag ion. It was concluded that the nucleotide derivatives were bonded through both the phosphate and the adenine group at the more positive potentials, but near the pzc the adenine group was responsible for the adsorption. However, the orientation of the adenine with respect to the electrode surface (the plane of the ring parallel or perpendicular to the surface) could not be inferred from the results.

3.2. POLYADENYLIC ACID AND DNA

Polyadenylic acid and DNA were also found to exhibit SERS on a Ag electrode.[61] In both polymers, the adenine modes dominated the surface spectrum. In DNA a band near 818 cm^{-1} attributed to the sugar phosphate symmetric stretch was found to decrease in intensity if the DNA was thermally denatured prior to the SERS studies. Bands near 734 and 1334 cm^{-1} were also sensitive to denaturation of the DNA, and became more intense with uncoiling of the double-stranded structure. It was concluded that some denaturation of the DNA was occurring on adsorption at the electrode on the basis of the greater intensity of these two bands in spectra from the surface as compared to solution spectra.

An interesting example of the application of SERS to the study of biopolymer structure at an electrode interface may be found in the isotope exchange studies by Koglin *et al.*[(62)] who undertook deuterium exchange experiments on adsorbed DNA. The advantage to this approach is that the adsorbed D_2O does not produce strong Raman signals (or SERS) as contrasted to D_2O solution exchange studies, wherein the Raman spectrum of the deuterium-exchanged DNA is obscured by the large D_2O background. As in the previously cited studies by these researchers, the adenine components were the most strongly enhanced in the Raman spectrum of surface-bound DNA. Assignments were made for the C—H, N—H, and O—H stretching vibrations in the DNA and their shifts on deuteration were monitored. Three classes of protons could be discerned: the OH protons of ribose; the CH protons of ribose; and CH protons at positions 2 and 8 in adenine. These were clearly resolvable and the authors concluded that it should be possible to determine the isotope exchange kinetics by following the intensity changes in these bands with time. However, based upon their previous results with DNA, it might be expected that these kinetics will not necessarily correspond to those measured in solution by other techniques, since the DNA appears to be partially denatured at the electrode surface.

3.3. LIPID MONOLAYERS

Lipid monolayers have often been used to model biological membranes. Thus, a recent study of calcium arachidate monolayers on Ag substrates provides an important background for future studies on natural membrane systems.[(63)] The Raman spectra obtained for one to twelve monolayers of this lipid deposited by the Langmuir-Blodgett technique on two different categories of rough substrates—gratings and Ag islands—were compared to IR spectra obtained from the same samples. It was found that in those cases where a SERS mechanism appeared to be operative, i.e., Ag island films either under or overlayering the monolayers, Raman spectra were observed for the hydrocarbon modes of the lipids in the vicinity of the Ag islands only. Moreover, based upon an analysis of the intensity of the 2880 cm^{-1} asymmetric CH stretch intensity relative to that of the symmetric CH stretch at 2850 cm^{-1}, which is maximal for the all-*trans* conformation of the hydrocarbon chains, it could be deduced that disorder was introduced near the Ag island regions. On the other hand, IR spectroscopy was found to sample the bulk of the monolayers and did not selectively detect the disordered regions. From an analysis of the IR signal-to-noise ratio, it was determined that *ca.* 10% of the monolayer was disordered in the presence of the Ag overlayer. In the Raman case, no enhancement was found for monolayers adsorbed on Ag films which were deposited on gratings with very small modulation periodicity ratios, whereas an enhancement factor of four was determined for monolayers in contact with island films. These results are promising in that they indicate the SERS effect detects only those molecules close to the Ag islands and suggest that overlayering a natural membrane with Ag islands may be a means of determining surface groups on the membrane. However, the results also indicate that the overlayering process may in itself disrupt the membrane. If this is the case for complex membranous systems, the technique may be limited to the identification of the surface groups and prove insensitive to structural properties of the membrane.

3.4. REDOX PROTEINS

Our interest in surface enhanced Raman and resonance Raman methods is concerned with development of techniques for studying relatively small biological molecules (porphyrins, lipids, etc.) and applying these to the study of larger more complex systems (proteins, membrane complexes). A practical goal of this work is the development of analytical methods for the detection of low concentrations of biomolecules in solution or in biological fluids. In addition, it is anticipated that these studies will also contribute to a fundamental understanding of how such systems behave at surfaces. Finally, a long-range goal is the development of SERS and SERRS methods for studying not only isolated biomolecules, but intact cells and tissue surfaces as well.

The types of information which we would like to obtain from the surface enhanced Raman studies include the following:

(1) identification of surface groups on large molecules;
(2) identification of bonding interactions between adsorbed molecules and the electrode surface;
(3) determination of molecular structure in the adsorbed state (i.e., native or denatured);
(4) determination of chemical reaction products when such are produced on the surface;
(5) determination of the redox state of systems which undergo electron transfer reactions;
(6) determination of potential dependence and the effect of potential on the above mentioned properties;
(7) determination of the excitation wavelength dependence (this is needed for molecules which exhibit normal Raman scattering as well as those which exhibit resonance Raman scattering in addition to surface enhancement);

(8) determination of the surface coverage (this is needed for accurate measurements of the enhancement factors);
(9) determination of distance effect on enhancement factor;
(10) determination of orientational effects on enhancement factors.

Of course, all of the above information is needed for SERS studies in general and it is likely that much of the fundamental information concerning distance and orientational effects will result from studies of small molecules in highly defined structures. In summarizing some of our own research results, examples will be given of information obtained in categories 2 through 7.

3.4.1. Cytochromes

Our first experiments involving the application of SERS to a redox active protein utilized horse heart cytochrome *c* and sperm whale myoglobin, both of which were found to adsorb spontaneously onto an anodized Ag electrode.[64] These proteins were chosen for study because their three-dimensional structures have been determined and they have also been subjected to numerous electrochemical studies. Initially, it was expected that the spectra of these proteins in the adsorbed state would consist of vibrations from the amino acid side chains on the surface of the proteins. In the case of Cyt *c*, for example, these side chains might include the lysines which surround the heme crevice. Contrary to our expectations, however, the surface spectrum which we observed consisted solely of modes assignable to the heme prosthetic group, which is known to be buried within the protein matrix. The spectrum was also found to be potential dependent, with the greatest intensity manifested at −0.6 V vs. SCE and a much weaker spectrum observed at potentials more positive than this value. The negative potential is near the potential of zero charge, suggesting that the bonding interactions are most likely nonpolar and do not involve ionic groups on the protein surface. In addition, it was noted that the redox state of the heme group was sensitive to the electrode potential (Figure 1A and 1B). Based upon the previously assigned oxidation state marker band at 1376 cm^{-1} for low spin ferriCyt *c*, it appeared that the surface-bound Cyt was in the oxidized state at the more positive potentials. A shift in this band to 1362 cm^{-1} at negative electrode potentials indicated that the heme was undergoing reduction. In general, the vibrational frequencies for both the ferroCyt *c* and the ferriCyt *c* were found to correspond closely with those observed for the analogous species in solution. Small frequency shifts and differences in the relative band intensities were observed, but these effects are characteristic of SERS for small molecules as well.

In this study, sperm whale myoglobin was also examined upon spontaneous adsorption onto a Ag electrode. As in the case of Cyt *c*, only the

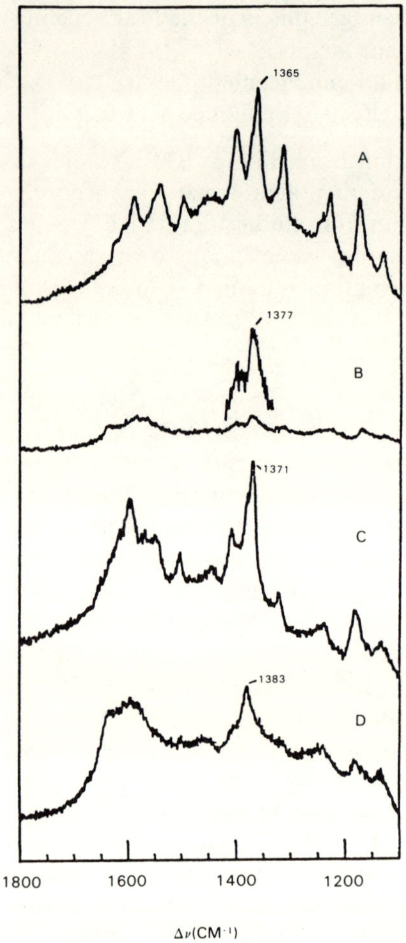

FIGURE 1. Surface enhanced resonance Raman spectra of cytochrome *c* and sperm whale myoglobin adsorbed onto an anodized Ag electrode. (A) Cytochrome *c* at −0.6 V vs. SCE; (B) cytochrome *c* at −0.2 V vs. SCE; (C) myoglobin at −0.6 V vs. SCE; (D) myoglobin at −0.2 V vs. SCE. Laser excitation wavelength = 514.5 nm; power = 40 mW; scan rate = 0.3333 Å/s; resolution = 2 cm^{-1}; counting interval = 1 s. Solution concentrations = 1×10^{-6} *M* for both proteins. From T. M. Cotton, S. G. Schultz, and R. P. Van Duyne, *J. Am. Chem. Soc.* **102**, 7960 (1980).

heme prosthetic group was observed for the adsorbed Mb. It was found to exhibit potential behavior similar to that of Cyt *c* in that the strongest spectrum was observed at negative potentials (−0.6 V, Figure 1C) but the decrease in signal intensity at more positive potentials (−0.2 V, Figure 1D) was not as dramatic as in the case of Cyt *c*. The oxidation state marker bands indicated a reduced heme at the negative potentials, which became oxidized at −0.2 V. In contrast to Cyt *c*, the band frequencies were quite different from those of Mb in solution, suggesting that some denaturation of the Mb occurred on the electrode surface.

Since both Cyt *c* and Mb surface enhanced Raman spectra were those of the heme groups, it is likely that some combination of surface and resonance enhancement is responsible (SERRS). An estimate of the enhancement factor due only to the surface effect was made by comparing the intensity of the Cyt *c* spectrum observed on Ag with that observed in solution. If monolayer coverage is assumed, an enhancement factor of five orders of magnitude was calculated. However, since the surface coverage measurements were not made, this factor could easily be off by an order of magnitude. Enhancements on the order of 10^3–10^4 are, nonetheless, significant, since very strong spectra can be observed under these conditions for solutions as dilute as 10^{-6} to 10^{-7} *M*.

The above study suggests that the enhancement mechanism responsible for producing RR spectra of the hemes in Cyt *c* and Mb is primarily electrodynamic in origin. In both proteins the hemes are buried within a protein matrix and, provided the protein is not denatured, would not contact the electrode directly. Murray[(65)] has examined both the theoretical and experimental predictions for the enhancement factor as a function of molecule separation from the Ag surface. Electromagnetic theory predicts that the decay length for molecules near the surface is on the order of 1000 Å. On highly roughened surfaces, where the surface roughness is smaller than or equal to the wavelength of light, the optical properties of the metal are governed by optical conduction resonances or collective electron resonances of the localized protrusions of the metal. These resonance can exist throughout the visible and depend strongly upon the surface morphology. Experimental systems which were examined and shown to adhere to the theroetical predictions include vacuum-deposited Ag films, which were roughened with I_2 vapor, Ag island films, and multilayered structures. In the latter, *p*-nitrobenzoic acid (PNBA) was deposited on Al films and coated with polymer layers of varying thickness. CaF_2 was deposited on top of the polymer layer to provide surface roughness, and Ag was then deposited on top of the CaF_2 layer. The intensity of the PNBA was monitored as a function of the polymer layer thickness (or distance from the Ag surface) and found to decrease by only a factor of 10 for an increase of 50 Å in the spacer thickness.[(65)] Thus, the results of these experiments show that an enhancement of the heme in Cyt and Mb is not unreasonable even though it is some distance from the electrode. The additional resonance enhancement undoubtedly contributes significantly to the overall intensity and precludes the detection of surface groups on the protein.

We have also examined another cytochrome complex, cytochrome cd_1, which is an oxidation–reduction enzyme composed of two subunits.[(66)] Each subunit contains one *c*-type heme and one *d*-type heme (a chlorin-type structure). The overall molecular weight of the protein is 120,000 daltons. The objective of this study was, first, to determine whether a protein of this size would exhibit surface enhancement and, second, to utilize the enhancement advantage, if observed, to obtain a Raman or resonance Raman spectrum of the protein for the first time. No previous spectrum had been reported for this protein, which is quite difficult to isolate in large quantities. Both objectives were met and, as in the case of the previously studied proteins, only the heme prosthetic group was observed for the surface-bound protein. By selective excitation of either the heme *c* or *d* it was possible to determine their vibrational spectra individually. This was easily achieved since the chromophores have different electronic absorption spectra. Data obtained by varying the electrode potential suggested that the heme *d* is closer to the protein surface in the complex than is heme *c*. Finally, another important advantage noted in studying this protein from two bacterial sources was the fluorescence quenching properties of the Ag electrode. One of the preparations was intensely fluorescent (probably due to impurities) and its solution RR spectrum could not be recorded, whereas the SERR spectrum was readily obtained. As noted previously in the case of dyes adsorbed on thin metal films, quenching probably results from lifetime shortening. This property of the metal surface will undoubtedly prove very useful in Raman and RR studies of biomolecules.

3.4.2. Reaction Center Complexes from Photosynthetic Bacteria

The reaction center complex is the smallest entity isolatable from photosynthetic organisms that is capable of carrying out the primary photochemistry of photosynthesis. This definition implies charge separation between the primary electron donor molecules (two bacteriochlorophyll molecules, the special pair) and the primary electron acceptor (a ubiquinone or menaquinone). The complex is hydrophobic and is located within the chromatophore membrane of these organisms. It has been isolated in relatively pure form from a number of different bacterial sources and the protein and pigment components have been characterized. The overall molecular weight is in excess of 100 kdaltons and the complex contains three polypeptides, four bacteriochlorophyll (BChl) molecules, two bacteriopheophytin (BPheo) molecules, and one or two quinone molecules. Two BChls function as the primary electron donor, one BPheo is believed to function as an intermediate electron acceptor, and the quinones function as the primary and secondary electron acceptor. The role of the remaining pigments is not clear at present. One or possible both of the BChls may be involved in some way as intermediate electron acceptors.

The primary events of photosynthesis as presently understood have been delineated by the application of several different techniques, including especially electronic absorption and ESR spectroscopy. Due to the short time scale of the initial electron transfer from the special pair BChls to an intermediate, only those techniques capable of picosecond time resolution can be expected to complete our understanding of reaction center photochemistry. Electronic absorption spectroscopy possesses the necessary time resolution, but lacks molecular specificity. Optical spectra of transient electron acceptors do not provide positive identification of these molecules. Resonance Raman spectroscopy has both the required time resolution capabilities and the molecular specificity needed to identify the transients. Therefore, we have been investigating electrochemical methods for characterizing the RR spectra of BChl and BPheo anion radicals in solution.(67,68) The information obtained should eventually prove useful in the identification of the intermediates in reaction center complexes. Recently, we have investigated the possibility of using surface enhanced Raman methods for this purpose.(69)

Our initial surface enhanced Raman study of reaction centers addressed several questions. First, would it be possible to observe spectra of these large complexes adsorbed onto the surface of a Ag electrode? Second, would the spectra be interpretable, or would they be an unintelligible superposition of the vibrational modes associated with the surface groups, the BChls, and the BPheos of the complex? Third, what effect would the various experimental parameters have upon the spectra, including excitation wavelength and electrode potential? Answers to these questions were sought to determine if SERS or SERRS has the potential for providing either structural information about the reaction center (i.e., the proximity of the pigments to the electrode surface) and/or identification of transient electron acceptors (following photoexcitation or reduction at negative electrode potentials). Our results provided encouraging answers to these questions. We were able to observe spectra of BChl or BPheo selectively by changing the excitation wavelength. The spectra of the reaction center complexes adsorbed onto Ag contained vibrations assignable entirely to BChl, BPheo, and LDAO, the detergent used in isolating the complex. Moreover, variation of the electrode potential showed opposite signal intensity dependence for BChl as compared to BPheo. The strongest BChl spectrum was observed at relatively negative potentials (−0.7 V), whereas the strongest BPheo spectrum was observed at potentials near 0 V. These results suggest a reorientation of the complex occurs on the electrode surface as a function of potential. At negative potentials, one or more BChl molecules is closest to the electrode surface and experiences both RR and surface enhanced Raman effects. At potentials near 0 V, one or both of the BPheo molecules is closest to the electrode surface and experiences both types of enhancement. We plan to continue these studies using a broader range of excitation wavelengths and electrode potentials.

Another important advantage observed in the above investigation was the reduction of background fluorescence from reaction centers adsorbed onto a Ag electrode. Normally, fluorescence is intense at room temperature and precludes the detection of RR scattering. Spectra were acquired from bulk solution concentrations of only 1×10^{-6} *M* in reaction centers.

3.5. PORPHYRINS

SERRS studies of isolated porphyrins and prosthetic groups of redox proteins are needed for interpretation of spectra obtained on intact proteins. The effects of the many experimental parameters on the spectra are more readily characterized with these relatively simpler molecules. Unfortunately, many of the prosthetic groups are not water soluble and cannot be used directly for surface enhanced Raman experiments. We have used two approaches to overcome this difficulty. The first is the use of water-soluble porphyrins as model compounds for hemes and chlorophylls. The second is the incorporation of water-insoluble compounds into phospholipid vesicles, which can then be studied by SERS or SERRS.

The two water-soluble porphyrins, tetrasodium *meso*-tetra(4-sulfanotophenyl-porphine (TSPP) and *meso*-tetra(4-carboxyphenyl)porphine (TCPP) were found to adsorb spontaneously onto anodized Ag electrodes from bulk solution concentrations as dilute as 10^{-7} *M*.[70] Both surface and resonance enhancement mechanisms contributed to the high sensitivity which we observed. A surface enhancement factor of approximately 10^5 was estimated by assuming monolayer coverages of the porphyrins. These results conflict with the colloid study[51] cited previously in which methyl red exhibited no surface enhancement, but only resonance enhancement. We cannot explain this lack of agreement at the present time.

An interesting finding in the SERRS studies of the water-soluble porphyrins is that both incorporated Ag ions into the porphyrin macrocycle during the anodization step. The reaction was easily characterized by the appearance of low-frequency modes typical of metallated porphyrins. The detection of chemical reactions involving molecules adsorbed at metal surfaces is another important advantage of the SERRS or SERS techniques. It would have been difficult for us to identify the Ag–porphyrin reaction by other spectroscopic techniques. The results also point out another important aspect of surface enhanced Raman experiments which involve electrode anodizations in the presence of the adsorbate. This is the propensity of Ag ions to form complexes with a large variety of compounds. Complex formation may be important in the enhancement process, especially in those cases where PMM are operative.

The two water-soluble porphyrins were found to behave quite differently as a function of electrode potential. TSPP showed the strongest SERRS spectra at potentials near 0 V and desorbed at more negative potentials, apparently due to coulombic repulsion by the sulfonate groups. TCPP, on the other hand, produced strong SERRS spectra at quite negative potentials (−0.6 V). Changes in the vibrational frequencies and the appearance of a new band at negative potentials suggested that reorientation of this porphyrin occurred as a function of potential.

The vibrational frequencies of several bands in the water-soluble porphyrins showed shifts on the order of $10\,cm^{-1}$ when adsorbed onto the electrode as compared to in solution or in the solid state. Relative enhancement of several bands was also noted. These results can be contrasted with the previously cited reaction center data. No significant shifts were observed in BChl or BPheo bands on adsorption of the reaction center complex onto the electrode when compared to previous spectra recorded for the complexes in solution. The lack of frequency shifts suggests that the reaction center porphyrins are not in direct contact with the electrode, but are buried within the polypeptide–detergent complex.

3.6. PHOSPHOLIPID VESICLES

Figure 2a and 2b depict SERRS spectra of protoporphyrin IX (PP-IX) in phosphatidylcholine (PC) vesicles adsorbed onto a Ag electrode at two different electrode potentials (−0.6 V and −0.2 V, respectively).[71] The molar ratio of PP-IX to PC is 1:100 in these preparations. A comparison of the SERRS spectra with that of solid PP-IX (Figure 2c) shows that the surface

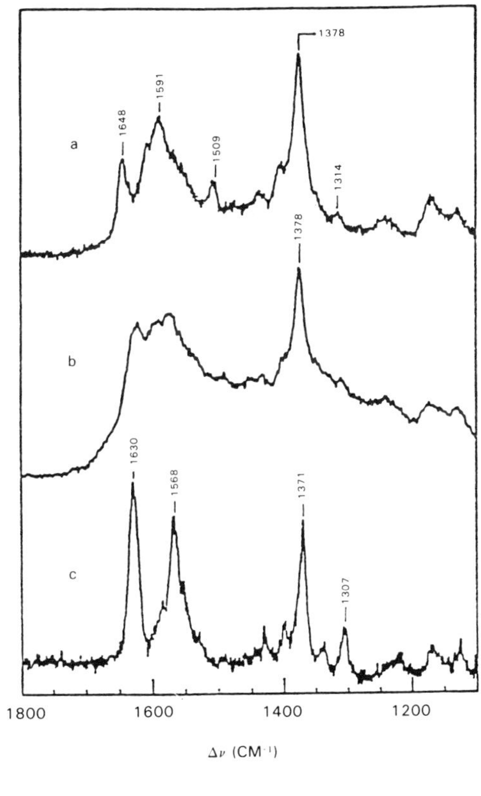

FIGURE 2. Surface enhanced resonance Raman spectra of protoporphyrin IX in phosphatidylcholine vesicles compared with the resonance Raman spectrum of protoporphyrin IX in the solid state. (a) PP-IX in phosphatidylcholine vesicles adsorbed onto an anodized Ag electrode. The mole ratio of PP-IX to PC was 1:100 and the bulk concentration of PP-IX in solution was 1×10^{-6} *M*. Electrode potential was −0.6 V vs. SCE; laser power was 130 mW; (b) same as in (a) except the electrode potential was −0.2 V vs. SCE; (c) solid PP-IX in a glass tube. Laser power was 200 mW.

spectra are due to PP-IX; no PC vibrations are observed. The surface spectra show frequency shifts and intensity variations which are typical of molecules adsorbed onto Ag. An interesting feature of these spectra is the shift in the spin state marker band near 1648 cm^{-1} (Figure 2a) to 1630 cm^{-1} (Figure 2b) with a change in electrode potential from −0.6 V to −0.2 V. This shift is characteristic of low-spin ferric hemes on reduction. However, the oxidation state marker band at 1378 cm^{-1} is unchanged with potential. It does not appear that the heme Fe is undergoing reduction. The band shift may represent a change in distance or orientation of the PP-IX in the vesicles as a function of electrode potential. These results are important with respect to previous SERRS experiments with cytochromes and myoglobin. The lack of reduction of PP-IX heme supports the interpretation that the porphyrin is located primarily within the hydrocarbon region of the PC vesicles. Only the polar propionic acid chains are extended into the head-group region of the vesicles. It does not appear, therefore, that the vesicles rupture near the electrode surface or that the PP-IX is pulled out of the vesicles. The cytochromes and myoglobin, on the other hand, did undergo reduction on the electrode surface. The orientation of the heme group in these proteins is no doubt optimal for electron transfer reactions.

3.7. HEMOGLOBIN

A preliminary examination of hemoglobin indicates that it will be possible to utilize SERRS to study this protein as well. Figure 3 shows the spectrum obtained at −0.6 V using 457.9 nm excitation. The hemoglobin was adsorbed from a μM solution onto an anodized Ag electrode. The spectrum is very intense and is due entirely to the heme groups, as was observed for cytochromes. A comparison of the surface spectrum with spectra reported in the literature for hemoglobin in solution shows the vibrations are within 10 cm^{-1} of the solution values.

4. CONCLUSIONS AND FUTURE DIRECTIONS

In this review, an attempt was made to summarize the experimental and theoretical aspects of the field of surface enhanced Raman scattering as presently understood. Both aspects are currently under intense scrutiny and new interpretations will undoubtedly be available when this review is published. At present, however, I do not believe it is possible for an unbiased researcher to conclude that any of the proposed mechanisms can explain this remarkable effect when considered alone. Therefore, I have attempted to include some examples from the studies of small molecules which support either or both the PMM and PEDM.

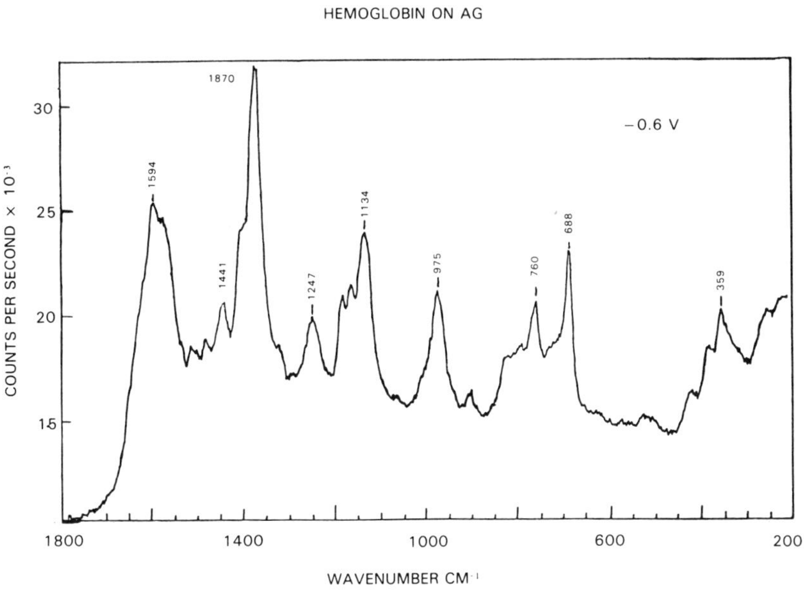

FIGURE 3. Surface enhanced resonance Raman spectrum of hemoglobin on a Ag electrode. The electrode potential was −0.6 V vs. SCE. Hemoglobin was adsorbed onto the electrode from a 1×10^{-6} *M* solution. Instrumental parameters were the same as in Figure 1, except the laser excitation wavelength was 457.9 nm.

The cited examples of SERS and SERRS studies of biologically important molecules show that, in spite of the limitations of our present understanding of the surface enhancement phenomenon, useful information can be obtained by this approach. Examples of the kinds of information and useful applications which we have sought in our own work include the detection of very low concentrations of porphyrins and proteins in aqueous solution, the determination of SERRS spectra of highly fluorescent biomolecules, and the determination of chemical and electron transfer reactions in molecules adsorbed onto Ag electrodes. We have deliberately examined a number of different macromolecular systems in order to determine the generality of this technique. Our future research will be concerned with quantitative measurements and structural studies of the proteins on metal surfaces.

The future application of surface enhanced methods for the study of biological systems is very promising. The full range of information available cannot be predicted, however, without quantitative determinations of surface coverages and enhancement factors for a variety of molecular species. These studies will also require more careful characterization of the surface roughness of the metal than is possible with electrodes. Thin metal films, or island structures, should be better suited for this purpose. The preparation of well-defined model systems on surfaces, e.g., monolayer structures incorporating dyes or naturally occurring prosthetic groups, will be useful for determining the effects of molecular orientation and distance from the metal surface on enhancement factors. These types of information may then prove useful for studying cells and tissues. When coupled with Raman microprobe analysis, SERRs and SERS may provide previously unattainable structural and molecular information about biological surfaces.

REFERENCES

1. R. L. McCreery, R. Pruiksma, and R. Fagan, Bioelectrochemistry—Examination of some examples, *CRC Crit. Rev. Anal. Chem.* **7**, 89–119 (1978).
2. R. Burny, E. Herbst, and M. Hinsenkamp, eds., *Electric Stimulation of Bone Growth and Repair*, Springer-Verlag, New York (1978).
3. P. T. Kissinger, J. B. Hart, and R. N. Adams, Voltammetry in brain tissue—A new neurophysiological measurement, *Brain Res.* **55**, 209–219 (1973).
4. M. Kessler, L. C. Clark, D. W. Lubbers, I. A. Silver, and W. Simon, eds., *Ion and Enzyme Electrodes in Biology and Medicine*, University Park Press, Baltimore (1976).
5. I. R. Miller and J. Rishpon, in: *Electrical Phenomena at the Biological Membrane* (E. Roux, ed.), pp. 93–110, Elsevier Scientific, Amsterdam (1977).
6. P. Bianco and J. Haladjian, Electrode-reaction of cytochrome *c*-551.5(C7) from *Desulfuromonas acetoxidans*, *J Bielectrochem. Bioenerg.* **8**, 239–245 (1981).
7. D. G. Davis, in: *The Porphyrins*, Vol. 5 (D. Dolphin, ed.), pp. 127–152, Academic Press, New York (1978).
8. M. Fleischmann, P. J. Hendra, and A. J. McQuillan, Raman spectra from electrode surfaces, *J. Chem. Soc., Chem. Commun.*, 80–81 (1973).
9. N. T. Yu, Raman spectroscopy—Conformational probe in biochemistry, *CRC Crit. Rev. Biochem.* **4**, 229–280 (1977).
10. M. L. Hair, *Infrared Spectroscopy in Surface Chemistry*, Dekker, New York (1967).
11. M. Fleischmann, P. J. Hendra, and A. J. McQuillan, Raman spectra of pyridine adsorbed at a silver electrode, *Chem. Phys. Lett.* **26**, 163–166 (1974).
12. D. L. Jeanmaire and R. P. Van Duyne, Surface Raman spectroelectrochemistry, *J. Electroanal. Chem.* **84**, 1–20 (1977).
13. M. G. Albrecht and J. A. Creighton, Anomalously intense Raman spectra of pyridine at a silver electrode, *J. Am. Chem. Soc.* **99**, 5215–5217 (1977).
14. G. Hagen, B. Simic-Glavaski, and E. Yeager, Raman spectrum of an adsorbed species on electrode surface, *J. Electroanal. Chem.* **88**, 269–275 (1978).
15. B. Pettinger and U. Wenning, Raman spectra of pyridine adsorbed on silver (100) and (111) electrode surfaces, *Chem. Phys. Lett.* **56**, 253–257 (1978).
16. H. Metiu, in: *Surface Enhanced Raman Scattering* (R. K. Chang and T. E. Furtak, eds.), pp. 1–34, Plenum Press, New York (1982).
17. G. C. Schatz, in: *Surface Enhanced Raman Scattering*, (R. K. Chang and T. E. Furtak, eds.), pp. 35–50, Plenum Press, New York (1982).
18. B. Pettinger and H. Wetzel, in: *Surface Enhanced Raman Scattering*, (R. K. Chang and T. E. Furtak, eds.), pp. 293–314 and references cited within, Plenum Press, New York (1982).

19 (a) L. A. Sanchez, R. L. Birke, and J. R. Lombardi, Surface enhanced Raman spectrum from pyridine on mercury, *Chem. Phys. Lett.* **79**, 219-223 (1981). (b) R. Naaman, S. J. Buelow, O. Cheschnovsky, and D. R. Herschbach, Surface enhanced Raman scattering from molecules adsorbed on mercury, *J. Phys. Chem.* **84**, 2692-2694 (1980).

20. W. Krasser and A. J. Renouprez, Enhanced Raman spectra of coadsorbed carbon monoxide and hydrogen on small nickel particles, *J. Raman Spectrosc.* **11**, 425-429 (1981).
21. H. Yamada, Y. Yamamoto, and N. Tani, Surface enhanced Raman Scattering (SERS) of adsorbed molecules on smooth surfaces of metals and a metal oxide, *Chem. Phys. Lett.* **86**, 397-400 (1982).
22. W. Krasser and A. J. Renouprez, Enhanced Raman scattering of benzene chemiadsorbed on small platinum clusters, *Solid State Commun.* **41**, 231-235 (1982).
23. M. Moskovits and D. P. Dilella, in: *Surface Enhanced Raman Scattering* (R. K. Chang and T. E. Furtak, eds.), pp. 243-274, Plenum Press, New York (1982).
24. H. Yamada and Y. Yamamoto, *Proc. 8th Intl. Conf. Raman Spectroscopy*, Bordeaux, France (1982), pp. 73-74.
25. T. Lopez-Rios, C. Pettenkofer, I. Pockrand, and A. Otto, Enhanced Raman scattering from aluminium films, *Surface Sci.* **121**, L541-L544 (1982).
26. Spectra have not been published in indium, but claimed in a communication: R. Aroca and R. O. Loufty, Enhanced Raman scattering from molecular dye films on silver, *J. Raman Spectrosc.* **12**, 262-265 (1982).
27. A. Otto, I. Pockrand, J. Billmann, and C. Pettenkofer, in: *Surface Enhanced Raman Scattering* (R. K. Chang and T. E. Furtak, eds.), pp. 147-172, Plenum Press, New York (1982).
28. S. G. Schultz, M. Janik-Czachor, and R. P. Van Duyne, Surface enhanced Raman spectroscopy—A reexamination of the role of surface roughness and electrochemical anodization, *Surface Sci.* **104**, 419-434 (1981).
29. R. P. Duyne, in: *Chemical and Biochemical Applications of Lasers*, Vol. 4 (C. B. Moore, ed.), pp. 101-185, Academic Press, New York (1978).
30. J. A. Creighton, C. G. Blatchford, and M. F. Albrecht, Plasma resonance enhancement of Raman scattering by pyridine adsorbed on silver or gold sol particles of size comparable to the excitation wave length, *J. Chem. Soc., Faraday Trans. 2* **75**, 790-798 (1979).
31. M. L. Lippitsch, Observation of surface enhanced Raman spectra by adsorption to silver colloids, *Chem. Phys. Lett.* **74**, 125-127 (1980).
32. M. Kerker, O. Siiman, L. A. Bumm, and D. S. Wang, Surface enhanced Raman scattering (SERS) of citrate ion adsorbed on colloidal silver, *Appl. Opt.* **19**, 3253-3255 (1980).
33. K. U. Von Raben, R. K. Chang, and B. L. Laube, Surface enhanced Raman scattering of dicyanoaurate (I) ions adsorbed on gold colloids, *Chem. Phys. Lett.* **79**, 465-469 (1980).
34. H. Wetzel and H. Gerischer, Surface enhanced Raman scattering from pyridine and halide-ions adsorbed on silver and gold sol particles, *Chem. Phys. Lett.* **76**, 460-464 (1980).
35. D. S. Wang and M. Kerker, Enhanced Raman scattering by molecules adsorbed at the surface of colloidal spheroids, *Phys. Rev. B.* **24**, 1777-1789 (1981).
36. A. J. McQuillan and C. G. Pope, Raman spectra of formate and acetate ions adsorbed on silver particles in aqueous solution, *Chem. Phys. Lett.* **71**, 349-352 (1980).
37. W. Krasser, U. Kettler, and P. S. Bechthold, Luminescence and enhanced Raman spectra of pure and pyridine-bonded small silver clusters, *Chem. Phys. Lett.* **86**, 223-226 (1982).
38. K. U. Von Raben, R. K. Chang, P. B. Dorian, and B. L. Laube, SERS from oxy anions and oxides adsorbed on silver powder, *Proc. 8th Intl. Conf. Raman Spectroscopy*, Bordeaux, France (1982), pp. 81-82.
39. R. M. Hart, J. G. Bergman, and A. Wokaun, Surface enhanced Raman scattering from silver particles on polymer-replica substrates, *Opt. Lett.* **7**, 105-107 (1982).

40. A. Girlando, M. R. Philpott, D. Meitmann, J. D. Swalen, and R. Santo, Raman spectra of thin organic films enhanced by plasmon surface-polaritons on holographic metal gratings, *J. Chem. Phys.* **72**, 5187–5191 (1980).
41. (a) M. Moskovits and D. P. DiLella, Surface enhanced Raman spectroscopy of benzene and benzene-*d*6 adsorbed on silver, *J. Chem. Phys.* **73**, 6068–6075 (1980). (b) H. Seki and M. Philpott, Surface enhanced Raman scattering on silver island films in an ultra-high vacuum, *J. Chem. Phys.* **73**, 5376–5379 (1980).
42. C. Y. Chen, I. Davoli, G. Ritchie, and E. Burstein, Giant Raman scattering and luminescence by molecules adsorbed on Ag and Au metal island films, *Surface Sci.* **101**, 363–366 (1980).
43. K. Sakoda, K. Ohtaka, and E. Hanamura, Surface enhanced Raman scattering in attenuated total reflection arrangement, *Solid State Commun.* **41**, 393–397 (1982).
44(a). P. N. Sanda, J. M. Warlaumont, J. E. Demuth, J. C. Tsang, K. Christmann, and J. A. Bradley, Surface-enhanced Raman scattering from pyridine on Ag (111), *Phys. Rev. Lett.* **45**, 1519–1523 (1980). (b) I. Pockrand and A. Otto, Raman scattering from silver/vacuum interfaces, *Applic. Surf. Sci.* **6**, 362–371 (1980).
45. J. C. Tsang, J. R. Kirtley, and J. A. Bradley, Surface-enhanced Raman spectroscopy and surface-plasmons, *Phys. Rev. Lett.* **43**, 772–775 (1979).
46. G. Placzek, Rayleigh-Streuung und Raman-Effekt, *Handbuch der Radiologie*, VI, 2, p. 209, Akademische Verlagsgesellschaft, Leipzig (1934).
47. P. R. Carey and H. Schneide, Resonance Raman spectroscopy in biochemistry and biology, *Quant. Rev. Biophys.* **11**, 309–370 (1978).
48. M. Fujihira and T. Osa, Internal-reflection resonance Raman spectroscopy for studies of adsorbed dye layers of electrode–solution interface, *J. Am. Chem. Soc.* **98**, 7850–7851 (1976).
49. H. Yamada, T. Amamiya, and H. Tsubormura, Resonance Raman spectra of rose-bengal adsorbed on a ZND electrode, *Chem. Phys. Lett.* **56**, 591–594 (1978).
50. C. A. Melendres and F. A. Cafasso, Electrochemical and *in situ* laser Raman spectroscopy studies on carbon supported iron phthalocyanine electrodes, *J. Electrochem. Soc.* **128**, 755–760 (1981).
51. A. Bachackashvilli, B. Katz, Z. Priel, and S. Efrima, *J. Phys. Chem.* **88**, 6185–6190 (1984).
52. R. Kotz and E. Yeager, Raman spectroscopy of cobalt phthalocyanine adsorbed on a silver electrode, *J. Electroanal. Chem.* **113**, 113–125 (1980).
53. R. Aroca and R. O. Loufty, Enhanced Raman scattering from molecular dye films on silver, *J. Raman Spectrosc.* **12**, 262–265 (1982).
54. M. L. Lippitsch, Observation of surface enhanced Raman spectra by adsorption to silver colloids, *Chem. Phys. Lett.* **74**, 125–127 (1980).
55. J. P. Heritage and A. M. Glass, in: *Surface Enhanced Raman Scattering* (R. K. Chang and T. E. Furtak, eds.), pp. 391–412 and references cited within, Plenum Press, New York (1982).
56. G. Ritchie and C. Y. Chen, in: *Surface Enhanced Raman Scattering* (R. K. Chang and T. E. Furtak, eds.), pp. 361–378, Plenum Press, New York (1982).
57. A. M. Glass, P. F. Liapo, J. G. Bergman, and D. H. Olson, Interaction of metal particles with adsorbed dye molecules: Adsorption and luminescence, *Opt. Lett.* **5**, 368–370 (1980).
58. S. Garoff, D. A. Weitz, and C. J. Sandroff, in: *Proc. 8th Intl. Conf. Raman Spectroscopy*, Wiley, New York (1982), pp. 55–56.
59. I. R. Nabiev, S. C. Trakhanov, E. S. Efremov, V. V. Marinyuk, and R. M. Lasorenko-Manevich, Anomalously intense Raman spectra of some biological molecules adsorbed on silver electrodes, *Bioorganicheskaya Khimiya* **7**, 941–945 (1981).
60. (a) E. Koglin, J. M. Sequaris, and P. Valenta, Surface enhanced Raman spectra of nucleic acid components adsorbed at a silver electrode, *J. Mol. Struct.* **60**, 421–425 (1980). (b) K. M. Koglin, E. Koglin, J. M. Sequaris, P. Valenta, and H. W. Nurnberg, Surface enhanced Raman spectra of nucleic acid components adsorbed at a silver electrode, *J. Electroanal. Chem.* **114**, 179–194 (1980).

61. J. M. Sequaris, E. Koglin, P. Valenta, and H. W. Nurnberg, Surface enhanced Raman scattering (SERS) spectroscopy of nucleic acids, *Ber. Bunsenges. Phys. Chem.* **85**, 512–513 (1981).
62. E. Koglin, J. M. Sequaris, and P. Valenta, Hydrogen–deuterium exchange in adenosine 5-monophosphate detected by surface enhanced Raman scattering, *Z. Naturforsch.* **36**, 809–812 (1981).
63. W. Knoll, M. R. Philpott, and W. G. Golden, Surface infrared and surface enhanced Raman vibrational spectra of monolayer assemblies in contact with rough metal surfaces, *J. Chem. Phys.* **77**, 219–225 (1982).
64. T. M. Cotton, S. G. Schultz, and R. P. Duyne, Surface-enhanced resonance Raman scattering from cytochrome c and myoglobin adsorbed on a silver electrode, *J. Am. Soc.* **102**, 7960–7926 (1980).
65. C. A. Murray, in: *Surface Enhanced Raman Scattering* (R. K. Chang and T. E. Furtak, eds.), pp. 203–221, Plenum Press, New York (1982).
66. T. M. Cotton, R. Timkovich, and M. S. Cork, Resonance Raman and surface-enhanced resonance Raman studies of cytochrome *cd*, *FEBS Lett.* **133**, 39–44 (1981).
67. T. M. Cotton, R. P. Van Duyne, Resonancd Raman spectoelectrochemistry of bacteriochlorophyll cation radical, *Biochem. Biophy. Res. Commun.* **82**, 424–433 (1978).
68. T. M. Cotton, K. D. Parks, and R. P. Van Duyne, Resonance Raman spectra of bacteriochlorphyll and its electrogenerated cation radical excitation of the Soret bands by use of stimulated Raman scattering from H2 and D2, *J. Am. Chem. Soc.* **102**, 6399–6407 (1980).
69. T. M. Cotton and R. P. Van Duyne, Resonance Raman scattering from *Rhodopseudomonas-*sphaeroids reaction centers adsorbed on a silver electrode, *FEBS Lett.* **147**, 81–84 (1982).
70. T. M. Cotton, S. G. Schultz, and R. P. Van Duyne, SERS from water-soluble porphyrins, *J. Am. Chem. Soc.*, **104**, 6528–6532 (1982).
71. T. M. Cotton, in preparation.

5

The Study of Interfacial Proteins and Biomolecules by X-ray Photoelectron Spectroscopy

Royston W. Paynter and Buddy D. Ratner

1. INTRODUCTION

Protein films adsorbed on surfaces have been shown to influence a number of biological processes including the biocompatibility of synthetic materials, opsonization of bacteria, biofouling in marine environments, and cell attachment and growth. Review articles describing the nature of the protein adsorption phenomenon and its biological consequences are available.[1-3] Protein thin films have presented problems to investigators exploring them because of the small amounts of material involved, often less than one microgram of protein per square centimeter of surface. In addition, a great deal of information is necessary in order to fully describe protein films, thereby further compounding the problems involved in studying these systems. X-ray photoelectron spectroscopy (XPS) is a relatively new technique that has shown great promise for exploring adsorbed protein films. This article will briefly consider the factors of importance in describing adsorbed protein films, and will discuss the limitations of other available methods for investigating certain of these factors. The advantages of XPS for studying these films will be expounded upon along with a consideration of what can and cannot be learned using XPS. A literature review will then be presented outlining how XPS has been used to investigate biological molecules. Finally, new developments in our laboratories involving the application of XPS to explore protein films will be described along with a

Royston W. Paynter • Laboratoire de Chirurgie Expérimentale, Pavillon Agriculture Service, Université Laval, Québec, P.Q. G1K 7P4, Canada. *Buddy D. Ratner* • Department of Chemical Engineering, University of Washington, Seattle, Washington 98195.

TABLE 1
Some Common Amino Acids

Name	Structure	Molar composition (omitting H) (%)			
		C	N	O	Other
Glycine	$NH_2{-}CH_2{-}COOH$	40.0	20.0	40.0	—
Alanine	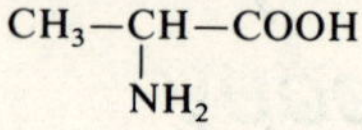	50.0	16.7	33.3	—
Leucine	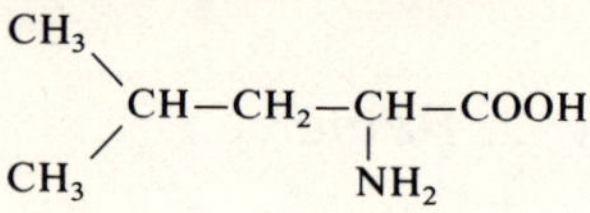	66.7	11.1	22.2	—
Asparagine	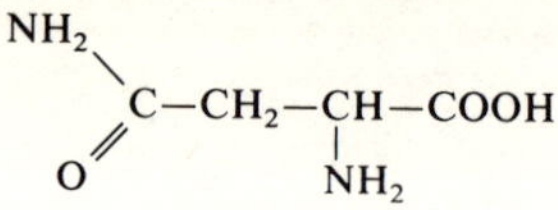	44.4	22.2	33.3	—
Lysine	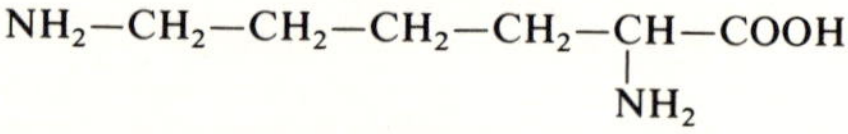	60.0	20.0	20.0	—
Serine	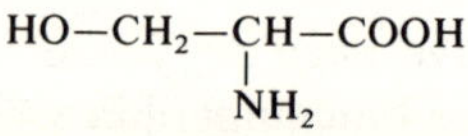	42.9	14.3	42.9	—
Tyrosine	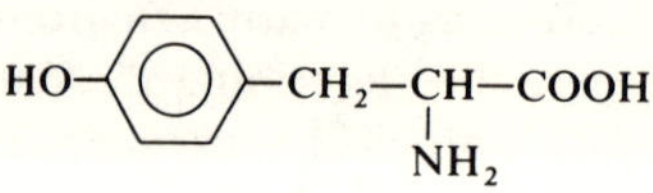	69.2	7.7	23.1	—
Arginine	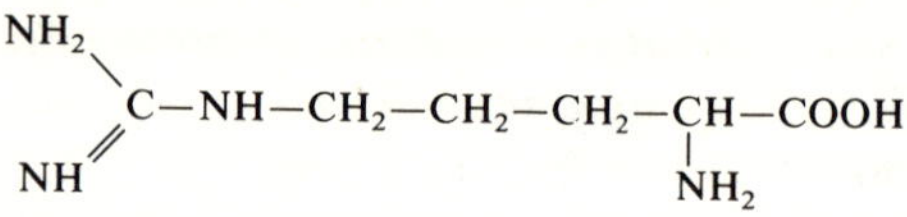	50.0	33.3	16.7	—
Proline	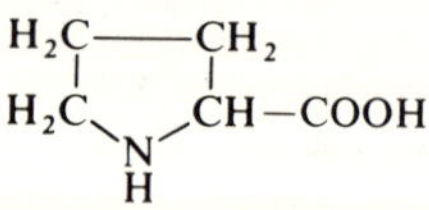	62.5	12.5	25.0	—
Tryptophan	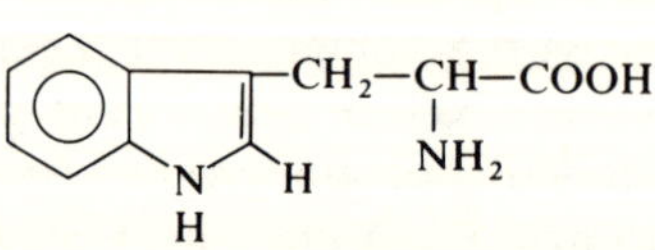	73.3	13.3	13.3	—
Cysteine	$HS{-}CH_2{-}CH(NH_2){-}COOH$	42.9	14.3	28.6	14.3 S
Thyroxine	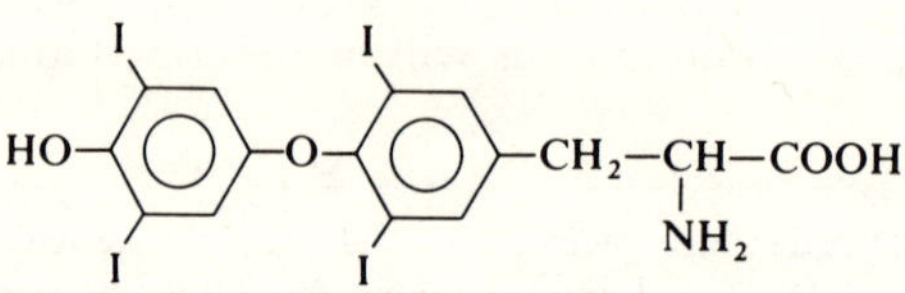	62.5	4.2	16.7	16.7 I

consideration of new developments in instrumentation and data analysis which are potentially applicable to increasing our understanding of adsorbed protein layers.

1.1. PROTEINS AND PROTEIN FILMS

Proteins are long-chain compounds composed of many different amino acids. The sequence of these amino acids in the polymer chain is referred to as the primary structure. All of these amino acids contain carbon, hydrogen, nitrogen, and oxygen. Many incorporate sulfur-containing functional groups. Thyroxine contains iodine. A number of the common amino acids are illustrated in Table 1. A surprisingly wide range of elemental compositions exists for amino acids. Also incorporated in some protein molecules are metals such as Cu, Fe, and Zn. Finally, cofactors (which often contain phosphorus), water, inorganic and organic ionic species, and lipids (e.g., cholesterol) are often associated with proteins. The elemental compositions of a few common proteins are presented in Table 2.

A compositional description of proteins is insufficient to completely characterize these molecules. The protein chains can organize themselves into ordered structures (helices, random coils, pleated sheets; the "secondary structure"), and these secondary structures further organize into complex units, often of a globular shape (the tertiary structure). Finally, these globular units can associate with each other in various well-defined stoichiometric and geometric units (quaternary structure). Partial descriptions of the tertiary and quaternary structure of a few proteins are presented in Table 3. An understanding of these levels of organization in proteins will be important in the interpretation of some of the XPS data on these molecules to be presented in the next sections of this article.

TABLE 2
Elemental Compositions of Selected Proteins[a]

		Molar %				
Protein	Reference	C	N	O	S	Fe
Albumin (human)	110	63.0	16.0	20.1	0.95	—
γ-Globulin (human)	110	62.8	15.8	21.0	0.37	—
Fibrinogen (human)	110	61.8	16.5	21.0	0.60	—
Hemoglobin (human)	111	64.5	17.4	17.6	0.41	0.09
Fibronectin (human)	112	63.2	15.2	21.1	0.48	—
Insulin (bovine)	113	63.5	15.0	20.6	0.85	—
Lysozyme (dog spleen)	113	61.7	17.4	19.9	1.1	—

[a] Ignoring hydrogen, "amide nitrogen," and prosthetic groups of conjugated proteins (except hemoglobin).

TABLE 3
Tertiary and Quaternary Structures of Selected Proteins

Protein	Molecular weight[a]	% in helical conformation[b]	Major axis[c] (Å)	Minor axis[c] (Å)	Subunit[d] No.	Subunit[d] Mol. wt.
Albumin	68,460	56	150	38	4	19,000
γ-Globulin	153,100	47	235	44	4	20,000 (L) 50,000
Fibrinogen	339,700	32	700	38	—	—
Hemoglobin	64,500	—	—	—	4	16,000
Fibronectin[e]	440,000	—	85	85	—	—
Insulin	11,466	65	102	31	2	5,733
Lysozyme	13,930	46	—	—	—	—

[a] From Reference (114).
[b] From Reference (115).
[c] From Reference (116).
[d] From Reference (117).
[e] For spherical molecule, from Reference (95).

Protein molecules will adsorb onto all surfaces. However, the precise nature of the film will be dependent upon the nature of the surface. A consideration of what might be learned about these protein films is important in understanding biological interactions with them and, with particular relevance to this article, in describing the limitations and advantages of the XPS technique for their study. Thus, a complete characterization of the protein layer might include the following:

(1) The total amount of protein on a surface.
(2) The amount of each specific protein in a protein film adsorbed from a complex protein mixture such as blood plasma.
(3) The rate at which each individual protein (and the complete complex mixture) adsorbs onto the surface.
(4) The strength of binding of each protein to the surface (i.e., exchange or reversibility of adsorption).
(5) Chemical reactions of proteins on the surface (e.g., cleavage, complexation, modification).
(6) Configurational alterations of proteins upon adsorption (denaturation, partial or complete alteration of secondary or tertiary structure).
(7) Orientation of the protein molecules at the surface (i.e., for molecules such as IgG, pointing up or down?)
(8) The thickness of the protein layer.
(9) The uniformity of the protein layer (e.g., degree of surface coverage).

TABLE 4
Techniques Which Have Been Used To Explore the Configuration and/or Organization of Adsorbed Proteins

Technique	Information	Reference	Problems
Electron microscopy	(a) Triad structure of fibrinogen (b) Distribution on surface	3, 118, 119	Dehydrated films; preparation artifacts; low solution concentration of protein
Antisera binding	Organization	120–121	Mainly accessibility measured; specificity problems
Infrared spectroscopy	(a) Fraction of carbonyls bound (b) Shift of amide I band	122–124	Low specificity
Ellipsometry	Thickness of adsorbed protein film	3, 118–126	Many possible interpretations of data; limited number of substrates
Microcalorimetry	Heat released upon adsorption	127	Nonspecific; many interpretations of data possible
Electrophoretic mobility	Charge at particle surface	128	Interpretation complex
Circular dichroism	Conformational change	129, 130	Low sensitivity; transparent substrates

(10) The organization of a complex multilayer protein film (i.e., the position of each protein molecule in the horizontal and vertical direction, and the accessibility of each protein molecule at the surface).

Many techniques have been used successfully to study the features of protein films described in (1)–(5), above. These are reviewed in other chapters in this volume. However, attempts to study factors (6)–(10) above have often met with experimental or interpretational difficulty. Problems with the experimental techniques used for factors (6)–(10) are summarized in Table 4. XPS can provide new insights into various aspects of those factors and give certain types of information inaccessible by other techniques. In the next section XPS and its utility in protein studies will be discussed.

1.2. XPS FOR PROTEIN STUDIES

XPS (also referred to as electron spectroscopy for chemical analysis, ESCA) is a technique in which the energy and flux of electrons ejected

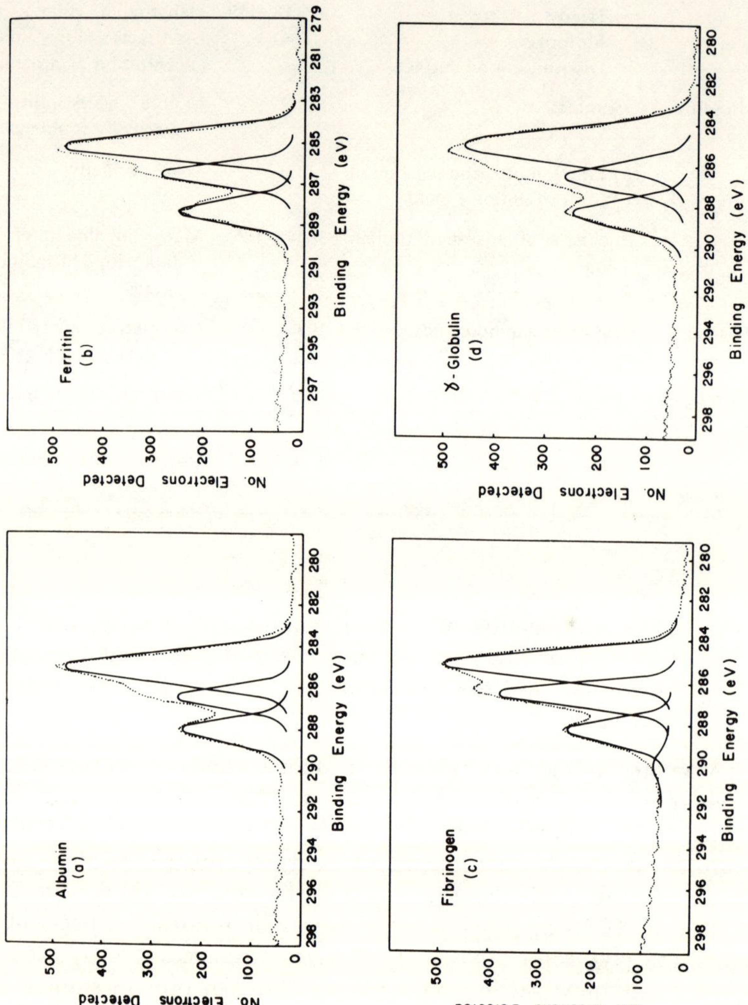
Albumin
(a)
Ferritin
(b)
Fibrinogen
(c)
γ - Globulin
(d)
No. Electrons Detected
Binding Energy (eV)

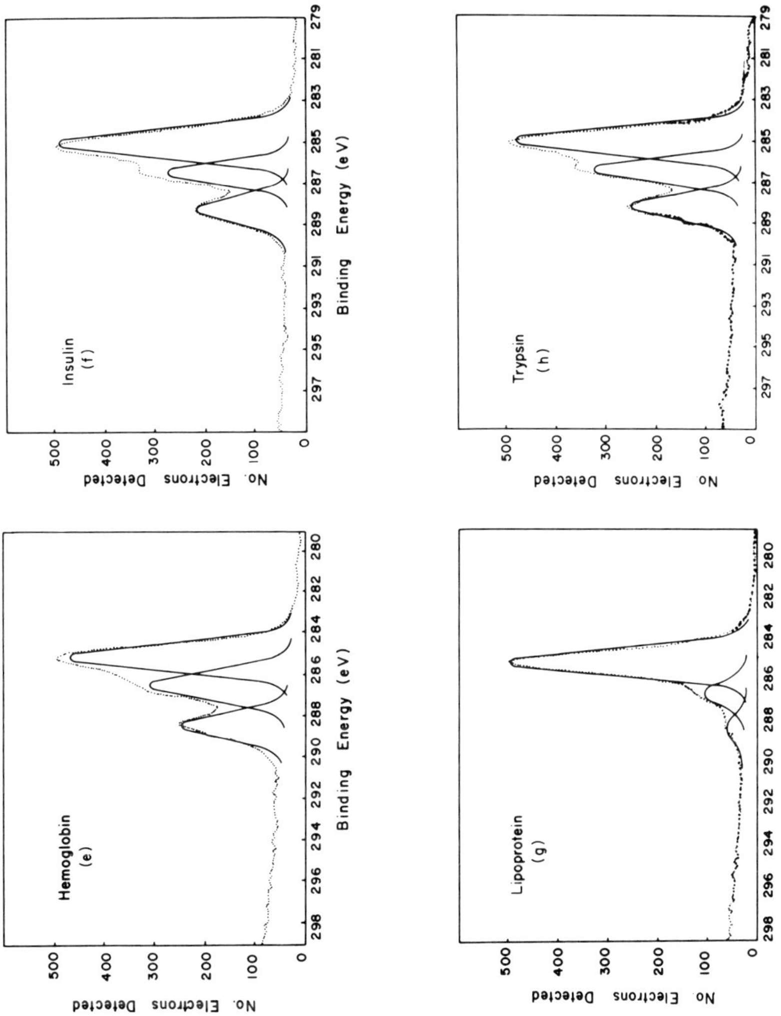

FIGURE 1. High-resolution C 1*s* spectra of selected proteins deposited as thick films.

from a surface by X-ray stimulated photoemission is measured. This technique is reviewed in Volume 1 of this series. The features and analytical capabilities of XPS which are of particular relevance for protein studies are the following:

(1) The ability to measure all elements present (except H) is potentially useful since a wide range of elements can be associated with specific proteins and other biomolecules.
(2) XPS typically samples to a depth of 50–100 Å. This is, in many cases, slightly more than the thickness of adsorbed protein films. This sampling depth is also somewhat greater than the thickness of individual protein molecules on a surface (see Table 3). Thus, both substrate and overlayer can be simultaneously studied.
(3) XPS studies can be performed on frozen, hydrated protein films.[(4)] This is potentially important since protein molecules probably denature in response to dehydration under high vacuum. Although XPS is not directly sensitive to the secondary or tertiary structure of protein molecules, such denaturation would be expected to alter the dimensions of the protein molecule and, consequently, may change the protein film thickness.
(4) XPS permits the examination of molecular bonding environments as well as the determination of which elements are present. This information can be useful for determining the oxidation state of various elements (e.g., S, Fe) or for learning about the specific carbon compounds present.
(5) Protein film thickness, homogeneity, and organization can be investigated through angular-dependent XPS studies.
(6) Sample preparation is simple.
(7) The XPS technique is, in most cases, nondestructive.

The specific advantages and disadvantages of XPS will now be considered by presenting a few examples. In Figure 1(a)–1(h), the carbon 1*s* spectra of a series of proteins deposited as thick (>200 Å) films are presented. Data derived from a complete analysis of the high resolution spectra of each protein are presented in Table 5. A few important conclusions can be drawn from these spectra and data. First, although in some cases substantial differences are noted in C 1*s* and elemental ratio data between proteins, it is unlikely that these data would be useful in an analytical sense in distinguishing which proteins are present in a complex film adsorbed from plasma or from other biological fluids. The exception to this might be if a particular unique signal (e.g., Zn for insulin, Fe for ferritin) were detected. On the other hand, distinctive C 1*s* peak shapes and elemental ratios might enable identification of a protein on a surface if that protein alone were present. The magnitude of the differences between proteins which might be expected can be estimated from Tables 2 and 5. However, the possibility of surface hydrocarbon contamination and its potentially deleterious effect on the ability to distinguish different proteins must always be considered.

Three areas in which XPS should be of particular value are in the measurement of the total amount of protein adsorbed, in the study of the modification, complexation, or reaction of proteins, and in the analysis of the structural organization of protein films. Each of these applications will be considered individually.

XPS is a quantitative technique. Therefore it can, under appropriate conditions, be used to estimate the level of protein on a surface. There are a number of factors which should be considered for this analysis. First the protein should contain an element not present in the substrate which will generate a unique XPS signal indicative of protein. With substrate materials which do not contain nitrogen, the nitrogen signal is useful to distinguish the protein. On Teflon the complete protein carbon signal can be used to indicate the protein since the C 1*s* signal from CF_2 groups in the Teflon is shifted to a higher binding energy than the C 1*s* signals generated by the functional groups in the protein.[4] Second, the protein film should not be too thick (<100 Å). Since XPS is a surface-sensitive technique, signals emanating from deep within a thick protein film are not detected. Thus, this method will only work well in the low range of adsorption levels. Finally, XPS will work best for measuring the amount of protein on a surface if a calibration curve is developed using data derived from an independent protein measurement method (e.g., radiolabeling). The advantages of the XPS method for such measurements are its high sensitivity

TABLE 5
Data Derived from XPS Spectra of Thick Protein Films[a]

	C 1*s* curve resolution (% area)			Molar % (omitting H)				
Protein	~288.3 eV	~286.5 eV	~285.0 eV	C	N	O	S	Other
Albumin	22.2	21.4	56.4	65.3	14.2	18.1	2.4	—
γ-Globulin	24.1	21.7	54.2	63.3	14.1	22.6	[b]	—
Fibrinogen	22.2	30.3	47.5	67.1	11.6	20.3	1.1	—
Hemoglobin	21.8	31.7	46.4	65.3	13.6	18.1	3.0	—
Insulin	20.7 ± 1.7	24.4 ± 3.1	55.4 ± 2.6	67.7	13.8	18.3	[c]	0.12 Zn
Ferritin	24.1	26.1	49.8	60.3	12.8	26.2	[c]	0.59 Fe
Trypsin	22.8	28.7	48.6	67.1	13.7	19.2	[c]	—
Lipoprotein	6.3	14.4	79.3	80.8	4.3	12.6	2.3	—

[a] Proteins prepared as 1 mg/ml solutions in water and air-dried on clean glass disks.
[b] Trace of sulfur detected.
[c] Not measured.

towards extremely low levels of adsorbed protein, its freedom from the necessity of using radiolabeled proteins (once a calibration curve is established), its specificity (XPS will not be "deceived" by adsorbed radioactive iodide ion, for example), and its ability to give other useful information simultaneously with the determination of protein levels.

XPS can be particularly useful for observing changes in the chemical nature of a protein film. Proteins on surfaces can be readily derivatized, and derivatization reagents can be chosen such that they contain unique atoms which are readily detected by XPS (e.g., halides, metals). Derivatization reactions which are useful in conjunction with XPS have been explored.[5,6] Complexation with inorganic or organic ions or even with other proteins is also potentially amenable to study. High sensitivity in the surface region coupled with the many levels of information obtainable from an XPS experiment make XPS a promising method for studying such reactions.

Three types of information might be derived in XPS studies of the organization of protein films. First, for uniform, relatively thin films, film

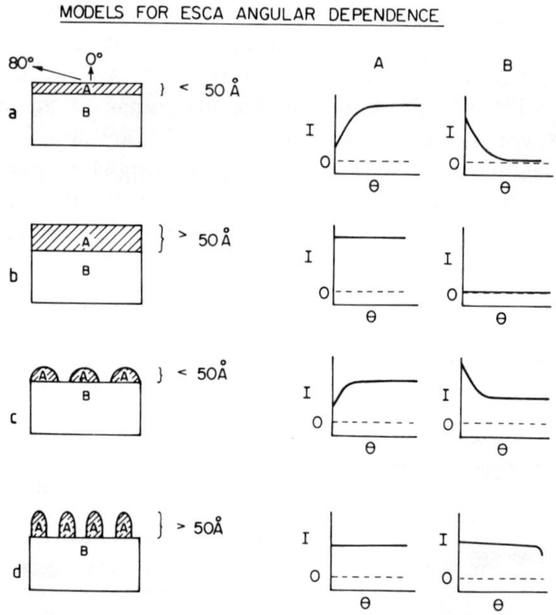

FIGURE 2. Qualitative angular-dependent XPS plots for various substrate-overlayer geometries. (a) Contiguous overlayer, thinner than the sampling depth. Surface sensitivity is enhanced at high photoelectron take-off angles. (b) Contiguous overlayer, thicker than the sampling depth. Substrate signal is not detected. (c, d) Patchy overlayer, thinner than the sampling depth and thicker than the sampling depth, respectively. The angular dependence of the XPS signals will depend upon the geometry of the system.

thickness might be estimated. Second, based upon the changes in intensity of a series of XPS peaks as the angle of the sample is varied with respect to the analyzer, the uniformity of coverage of the protein film might be ascertained. This principle is illustrated in Figure 2, and will be discussed in more detail when studies on protein films from our laboratory are presented. Finally, if a protein molecule has been tagged wth a unique atom to which XPS is particularly sensitive, its position within the protein film (vertical localization) should be amenable to analysis by angular-dependent studies.

It is interesting to speculate on the relationship between the data presented in Table 3, and XPS data interpreted using models such as those shown in Figure 2. If protein films on surfaces were found to be significantly thinner than the minimum globular dimension, the only reasonable explanation would be that denaturation has taken place. On the other hand, a discontinuous film in which the thickness of the protein patches was of approximately the same magnitude as the globular dimension of the protein could be interpreted as being indicative of adsorption in a relatively unaltered state. These situations have been observed in a previous study,[4] and this type of information can be used for generating corroborative evidence concerning alterations in conformation upon adsorption.

2. LITERATURE REVIEW—XPS OF BIOLOGICAL MOLECULES

In this section, the rapidly growing literature concerning the study of biological molecules by XPS will be sampled. This is useful to gain perspective on the wide range of applications of the technique in biology and in the examination of proteins at solid interfaces.

2.1. VALENCE BAND STUDIES

Ultraviolet photoelectron spectroscopy (UPS),[7] a technique related to XPS which probes the valence band of molecules in the gaseous state using a He(II) source, deserves to be mentioned as a powerful tool for studying the electronic structure of biological molecules.[8,9] This is of special interest in elucidating the activity of drugs,[10–14] and molecular orbital calculations are frequently used in assigning the spectral bands. An example is the UPS study of thalidomide,[15] a two-part molecule. It was found that the interaction of the nonteratogenic glutarimide and phthalimide subunits was "through-bond" rather than "through-space," and it was proposed that the resulting changes in the electronic structure may allow specific interactions with DNA. In the study of four opiate narcotics,[16] however, it was concluded that the molecular, rather than the electronic, structure was the prime determinant of analgesic activity.

A range of biological molecules have been ranked with respect to electron-donor ability by UPS.[17-18] Amino acids[19-21] and other biomolecules[22] have also been studied.

2.2. XPS OF SMALL MOLECULES

XPS studies have been made to correlate the electronic structure of biological molecules, as reflected by the core orbitals, with that calculated theoretically. Early comparisons of the C 1*s* and N 1*s* binding energies measured by XPS and calculated orbital energies for adenine,[23] cytosine, and thymine[24] were criticized on the basis of the inadequate spectral resolution of the technique.[25] Subsequent work[26,27] and the refinement of the theoretical calculations[28,29] have shown that such correlations are indeed valid.

Information obtained from XPS has been used to supplement that from UPS in a study of morphine and nalorphine.[14] The N 1*s* peaks of the two molecules were found to have identical energies, contradicting the popular belief that the differences in their activities are due to different charges on the nitrogen atoms.

XPS combined with CNDO/2 molecular orbital calculations has been applied to the study of amino acids, dipeptides, and polypeptides,[30] in which the N 1*s* peak was shown to reveal zwitterion structures. Uracil, thiouracil,[31] and pyrimidine and purine bases[32] have also been investigated in this manner.

2.3. XPS OF ORGANIC SULFUR

Initial concern about an anomalously high (4 eV) chemical shift reported for sulfur when complexed with Cu(II) in bean plastocyanin[33] emphasized the need for care in the preparation of samples for XPS, in this case to avoid sulfate impurities. The assignment of the high binding energy (h.b.e.) sulfur to a contaminant,[34] however, did not explain its apparent dependence on the presence of Cu(II).[35] Further investigation of the h.b.e. component[36] did not support another suggestion that it was a charge-transfer satellite,[37] but instead confirmed the occurrence of Cu-catalyzed radiation damage.[38] It appears that in the study of Cu(II) chelates by XPS, the nature of the material can be altered by interaction with the X-ray beam, possibly involving the internal redox reaction[39] $2Cu(II)SR \rightarrow Cu(I)_2RSSR$, or the Cu(II)-catalyzed conversion of sulfur to sulfone by residual water in the spectrometer vacuum.[40] Nonetheless, XPS has been successfully employed in the determination of sulfoxide in peptide molecules.[41]

2.4. XPS OF PORPHYRINS

Not surprisingly, the N 1*s* peak has been the center of attention in XPS studies of porphyrins, which are able to complex a metal ion via their four nitrogen atoms. Examination of the free base has revealed two distinct bonding environments for the nitrogen[42,43] implying that the two free-base protons are bound to one nitrogen atom each. In studies of metalloporphyrins, correlations have been found between the N 1*s* binding energy, metal electronegativity,[44] and nitrogen-to-metal atom bond length.[45] In addition, the N 1*s* peak was observed to possess a "shake-up" satellite at +2.8 eV,[46] and to have a line shape dependent on whether or not the metal ion was paramagnetic.[44] XPS has also been employed in the study of reduced porphyrins[47] and the binding of tetra(aminophenyl)porphyrins to glassy carbon.[48]

2.5. XPS OF METAL-CONTAINING PROTEINS

Studies of most metalloproteins have been complicated by the very low concentrations of metal ion present. Early XPS investigations of ferredoxins (non-heme iron proteins) were particularly subject to poor signal-to-noise ratios.[49] Nonetheless, it was possible to determine the oxidation state of the iron atoms[50] as well as information about magnetic state[51,52] from the exchange-splitting of the Fe 3*s* peak. XPS has been employed in conjunction with infrared and electron paramagnetic resonance spectroscopy to provide information on the structure of iron phthalocyanines and the spin multiplicity of the iron atoms.[53]

The chelation of Cd and Zn by sulfur in thionein[54] and of Cu and Zn by erythrocuprein[55,56] have been studied by XPS. In the latter work, a Cu:Zn ratio of 1:2.8 was observed, in contrast to a ratio of 1:1 obtained by atomic absorption spectrophotometry, from which it was deduced that the Cu atoms were located on the outer surface of the protein and the Zn atoms in the core. The Cu:Zn ratio observed by XPS indicates a burial depth of about 20 Å for the Zn atoms; however, the signal/noise ratios of the published spectra are rather poor. Studies have also been made which provide information on the redox behavior of Cu complexes in the spectrometer X-ray beam. In chelates of Cu(I) and Cu(II) with biological ligands,[57] photoreduction of Cu(II) was not observed, and it was found that the extent of the splitting of the Cu(II) 2*p* satellite was dependent on the ligand. In a Cu-penicillamine complex the metal atom was found in the Cu(I) state regardless of treatment with superoxide.[58] Cu(I)·SR was suggested as the stabilizing moiety. XPS has also featured as part of multitechnique characterizations of a Cu–AMP complex[59] and Cu(II) complexes of two newly synthesized tripeptides.[60]

In studies of DNA–Pt complexes,[61] XPS was able to differentiate between bound *cis*- and *trans*-$Pt(NH_3)_2Cl_2$. The N 1*s* peak shape changed upon the interaction of Pt(II) with the N^7(Gua) atoms in DNA. O^6(Gua) was found to participate in the binding of the *cis* (but not the *trans*) form of the Pt compound, affecting the specific hydrogen bonding characteristics of the guanine residues. This observation was related to the antitumor activity of *cis*-Pt compounds, which is not found in the *trans* forms.

XPS has also been employed in the study of selenium-containing amino acids,[62] Se-Glutathione peroxidase[63] and Hg-thionein.[64]

2.6. OTHER APPLICATIONS OF XPS IN BIOLOGY

The uses of XPS in biological studies have been reviewed by Millard.[65,66] To date, XPS studies of plasma proteins have not been used to provide new structural information, as the wide spectrum of bond types present can render a meaningful deconvolution of the carbon and oxygen peaks difficult.[67] However, the data in Figure 1 and Table 5 do indicate that, in some cases, such deconvolution can be useful. Also, it has been shown that the nitrogen peak consists of two major components, one characteristic of the backbone peptide linkages, the other of nitrogens in the side chains, which are capable of forming zwitterions and which show a chemical shift.[68,69] The quantitative nature of XPS has been demonstrated in the comparison of the observed and calculated elemental ratios for wool and plasma proteins,[65,66] and in the use of XPS for the estimation of protein grain,[70,71] utilizing the nitrogen and sulfur peaks. In the latter study, a good correlation was obtained between XPS and chemical analysis. XPS has also been used to quantitate the uptake of lysine-specific molecules by BSA,[65,72,73] and to monitor the loss of water on heating from chlorophyll *a* monohydrate.[74]

In the study of cells by XPS, *in situ* etching techniques have been employed to provide depth profiles. The presence of thallium in erythrocyte membranes has been reported in one study[75] using argon ion etching. Others have utilized oxygen plasma etching, which has the advantage of concentrating trace metals in an ash, although the technique therefore becomes less quantitative as etching proceeds. Examples include the study of bacterial[76] and mouse fibroblast[77] cells, in which polystyrene microspheres were used to calibrate the etching process, and the depth profiling of calcium in hepatoma cells during pyruvate-induced DNA synthesis.[78]

2.7. XPS OF SURFACE-ADSORBED BIOMOLECULES

To date little use has been made of XPS in evaluating proteins adsorbed on surfaces, which would be of interest in the biomaterials field. In one study, the results of *in vitro* blood compatibility testing of a series of polymeric materials have been correlated with the N 1*s* signal obtained from human *ex vivo* shunts (representing the protein adsorbed).[79] A study relevant to adsorbed films of biomolecules has been made of amino acid overlayers on silver,[80] in which the N 1*s* peak was used as an indicator of the adsorbed species. The total XPS signal was found to be proportional to the atomic density of the three amino acids studied, and values for the depths of the overlayers were estimated from the signal intensity as a function of the amino acid concentration in the solution from which it was adsorbed. A linear relationship was obtained, although island formation was inferred with adsorption from solutions of high concentration.

The N 1*s* peak was also utilized to monitor the elution of adsorbed plasma proteins from an 85%/15% poly(methyl methacrylate/acrylic acid) copolymer by a series of 20-min immersions in isotonic saline and alcohol/water mixtures of increasing strength.[81] Radiolabeled protein was used to provide a check on the ability of XPS to quantitate the adsorbed protein remaining after each wash (Figure 3).

The S 2*p* peak has been used as an indicator for heparin adsorbed on poly(amido amine)-grafted PVC.[67] The amount of adsorbed heparin was found to be independent of the graft levels used. A study of the anticoagulant effect of surface-immobilized heparin utilized the S 2*p* peak to monitor the ability of the heparin to penetrate a protein layer adsorbed from platelet-poor blood plasma.[82] The S 2*p* binding energy was used to differentiate

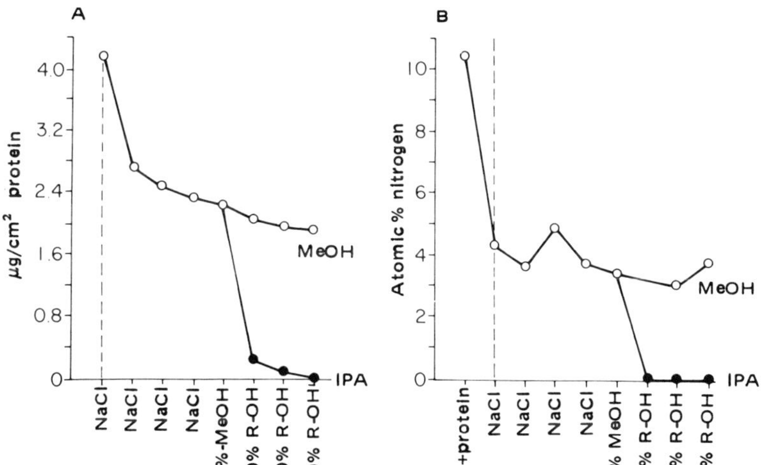

FIGURE 3. Elution of ^{125}I-labeled IgG from 85:15 poly(methyl methacrylate/acrylic acid), using a series of washes in isotonic saline followed by aqueous methanol or isopropanol solutions of increasing strength. (A) Determined by radiocounting, (B) by XPS.

between heparin (231.9 eV) and plasma protein (227.0 eV), and it was found that the adsorbed protein layer masked the heparin completely.

The contiguity of hemoglobin molecules in layers adsorbed on PTFE and platinum has been investigated by XPS,[4] several photoelectron take-off angles being employed. It was concluded that whereas a monolayer forms on platinum, patches of exposed substrate remained on PTFE. This observation that the coverage of polymeric substrates by proteins is incomplete, which is supported by XPS measurements of plasma proteins adsorbed on polyacrylics,[83] may have significance in the context of biomaterials research, and may open up new avenues for the uses of XPS in the study of protein films.

3. RECENT STUDIES ON ADSORBED PROTEIN FILMS

This section deals with recent applications of XPS in our laboratory in the study of the morphology of adsorbed protein layers. These data from angular-dependent XPS studies shed new light on the adsorption isotherms obtained using radiolabeled proteins.[3,84–91]

Previous XPS studies of the adsorption of hemoglobin on PTFE[4] and of other plasma proteins on various acrylic copolymers,[83] as well as investigations using transmission electron microscopy,[92,93] have indicated that the coverage of polymeric substrates by adsorbed protein is incomplete. In this section the results of preliminary XPS studies of the adsorption of hemoglobin and fibronectin on poly(vinyl fluoride) (PVF), poly(vinylidene fluoride) (PVF_2) and poly(tetrafluoroethylene (PTFE) are presented which further support that conclusion. Hemoglobin was selected as an example of a protein which inhibits the adhesion of cells to surfaces,[94] and fibronectin as an example of a protein which enhances cell adhesion and spreading.[95] The implications of these studies for cell adhesion will be the subject of future experimental investigations.

The proteins were radiolabeled with ^{125}I and diluted to the required concentrations. The polymer samples were then immersed for two hours in the protein solutions and rinsed using a solution displacement technique, to avoid passing the samples through an air interface. The amount of adsorbed protein was determined by radiocounting and converted to $\mu g/cm^2$ using the planar area of the sample. The specimens were then stored in buffer at 4 °C prior to examination by XPS.

To avoid the denaturation of the protein in the spectrometer vacuum, the samples with adsorbed hemoglobin were lightly blotted, then freeze-dried in the entry lock before examination at −50 °C. The condensation of carbonaceous contamination onto the samples was minimized by the use of a liquid nitrogen-cooled X-ray source.[96] Most of the fibronectin samples were dried in air and examined at ambient temperature. However, the results thereby obtained were consistent with those from fibronectin samples which were freeze-dried.

Of the four elements detected—carbon, nitrogen, oxygen, and fluorine—fluorine originates only from the substrate, nitrogen and oxygen almost entirely from the protein, and carbon from both. The fluorine signal is therefore a useful marker for the substrate, and its attenuation by a layer of protein (Figure 4) may be described by the Beer–Lambert law:

$$I_s = I_s^0[(1 - f) + f \cdot \exp(-d/\lambda \cos \theta)] \tag{1}$$

where I_s = observed intensity of the substrate signal, I_s^0 = the intensity of the signal from the clean substrate, f = the fraction of the substrate surface covered by protein, d = mean depth of the protein layer, λ = inelastic mean free path for the appropriate photoelectron, and θ = photoelectron take-off angle. The intensity of the signals characterizing the adsorbed protein will be governed by a similar relationship:

$$I_p = I_p^0 f[1 - \exp(d/\lambda \cos \theta)] \tag{2}$$

where I_p = observed intensity of the protein signal and I_p^0 = the intensity of the signal from a pure, thick sample of the protein. The carbon signal intensity will be described by a combination of Eqs. (1) and (2).

With I_s^0 and I_p^0 determined from standards, the equations may be solved for "f", the fraction of surface covered by protein, by substituting for "d" (in angstrom units):

$$d = 100(\mu g/cm^2)/1.4f \tag{3}$$

utilizing the value for $\mu g/cm^2$ protein obtained from radiocounting the labeled protein, and assuming a protein density of 1.4 g/cm^3.[97]

These calculations were performed on the XPS results obtained and average "f" values determined from the N, O, and F signals at two or three photoelectron take-off angles per sample. The signal intensities were internally standardized by expressing them as a percentage of the total for all

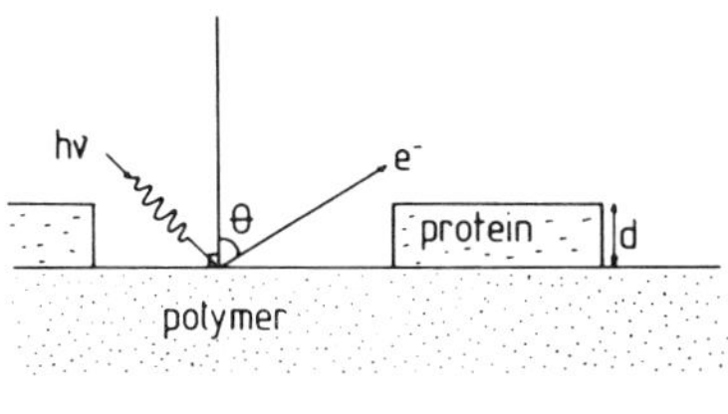

FIGURE 4. Model for adsorbed protein on a polymeric substrate. The protein is assumed to be of uniform thickness and to cover a fraction "f" (which may be 1) of the polymer surface. Shadowing effects are ignored.

four elements, and the predictive calculations outlined above developed accordingly.

As an example, the case of 0.42 $\mu g/cm^2$ of fibronectin adsorbed on poly(vinylidene fluoride) will be considered. At $\theta = 0°$, the XPS analysis of this sample gives an apparent surface composition of 62.9% C, 7.2% N, 16.2% O, and 13.6% F. Under these conditions the XPS analysis of the clean polymer was 54.3% C, 0.7% N, 3.8% O, and 41.3% F, and the analysis of a thick film of fibronectin on platinum gave 64.3% C, 12.5% N, and 23.2% O. Using this information for the I_s^0 and I_p^0 terms in Eqs. (1) to (3), it is possible to predict the apparent surface composition of a sample of the polymer with 0.42 $\mu g/cm^2$ of adsorbed fibronectin as a function of "f", employing λ values calculated according to Seah and Dench[98]:

f	%C	%N	%O	%F
1.0	63.6	10.2	20.3	5.9
0.8	62.1	8.8	17.8	11.3
0.65	60.8	7.5	15.7	15.9
0.6	60.4	7.1	14.9	17.6
0.55	59.9	6.6	14.1	19.4

The value of "f" which gives the figure of 7.2% N as observed experimentally is $f = 0.62$; to give agreement with %O = 16.2, $f = 0.69$, and to give agreement with %F = 13.6, $f = 0.72$.

For the same sample with $\theta = 60°$, the experimentally determined surface composition is 69.9% C, 6.9% N, 16.2% O, and 7.1% F. The I_s^0 and I_p^0 values are now obtained from an examination of the pure samples of polymer and protein at 60°. The polymer values are 55.2% C, 0.8% N, 5.0% O, and 39.0% F; and the protein values are 68.8% C, 9.5% N, and 21.8% O. Using Eqs. (1) to (3) again, it is possible to calculate the apparent surface composition as a function of "f":

f	%C	%N	%O	%F
1.0	68.6	9.2	21.4	0.77
0.8	66.1	7.5	18.3	8.1
0.7	64.8	6.6	16.7	11.8
0.6	63.5	5.7	15.1	15.7

The "f" values corresponding to the experimental data are: 0.74 on the basis of %N, 0.70 on the basis of %O, and 0.83 on the basis of %F.

The mean "f" value is calculated by averaging those estimated at both angles, dropping the highest and lowest:

$$f \pm \text{s.d.} = 0.71 \pm 0.02$$

The results of these calculations are presented in Figures 5 and 6. The

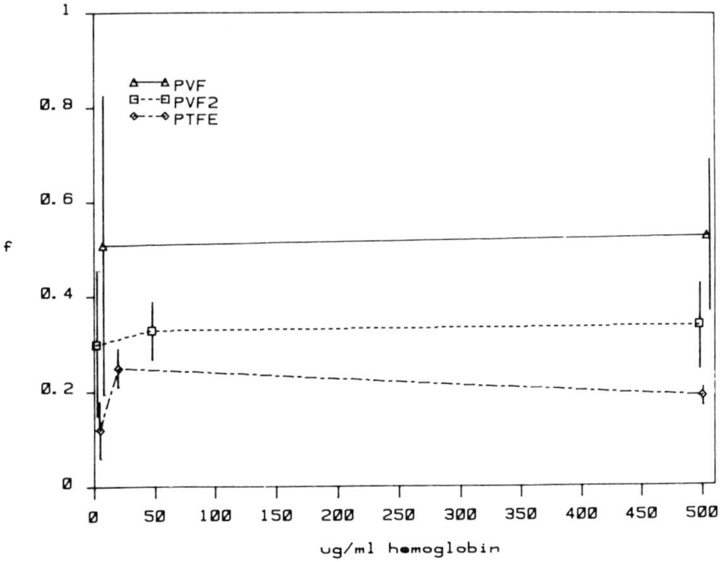

FIGURE 5. Fraction of surface covered by hemoglobin as determined by XPS versus protein concentration in solution. The error bars show ±2 s.d.'s.

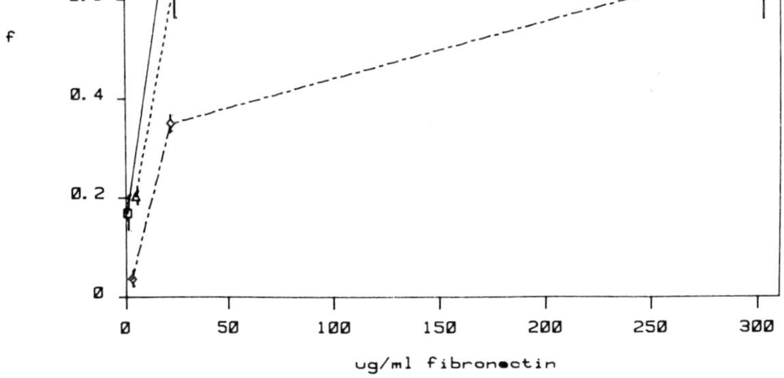

FIGURE 6. Fraction of surface covered by fibronectin as determined by XPS versus protein concentration in solution. The error bars show ±2 s.d.'s.

fraction of surface covered as determined by XPS is plotted against the protein concentration in solution.

Although the calculations of the coverage of the substrates by proteins are imprecise, it can be seen that the surface coverage rises, initially, as a function of protein concentration in solution, but that a continuous overlayer is not observed under the experimental conditions described here. In adsorption on the fluorine-containing substrates, the two proteins studied exhibit different surface coverage curves. Hemoglobin (Figure 5) reaches an early equilibrium with respect to surface coverage. The value of "f" at this equilibrium roughly correlates with the critical surface tension of the fluoropolymer (Figure 7), and lies well below 1. Fibronectin fails to reach such an equilibrium within the range of solution concentrations studied (Figure 6) and indeed appears to approach the formation of a continuous overlayer.

With best values for "f" thereby determined, the best values for "d", the mean protein layer depth, are given by Eq. (3). The values are listed in Table 6.

The data indicate that as the coverage of the surface by protein increases, so does the thickness of the protein layer, so that protein adsorbs to protein already bound as well as to fresh polymer surface. The observation of areas of bare polymer substrate after two hours in protein solution is indicative of a process whereby adsorbed protein may be desorbed again.

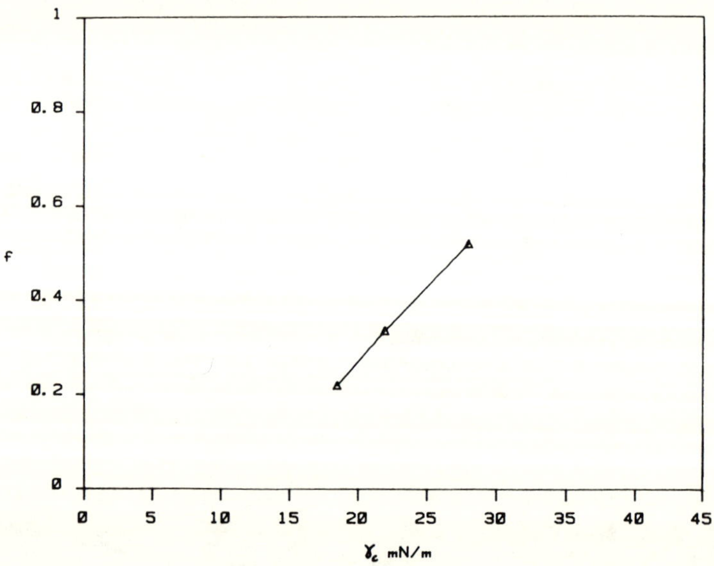

FIGURE 7. Fraction of surface covered by hemoglobin at 250 μg/ml solution concentration (from Figure 5) versus fluoropolymer critical surface tension.

TABLE 6
Estimated Mean Layer Depths (Å) of Protein Coverage on Fluoropolymer Substrates

	μg/ml Hemoglobin in solution				μg/ml Fibronectin in solution		
Substrate	10	20	50	500	1	20	300
PVF	18 ± 11			71 ± 22	18 ± 1	35 ± 4	73 ± 3
PVF_2	26 ± 13		50 ± 9	95 ± 25	26 ± 6	42 ± 1	137 ± 30
PTFE	65 ± 33	63 ± 10		177 ± 19	85 ± 33	108 ± 5	228 ± 37

This may involve a "configurational relaxation time,"[(99)] necessary to accomplish the conformational changes required for the adhesion of the protein to the substrate. Alternatively, it may be that adsorbed protein is not stabilized until noncovalent cross-links are formed with other adsorbing molecules.

The difference observed in the thickness of the fibronectin layer at very low levels of coverage on the three fluoropolymers in Table 6 may reflect conformational differences in the adsorbed fibronectin, if it is assumed that multilayer adsorption has not taken place from the most dilute protein solution. PVF and PVF_2 may adsorb fibronectin in its extended conformation and PTFE may adsorb the protein in its globular form.[(95)]

The foregoing data treatment is intentionally simple and does not address the question of the surface roughness of the polymer samples. Surface roughness or texture can have a significant effect on XPS data and has been considered from both theoretical and experimental standpoints in a number of papers.[(100–107)] TEM studies of plasma proteins adsorbed on Teflon[(92,93)] suggest that the dried films are extensively reticulated regardless of the polymer surface roughness. The data treatment for a noncontiguous protein film described above would still be valid on rough surfaces, except for a slight underestimation of the true surface area of the sample and an incomplete description of the photoelectron take-off angle. The importance of surface roughness in this application of XPS will be the subject of a future investigation.

In conclusion, a novel use of XPS as a tool for the quantitation of the coverage of various substrates by proteins has been described. It has been shown that the dried films of both hemoglobin and fibronectin adsorbed on fluoropolymers leave areas of substrate exposed. The extent of surface coverage has been shown to be a function of adsorbate, adsorbent, and protein concentration in solution. It is felt that the precision of the technique may be improved by the use of a photoelectron spectrometer with a better vacuum and higher count rate. With an instrument with a higher energy resolution, deconvolution of the C 1*s* peak may provide further information regarding the protein structure at the surface. New X-ray photoelectron spectrometers that also provide spatial resolution,[108] may be useful for observing the uniformity of protein coverage over biomedical devices and for estimating the size of the protein patches.[109]

ACKNOWLEDGMENTS

The authors wish to thank Barbara Pederson (University of Washington) and Richard Bigelow (Xerox Corporation) for assistance with the experimental work, as well as Dr. Thomas Horbett (University of Washington) and Dr. H. Ronald Thomas (Xerox Corporation) for valuable advice and discussion. Funding was provided by NHLBI grants HL 19419, HL 25951, and NIH DRR grant RR01296.

REFERENCES

1. T. A. Horbett, in: *Biomaterials: Interfacial Phenomena and Applications* (S. L. Cooper and N. A. Peppas, eds.), *Adv. Chem. Ser.* **199**, 233–244 (1982).
2. J. L. Brash, in: *Interaction of the Blood with Natural and Artificial Surfaces* (E. W. Salzman, ed.), pp. 37–60, Marcel Dekker Inc., New York (1981).
3. J. L. Brash and D. J. Lyman, in: *The Chemistry of Biosurfaces* (M. L. Hair, ed.), Vol. 1, pp. 177–232, Marcel Dekker Inc., New York (1971).
4. B. D. Ratner, T. A. Horbett, D. Shuttleworth, and H. R. Thomas, Analysis of the organization of protein films on solid surfaces by ESCA, *J. Colloid Interface Sci.* **83**, 630–642 (1981).
5. D. S. Everhart and C. N. Reilley, Chemical derivatization in electron spectroscopy for chemical analysis of surface functional groups introduced on low-density polyethylene film, *Anal. Chem.* **53**, 665–676 (1981).
6. D. Briggs and C. R. Kendall, Derivatization on discharge-treated LDPE: an extension of XPS analysis and a probe of specific interactions in adhesion, *J. Adhesion Adhesives, 2,* **1982** (January), 13–17.
7. D. Briggs, *Handbook of X-ray and Ultraviolet Photoelectron Spectroscopy*, Heyden, London (1977).
8. S. P. McGlynn, D. Dougherty, T. Mathers, and S. Abdulner, in: *Excited States in Organic Chemistry and Biochemistry* (B. Pullman and N. Goldblum, eds.), pp. 247–256, D. Reidel Publishing Company, Dordrecht, Holland (1977).
9. S. P. McGlynn, H. Aldrich, and D. Dougherty, The impact of photoelectron spectroscopy on biology, *J. Mol. Struct.* **45**, 119–138 (1978).
10. F. Kajfez, L. Klasinc, V. Sunjic, and T. Toth, Application of photoelectron spectroscopy to biologically active molecules and their constituent parts. II. (1). 1-4-Benzodiazepines, *J. Heterocycl. Chem.* **12**, 1175–1178 (1975).
11. F. Kajfez, L. Klasinc, and V. Sunjic, Application of photoelectron spectroscopy to biologically active molecules and their constituent parts. IV. Methylnitroimidazoles, *J. Heterocycl. Chem.* **16**, 529–531 (1979).

12. D. Ajo, I. Fragala, G. Granozzi, and E. Tondello, Electronic structure of xanthine and its biological methyl derivatives by UV photoelectron spectroscopy, *Spectrochim. Acta* **34A**, 1235–1238 (1978).
13. D. Ajo, M. Biagini Cingi, I. Fragala, and G. Granozzi, UV Photoelectron spectra of biological xanthines: theophylline, theobromine and caffeine, *Spectrosc. Lett.* **10**, 757–761 (1977).
14. H. E. Popkie, W. S. Koski, and J. J. Kaufman, *Ab-initio* LCAO-MO-SCF calculations of morphine and nalorphine and measurement of their photoelectron spectra, *J. Am. Chem. Soc.* **98**, 1342–1345 (1976).
15. L. Klasinc, N. Trinajstic, and J. V. Knop, Application of photoelectron spectroscopy to biologically active molecules and their constituent parts. VIII. Thalidomide, *Int. J. Quant. Chem. Quant. Biol. Symp.* **7**, 403–409 (1980).
16. L. Klasinc, B. Ruscic, A. Sabljic, and N. Trinajstic, Application of photoelectron spectroscopy to biologically active molecules and their constituent parts. 6. Opiate narcotics, *J. Am. Chem. Soc.* **101**, 7477–7482 (1979).
17. D. Dougherty, E. S. Younathan, R. Voll, S. Abdulnur, and S. P. McGlynn, Photoelectron spectroscopy of some biological molecules, *J. Electron Spectrosc. Relat. Phenom.* **13**, 379–393 (1978).
18. D. Dougherty and S. P. McGlynn, Photoelectron spectroscopy of carbonyls. Biological molecules, *J. Chem. Phys.* **67**, 1289–1290 (1977).
19. T. P. Debies and J. W. Rabalais, Electronic structure of amino acids and ureas, *J. Elect. Spectrosc. Rel. Phenom.* **3**, 315–322 (1974).
20. D. Dougherty, K. Wittel, J. Meeks, and S. P. McGlynn, Photoelectron spectroscopy of carbonyls. Ureas, uracils and thymine, *J. Am. Chem. Soc.* **98**, 3815–3820 (1976).
21. L. Klasinc, Application of photoelectron spectroscopy to biologically active molecules and their constituent parts. III. Amino acids, *J. Electron Spectosc. Relat. Phenom.* **8**, 161–164 (1976).
22. S. C. Khandelwal and J. L. Roebber, The photoelectron spectra of tetraphenylporphine and some metallotetraphenylporphyrins, *Chem. Phys. Lett.* **34**, 335–359 (1975).
23. M. Barber and D. T. Clark, Molecular core binding energies, carbon and nitrogen 1*s* levels for adenine, *J. Chem. Soc., Chem. Commun.* **1970**, 23–24.
24. M. Barber and D. T. Clark, Molecular core binding energies; carbon and nitrogen 1*s* levels for cytosine and thymine, *J. Chem. Soc., Chem. Commun.* **1970**, 24–25.
25. C. R. Brundle, Some recent advances in photoelectron spectroscopy, *Appl. Spectrosc.* **25**, 8–23 (1971).
26. R. Rein, A. Hartman, and S. Nir, On interpretation of ESCA spectra with application to thymine and other molecules, *Israel J. Chem.* **10**, 93–100 (1972).
27. A. Imamura, H. Fujita, and C. Nagata, Calculation of chemical shifts of inner-shell electrons for the DNA bases by CNDO/2 method including the charge redistribution effect, *J. Am. Chem. Soc.* **94**, 6287–6293 (1972).
28. K. Ishida, H. Kato, H. Nakatsuji, and T. Yonezawa, An MO interpretation of the chemical shifts of the binding energies of inner-shell electrons. II. A study of intramolecular shifts, *Bull. Chem. Soc. Japan* **45**, 1574–1575 (1972).
29. A. van der Avoird, On the relationship between molecular core binding energies and atomic charges, *J. Chem. Soc., Chem. Commun.* **1970**, 727–728.
30. D. T. Clark, J. Peeling, and L. Colling, An experimental and theoretical investigation of the core level spectra of a series of amino acids, dipeptides and polypeptides, *Biochim. Biophys. Acta* **453**, 533–545 (1976).
31. J. Peeling, F. E. Hruska, D. M. McKinnon, M. S. Chauhan, and N. S. McIntyre, ESCA studies of the uracil base. The effect of methylation, thionation and ionization on charge distribution, *Can. J. Chem.* **56**, 2405–2411 (1978).

32. J. Peeling, F. E. Hruska, and N. S. McIntyre, ESCA spectra and molecular charge distributions for some pyrimidine and purine bases, *Can. J. Chem.* **56**, 1555–1561 (1978).
33. E. I. Solomon, P. J. Clendening, H. B. Gray, and F. J. Grunthaner, Direct observation of sulfur coordination in bean plastocyanin by x-ray photoelectron spectroscopy, *J. Am. Chem. Soc.* **97**, 3878–3879 (1975).
34. J. Peeling, B. G. Haslett, I. M. Evans, D. T. Clark, and B. Boulter, Some observations on the ESCA spectra of plastocyanins, *J. Am. Chem. Soc.* **99**, 1025–1028 (1977).
35. J. A. Wurzbach, P. J. Grunthaner, D. M. Dooley, H. B. Gray, F. J. Grunthaner, R. R. Gay, and E. I. Solomon, Sulfur 2*p* photoelectron spectrum of *Limulus oxyhemocyanin.* Reply to observations on the ESCA spectra of plastocyanins, *J. Am. Chem. Soc.* **99**, 1257–1258 (1977).
36. V. Srinivisan, E. I. Stiefel, A. Elsberry, and R. A. Walton, X-ray photoelectron spectra of inorganic molecules. 21. Sulfur 2*p* chemical shifts associated with the binding of thiol and thioether groups to transition metal ions, *J. Am. Chem. Soc.* **101**, 2611–2614 (1979).
37. S. Larsson, Sulfur 2*p* photoelectron spectrum of blue copper proteins. Comment on papers by Solomon *et al.* and Peeling *et al.*, *J. Am. Chem. Soc.* **99**, 7708–7709 (1977).
38. M. Thompson, J. Whelan, D. J. Zemon, B. Bosnich, E. I. Solomon, and H. B. Gray, Sulfur 2*p* photoelectron spectrum of 1,8-bis(2-pyridyl)-3,6-dithiaoctane and its copper(II) complex. Possible interpretation of the S2*p* 168-eV peak in blue copper proteins, *J. Am. Chem. Soc.* **101**, 2482–2483 (1979).
39. R. A. Walton, X-ray photoelectron spectra of copper complexes considered as models for metalloproteins containing copper-sulfur bonds, *Inorg. Chem.* **19**, 1100–1101 (1980).
40. M. Thompson, R. B. Lennox and D. J. Zemon, Radiation-induced surface redox and shake-up structure in x-ray photoelectron spectra of copper(II) chelates, *Anal. Chem.* **51**, 2260–2263 (1979).
41. D. Jones, G. Distefano, C. Toniolo, and G. M. Bonora, A new method for determining sulfoxides in peptide molecules using x-ray photoelectron spectroscopy, *Biopolymers* **17**, 2703–2713 (1978).
42. M. V. Zeller and R. G. Hayes, X-ray photoelectron spectroscopic studies on the electronic structures of porphyrin and phthalocyanine compounds, *J. Am. Chem. Soc.* **95**, 3855–3860 (1973).
43. Y. Niwa, H. Kobayashi, and T. Tsuchiya, X-ray photoelectron spectroscopy of azaporphyrins, *Inorg. Chem.* **15**, 2336–2342 (1976).
44. D. H. Karweik and N. Winograd, Nitrogen charge distributions in free-base porphyrins and their reduced analogues observed by x-ray photoelectron spectroscopy, *Inorg. Chem.* **15**, 2336–2342 (1976).
45. D. K. Lavallee, J. Brace, and N. Winograd, X-ray photoelectron spectra of *N*-methyltetraphenylporphyrins: Evidence for a correlation of binding energies with metal–nitrogen bond distances, *Inorg. Chem.* **18**, 1776–1780 (1979).
46. A. J. Signorelli and R. G. Hayes, Studies of tetra-(*p*-fluorophenyl)porphyrins by x-ray photoelectron spectroscopy: effect of reduction on charge distribution in Fe compounds, *J. Chem. Phys.* **64**, 4517–4522 (1976).
47. Y. Niwa, X-ray photoelectron spectroscopy of reduced porphyrins, *J. Chem. Phys.* **62**, 737–738 (1975).
48. J. C. Lennox and R. W. Murray, Chemically modified electrodes. 10. Electron spectroscopy for chemical analysis and alternating current voltammetry of glassy carbon-bound tetra(aminophenyl)porphyrins, *J. Am. Chem. Soc.* **100**, 3710–3714 (1978).
49. L. N. Kramer and M. P. Klein, in: *Electron Spectroscopy* (D. A. Shirley, ed.), pp. 733–751, North-Holland, Amsterdam (1972).
50. D. Leibfritz, X-ray photoelectron spectroscopy of iron-containing proteins—the valency of iron in ferredoxins, *Angew Chem., Int. Ed.* **11**, 232–234 (1972).

51. P. T. Andrews, C. E. Johnson, B. Wallbank, R. Cammack, D. O. Hall, and K. K. Rao, X-ray photoelectron spectra of iron-sulfur proteins, *Biochem. J.* **149**, 471-474 (1975).
52. J. C. Chang, Y. M. Kim, and T. P. Das, Theory of exchange splitting of core levels in X-ray photoemission spectroscopy of hemin, *J. Chem. Phys.* **62**, 4350-4354 (1975).
53. S. Maroie, M. Savy, and J. J. Verbist, ESCA and EPR studies of monomer, dimer and polymer iron phthalocyanines: Involvements for the electrocatalysis of O reduction, *Inorg. Chem.* **18**, 2560-2567 (1979).
54. U. Weser, H. Rupp, F. Donay, F. Linnemann, W. Voelter, W. Voetsch, and G. Jung, Characterization of Cd, Zn-thionein (metalloprotein) isolated from rat and chicken liver, *Eur. J. Biochem.* **39**, 127-140 (1973).
55. G. Jung, M. Ottnad, W. Bohenkamp, and U. Weser, X-ray photoelectron spectroscopy (XPS) of bovine erythrocuprein, *FEBS Lett.* **25**, 346-348 (1972).
56. G. Jung, M. Ottnad, W. Bohenkamp, W. Bremser, and U. Weser, X-ray photoelectron spectroscopic studies of copper- and zinc-amino acid complexes and superoxide dismutase, *Biochim. Biophys. Acta* **295**, 77-86 (1973).
57. H. Rupp and U. Weser, Copper (I) and copper(II) in complexes of biochemical significance studied by X-ray photoelectron spectroscopy, *Biochim. Biophys. Acta* **446**, 151-165 (1978).
58. M. Younes and U. Weser, Superoxide dismutase activity of copper-penicillamine: Possible involvement of Cu(I) stabilized sulfur radical, *Biochem. Biophys. Res. Commun.* **78**, 1247-1253 (1977).
59. U. Weser, G.-J. Strobel and W. Voelter, ^{13}C-Nuclear magnetic resonance and x-ray photoelectron spectroscopy of Cu-AMP, *FEBS Lett.* **41**, 243-247 (1974).
60. Y. Sugiura, Newly synthesized sulfhydryl- and imidazole-containing tripeptides with a specific copper-binding site, *Inorg. Chem.* **17**, 2176-2182 (1978).
61. M. M. Millard, J. P. Macquet, and T. Theophanides, X-ray photoelectron spectroscopy of DNA·Pt complexes. Evidence of O^6(Gua)·N_7(Gua) chelation of DNA with *cis*-dichlorodiamine platinum (II), *Biochim. Biophys. Acta* **402**, 166-170 (1975).
62. H. Rupp and U. Weser, X-ray photoelectron spectroscopy of some selenium-containing amino acids, *Bioinorg. Chem.* **5**, 21-32 (1975).
63. D. Chiu, A. L. Tappel, and M. M. Millard, Improved procedure for x-ray photoelectron spectroscopy of selenium-glutathione peroxidase and application to the rat liver enzyme, *Arch. Biochm. Biophys.* **184**, 209-214 (1977).
64. G. Sokolowski, W. Pilz, and U. Weser, X-ray photoelectron spectroscopic properties of Hg-thionein, *FEBS Lett.* **48**, 222-225 (1974).
65. M. M. Millard, in: *Protein-Metal Interactions* (M. Friedman, ed.), pp. 589-619, Plenum Press, New York (1974).
66. M. M. Millard, Surface characterization of biological materials by x-ray photoelectron spectroscopy, *Contemp. Top. Anal. Clin. Chem.* **1978** (series 3), 1-55.
67. P. Ferruti, R. Barbucci, N. Danzo, A. Torrisi, O. Puglisi, S. Pignataro, and P. Spartano, Preparation and ESCA characterization of poly(vinyl chloride) surface-grafted with heparin-complexing poly(amido amine) chains, *Biomaterials* **3**, 33-37 (1982).
68. C. R. Cothern, W. E. Moddeman, R. G. Albridge, W. J. Sanders, P. L. Kelly, W. S. Hanley, and L. Field, Some bonding properties of D-penicillamine and related compounds measured by X-ray photoelectron spectroscopy, *Anal. Chem.* **48**, 162-166 (1976).
69. L. D. Hulett and T. A. Carlson, Analysis of compounds of biological interest by electron spectroscopy, *Clin. Chem.* **16**, 677-680 (1970).
70. M. P. Klein and L. N. Kramer, Determination of protein quantity and quality by x-ray photoelectron spectroscopy, *Symp. Seed Proteins* **1972**, 265-276.
71. J. Peeling, D. T. Clark, I. M. Evans, and D. Boulter, Evaluation of ESCA technique as a screening method for the estimation of protein content and quality in seed meals, *J. Sci. Food Agric.* **27**, 331-340 (1976).

72. M. M. Millard and M. S. Masri, Detection and analysis of protein functional group modification by X-ray photoelectron spectrometry, *Anal. Chem.* **46**, 1820–1822 (1974).
73. M. M. Millard and M. Friedman, X-ray photoelectron spectroscopy of BSA and ethyl vinyl sulfone modified BSA, *Biochem. Biophys. Res. Commun.* **70**, 445–451 (1976).
74. N. Winograd, A. S. Shepard, D. H. Karweik, V. J. Koester, and F. K. Fong, X-ray photoelectron spectroscopic studies of the thermal stability of chlorophyll *a* monohydrate, *J. Am. Chem. Soc.* **98**, 2369–2370 (1976).
75. R. G. Meisenheimer, J. W. Fischer, and S. J. Rehfeld, Thallium in human erythrocyte membranes: An X-ray photoelectron spectroscopy study, *Biochem. Biophys. Res. Commun.* **68**, 994–999 (1976).
76. M. M. Millard, R. Scherrer, and R. S. Thomas, Surface and depth profile composition of bacterial cells by X-ray photoelectron spectroscopy and oxygen plasma etching, *Biochem. Biophys. Res. Commun.* **72**, 1209–1217 (1976).
77. M. M. Millard and J. C. Bartholemew, Surface studies of mammalian cells grown in culture by x-ray photoelectron spectroscopy, *Anal. Chem.* **49**, 1290–1296 (1977).
78. L. Pickart, M. M. Millard, B. Beiderman, and M. M. Thaler, Surface analysis and depth profiles of calcium in hepatoma cells during pyruvate-induced DNA synthesis, *Biochem. Biophys. Acta* **544**, 138–143 (1978).
79. J. P. Fischer, P. Fuhge, K. Burg, and N. Heimburger, Methoden zur Herstellung und Charakterisierung von Kunststoffen mit verbesserter Blutverträglichkeit, *Angew. Makromol. Chem.* **105**, 131–165 (1982).
80. R. J. Colton, J. S. Murday, J. R. Wyatt, and J. J. DeCorpo, Combined XPS and SIMS study of amino acid overlayers, *Surface Sci.* **84**, 235–248 (1979).
81. D. K. Gilding, R. W. Paynter and J. E. Castle, Quantitative evaluation of water structuring and protein adsorption on the suface of hydrophilic polymers by ESCA, *Biomaterials* **1**, 163–168 (1980).
82. C. D. Ebert and S. W. Kim, Immobilized heparin: spacer arm effects on biological interactions, *Thromb. Res.* **26**, 43–57 (1982).
83. R. W. Paynter, *The Use of X-ray Photoelectron Spectroscopy to Study the Interaction of Biomaterials with Physiological Media*, PhD Thesis, University of Surrey, U.K. (1981).
84. W. H. Grant and R. E. Dehl, in: *Adhesion and Adsorption of Polymers* (L. H. Lee, ed.), Part B, pp. 827–835, Plenum Press, New York (1980).
85. S. W. Kim and R. G. Lee, in: *Applied Chemistry at Protein Interfaces* (R. E. Baier, ed.), *Adv. Chem. Ser.* **145**, 218–229 (1975).
86. R. G. Lee and S. W. Kim, Adsorption of proteins onto hydrophobic polymer surfaces: Adsorption isotherms and kinetics, *J. Biomed. Materials Res.* **8**, 251–259 (1974).
87. W. J. Dillman, Jr. and I. F. Miller, On the adsorption of serum proteins on polymer membrane surfaces, *J. Colloid Interface Sci.* **44**, 221–241 (1973).
88. W. Norde, in: *Adhesion and Adsorption of Polymers* (L. H. Lee, ed.), Part B, pp. 801–825, Plenum Press, New York (1980).
89. P. K. Weathersby, T. A. Horbett, and A. S. Hoffman, Fibrinogen adsorption to surfaces of varying hydrophilicity, *J. Bioeng.* **1**, 395–410 (1977).
90. T. A. Horbett, P. K. Weathersby, and A. S. Hoffman, The preferential adsorption of hemoglobin to polyethylene, *J. Bioeng.* **1**, 61–78 (1977).
91. B. K. Lok, Y.-L. Cheng, and C. R. Robertson, Protein adsorption on crosslinked polydimethylsiloxane using total internal reflection fluorescence, *J. Colloid Interface Sci.* **91**, 104–116 (1983).
92. R. C. Eberhart, M. E. Lynch, F. H. Bilge, and H. A. Arts, Effects of fluid shear and temperature on protein adsorption on Teflon surfaces, *Trans. Am. Soc. Art. Int. Org.* **26**, 185–192 (1980).

93. R. C. Eberhart, M. E. Lynch, F. H. Bilge, J. F. Wissinger, M. S. Muro, S. R. Ellsworth, and A. J. Quattrone, in: *Biomaterials: Interfacial Phenomena and Applications* (S. L. Cooper and N. A. Peppas, eds.), *Adv. Chem. Ser.* **199**, 293-315 (1982).
94. T. A. Horbett, unpublished data.
95. F. Grinnell and M. K. Feld, Adsorption characteristics of plasma fibronectin in relationship to biological activity, *J. Biomed. Materials Res.* **15**, 363-381 (1981).
96. D. T. Clark, H. R. Thomas, A. Dilks, and D. Shuttleworth, A method to reduce the hydrocarbon contamination of samples in X-ray photoelectron spectroscopy, *J. Electron Spectrosc. Relat. Phenom.* **10**, 455-460 (1977).
97. K. O. Pedersen, in: *Comprehensive Biochemistry* (M. Florkin and E. H. Stotz, eds.), Vol. 7, p. 212, Elsevier, Amsterdam (1963).
98. M. P. Seah and W. A. Dench, Quantitative electron spectroscopy of surfaces: A standard data base for electron inelastic mean free paths in solids, *Surface Interface Anal.* **1**, 2-11 (1979).
99. B. D. Fair and A. M. Jamieson, Studies of protein adsorption on polystyrene latex surfaces, *J. Colloid Interface Sci.* **77**, 525-534 (1980).
100. C. S. Fadley, R. J. Baird, W. Siekhaus, T. Novakov and S. A. L. Bergstrom, Surface analysis and angular distributions in x-ray photoelectron spectroscopy, *J. Electron Spectrosc. Relat. Phenom.* **4**, 93-137 (1974).
101. P. H. Holloway, The effect of surface roughness on Auger electron spectroscopy, *J. Electron Spectrosc. Relat. Phenom.* **7**, 215-232 (1975).
102. M. F. Ebel, Zur Bestimmung der Reduzierten Dicke d/λ dünner Schichten mittels XPS, *J. Electron Spectrosc. Relat. Phenom.* **14**, 287-322 (1978).
103. Y. M. Cross and J. Dewing, Thickness measurements on layered materials in powder form by means of XPS and ion sputtering, *Surface Interface Anal.* **1**, 26-31 (1979).
104. O. K. T. Wu, G. G. Peterson, W. J. LaRocca, and E. M. Butler, ESCA signal intensity dependence on surface area (roughness), *Appl. Surface Sci.* **11/12**, 118-130 (1982).
105. M. F. Ebel, Determination of reduced thickness by means of the variable take-off angle technique, *J. Electron Spectrosc. Relat. Phenom.* **22**, 333-346 (1981).
106. M. F. Ebel, Evaluating ratio data for the determination of reduced thickness by XPS, *Surface Interface Anal.* **3**, 149-152 (1981).
107. M. F. Ebel and J. Wernisch, Shading at different take-off angles in x-ray photoelectron spectroscopy, *Surface Interface Anal.* **3**, 191-193 (1981).
108. C. Bryson, *Impact of improved spatial resolution in ESCA*, presented at 4th Symposium on Applied Surface Analysis, Dayton, Ohio, June, 1982.
109. M. F. Ebel, G. Rittler, J. Wernisch and F. Primik, X-ray photoelectron spectroscopy applied to insular films, *J. Electron Spectrosc. Relat. Phenom.* **25**, 67-73 (1982).
110. F. W. Putnam, in: *The Proteins. Composition, Structure and Function*, 2nd edition (H. Neurath, ed.), Vol. 3, p. 180, Academic Press, New York (1965).
111. M. O. Dayhoff and R. V. Eck, *Atlas of Protein Sequence and Structure 1967-68*, pp. 124, 137, National Biomedical Research Foundation, Silver Spring, Md.
112. M. W. Mosesson, A. B. Chen, and R. M. Huseby, The cold insoluble globulin of human plasma: Studies of its essential structural features, *Biochim. Biophys. Acta* **386**, 509-524 (1975).
113. G. R. Tristam and R. H. Smith, in: *Advances in Protein Chemistry* (C. B. Anfinsen, Jr., ed.), Vol. 18, pp. 227-318, Academic Press, New York (1963).
114. H. A. Sober, *Handbook of Biochemistry*, pp. C10-C12, Chemical Rubber Co.. Cleveland (1970).
115. H. A. Scheraga, *Protein Structure*, p. 190, Academic Press, New York (1969).
116. E. G. Young, in: *Comprehensive Biochemistry* (M. Florkin, ed.), Vol. 7, p. 50, Elsevier, Amsterdam (1963).

117. H. E. Schultze and J. F. Heremans, *Molecular Biology of Human Proteins*, Vol. 1, p. 34, Elsevier, Amsterdam (1966).
118. R. R. Gorman, G. E. Stoner, and A. Catlin, The adsorption of fibrinogen. An electron microscope study, *J. Phys. Chem.* **75**, 2103–2106 (1971).
119. R. C. Eberhart, L. D. Prokop, J. Wissenger, and M. A. Wilkov, Observation of Albumin deposits on Teflon surfaces, *Trans. Am. Soc. Art. Int. Org.* **23**, 124–139 (1977).
120. J. E. Rubin, K. Fani, E. A. Friedman, and G. M. Berlyne, Uses of fluorescent antisera (FAS) to identify blood components deposited on dialyzer membranes, *Trans. Am. Soc. Art. Int. Org.* **24**, 471–473 (1978).
121. L. Vroman and A. L. Adams, Peculiar behavior of blood at solid interfaces, *J. Polym. Sci., Part C*, No. 34, 159–165 (1971).
122. B. W. Morrissey, The adsorption and conformation of plasma proteins: A physical approach, *Ann. N.Y. Acad. Sci.* **283**, 50–64 (1977).
123. R. J. Jakobsen and R. M. Gendreau, Blood plasma/implant interfaces. FT-IR studies of adsorption on polyethylene and heparin-treated polyethylene surfaces, *Art. Org.* **2**, 183–188 (1978).
124. S. I. Stupp, J. W. Kauffman, and S. H. Carr, Interactions between segmented polyurethane surfaces and the plasma protein fibrinogen, *J. Biomed. Materials Res.* **11**, 237–250 (1977).
125. L. Vroman and A. L. Adams, Identification of rapid changes at plasma–solid interfaces, *J. Biomed. Materials Res.* **3**, 43–67 (1969).
126. B. W. Morrissey, L. E. Smith, R. R. Stromberg, and C. A. Fenstermaker, Ellipsometric investigation of the effect of potential on blood protein conformation and adsorbance, *J. Colloid Interface Sci.* **56**, 557–563 (1976).
127. E. Nyilas, T. H. Chiu, and G. A. Herzlinger, Thermodynamics of native protein/foreign surface interactions. 1. Calorimetry of the human γ-globulin/glass system, *Trans. Am. Soc. Art. Int. Org.* **20**, 480–490 (1974).
128. D. Chattoraj and H. Bull. Electrophoresis of adsorbed protein, *J. Am. Chem. Soc.* **81**, 5128–5132 (1959).
129. C. R. McMillin and A. Walton, A circular dichroism technique for the study of adsorbed protein structure, *J. Colloid Interface Sci.* **48**, 345–349 (1974).
130. T. Akaike, Y. Sakurai, K. Kosuge, K. Kuwana, A. Katoh, T. Ohwaki, S. Migata, K. Katouka, and T. Tsuruta, Study of the interaction between plasma proteins and synthetic polymers by circular dichroism, *ACS Polym. Prepr., Am. Chem. Soc., Div. Polym. Chem.*, **20**(1), 581–584 (1979).

6

Chromatographic Analysis of Protein Adsorption

Bernard Koltisko and Alan Walton

1. INTRODUCTION

The interfacial adsorption of proteins is of fundamental importance to both the health sciences and certain biotechnology industries. In recent years, there has been an increase in study of adsorption mechanisms using flow or dynamic systems.[1-9] These types of systems more accurately model the protein–surface interactions experienced in vascular grafts, catheters, and *ex vivo* blood flow lines used in surgical procedures. In addition, it has been shown that the dynamics of an adsorbed protein layer, in terms of turnover and exchange, are influenced by flow rate, and differ from those measured under static conditions.[3,10] Thus, the use of flow systems may not only be desirable, but may be necessary to fully understand the complexities of the adsorption process under physiological conditions.

When a flow system is chosen to study protein adsorption, the theory and techniques of chromatographic analysis can be employed. Traditionally, this versatile analytical tool is thought of as a separation process. However, a wide range of other physicochemical aspects such as solution thermodynamics, intermolecular interactions, catalysis, and reaction kinetics have now been investigated by chromatographic techniques.[11,12]

The goal of this chapter is to acquaint the researcher with the principles of chromatographic analysis and show how these can be used to study the adsorption process. While some of the references quoted deal with protein adsorption directly, several are concerned with the adsorption of small

---

Bernard Koltisko • Department of Macromolecular Science, Case Western Reserve University, Cleveland, Ohio 44106. *Alan Walton* • University Genetics Company, 537 Newton Avenue, Norwalk, Connecticut 06852. Correspondence should be addressed to Bernard Koltisko.

molecules in both gas and liquid systems. The overall modeling in these latter examples can be generalized on a phenomenological basis to any system, and these studies have been cited to illustrate conceptually and experimentally the types of data available through chromatographic analysis.

2. GENERAL THEORY

For any flow system, the distribution of solute in solution and adsorbed can be described by the following mass balance equation:

$$D_{\mathrm{ax}}\frac{\partial^2 c}{\partial x^2} - v\frac{\partial c}{\partial x} - \frac{\partial q}{\partial t} = \frac{\partial c}{\partial t} \tag{1}$$

where D_{ax} is an average solute diffusion coefficient accounting for fluctuations in the diffusion process axially or across an adsorption column or tube, x is the position along the flow direction, c and q are the concentrations of the solute in solution and adsorbed, respectively, v is the flow velocity, and t is the time variable. This statement is general for any flow system, and assumes one solute and isothermal conditions. The equation describes the three principal factors of diffusion, convection, and adsorption which influence the final solute distribution function ($\partial c/\partial t$). It is this function which is detected experimentally by following either the depletion of protein from solution or the gain of protein on the adsorbent. The former case is more prevalent in chromatographic analysis, and usually employs a detector of moderate sensitivity, such as UV–vis or refractive index, placed downstream of a high surface area adsorbate. For lower surface area materials, radiolabeling of the protein can be employed. Great advances have been made in this area to improve detection capabilities, particularly in the use of fluorescence and IR spectroscopy, which can now allow rapid measurement of protein uptake, and also detection of conformational state of the adsorbed protein.[8,13]

The convective process is given by the second term of Eq. (1). This function defines concentration as a function of position in a flow stream, which is dependent on flow rate and time. In the case of tubing or other simple flow lines, the flow velocity v is the volumetric flow rate w divided by the cross-sectional area of the line. For a column packed with adsorbent material, the relationship

$$v = \frac{wL}{\alpha} \tag{2}$$

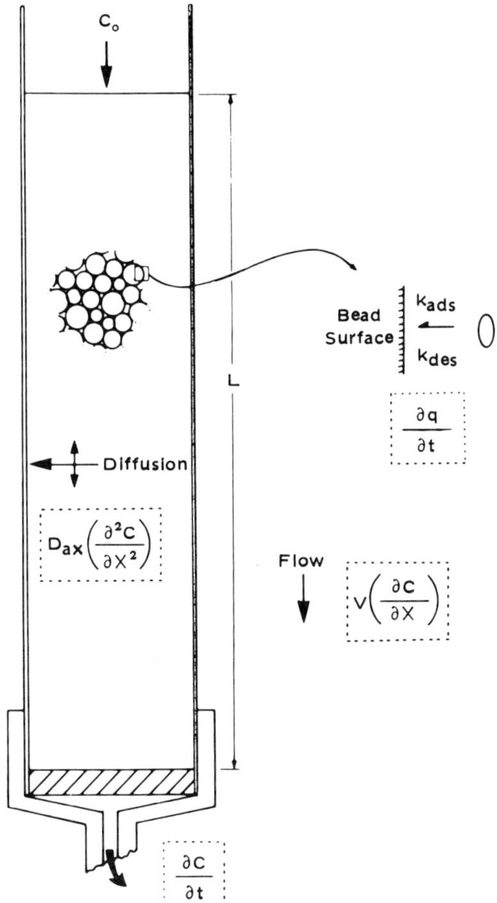

FIGURE 1. Typical column chromatographic system illustrating the contributing elements to the generalized effluent profile equation.

can be used, where L is the column length and α is the interstitial volume of the mobile phase, referred to as the void volume.

In choosing a flow rate, several factors must be considered. The type of flow pattern, either laminar or turbulent, can be predicted by evaluating the Reynolds number

$$Re = \frac{\rho v d}{\eta} \tag{3}$$

where ρ and η are the solution density and viscosity, respectively, and d is the flow path diameter. For a straight tube, laminar flow is maintained if Re is below 2100.[14] Due to the packing material, large fluctuations in flow velocity and direction are encountered in an adsorption column, and *Re* values of about 1–100 should be maintained to ensure laminar flow.[15] In this case, *Re* is approximated by substituting *dp*, the particle diameter, for *d*.[16] Whole blood in the body exhibits a wide range of *Re* values comprising both the laminar and turbulent regime.[17]

Evaluation of shear rate is also important due to the conformational sensitivity of proteins. Shear rate is defined by the change in flow velocity from the surface of the adsorbent to some point perpendicular to the surface in the flow stream. In a straight tube under laminar flow conditions, the wall shear rate, $\dot{y}$, can be evaluated by the following relation

$$\dot{y} = \frac{32w}{\pi d^3} \tag{4}$$

where d is the straight tube diameter. As a comparison, shear rates on the order of $10^2\ \mathrm{s}^{-1}$ are encountered by whole blood *in vivo*.[17] Due to the irregular geometry and range of flow velocities, shear rate cannot be readily determined in a packed column system.

While pure convection does not change the concentration profile of the solute, diffusion of the solute is a major contributor to the broadening of the elution profile. Evaluated as a Fick's second law process, these effects are given in the first term of Eq. (1). In comparison to unperturbed bulk solutions, molecular diffusion may be hindered by the presence of a solid packing material in a column. This packing also disturbs normal flow patterns across the column, altering the velocity of the molecules and creating random lateral and forward migration of molecules.[15] The process is coupled with normal molecular diffusion and has been termed convective diffusion, flow dispersion, or eddy diffusion. Thus, in an adsorption column, one measures an effective diffusion coefficient which reflects these additional elements. This is a factor of the column packing structure, determined by the size of the channels open for migration of the solute around the particles. In addition, a "wall effect" is present where perturbation in the packing structure occurs near the wall of a column resulting in a velocity profile dependent upon radial position in the column.[18]

Porous substrates allow inter-particle diffusion to occur. Rates of interaction in traditional adsorption or ion-exchange resins, such as silica gel and divinylbenzene–styrene beads, depend upon the inter-particle diffusion term, since the bulk of interaction sites occur on the inside of the particle.[19,20] Normally, most modeling procedures have treated this diffusional term separately from the external diffusion term by simply partitioning the total mobile phase volume into both inter- and intra-particle volume elements.[21,22]

The third and last term is a measure of the adsorption process. In evaluation of this term, the functional relationship between the amount adsorbed and the solution concentration, $q = f(c)$, must be considered. Plots of these functions, called isotherms, assume several forms dependent upon the adsorption mechanism. Several standard isotherm types have been used in predicting effluent solute distribution profiles from adsorption columns, including constant partition (linear),[23,24] Langmuir,[25,26] and Freundlich isotherms.[27,28]

3. FRONTAL ANALYSIS

One chromatographic technique which is well-suited to the study of the adsorption process is frontal analysis. In this technique, a continuous stream of solution containing adsorbate is applied to the adsorption column. After elution, the initial sharp boundary of the solute will be retarded as compared with the void volume and/or changed in shape, ranging from sharp to extremely diffuse. One may analyze the breakthrough profile as the protein solution begins to elute from the column or the tail profile produced after the protein solution is switched to pure solvent.

Originally, frontal analysis was developed for separation purposes, based on the relative adsorption affinities of different solutes in a mixture. This technique was applied to many substances including proteins,[29] but was limited as a separation technique due to the need for weakly sorbing compounds. However, it has been observed that the type of isotherm of the solute contributes to the shape of the effluent concentration profile. The slope of a Langmuir-type isotherm decreases with increasing concentration. Higher concentration regions "spill over" to maintain the sharp front. In contrast, the tail profile will be diffuse. An S-type isotherm (convex) in which the isotherm slope increases with concentration produces a diffuse front and a sharp tail profile.[30] A linear isotherm at low concentration exhibits equivalent front and rear boundaries.[12]

Two methods can be used to construct isotherms using frontal analysis. The first is termed the integral method, and can be applied to either a sharp or diffuse front. The bounded area in Figure 2 is simply integrated to obtain q at the applied c value. This result will give one point on the adsorption isotherm. Irreversibly bound material does not interfere with the analysis, and the data obtained by this dynamic method can agree with static measurements.[31]

A second technique, also illustrated in Figure 2, is the point technique in which a diffuse front or diffuse tail is analyzed to obtain the whole isotherm of a system up to the applied value of c.[32] This is accomplished

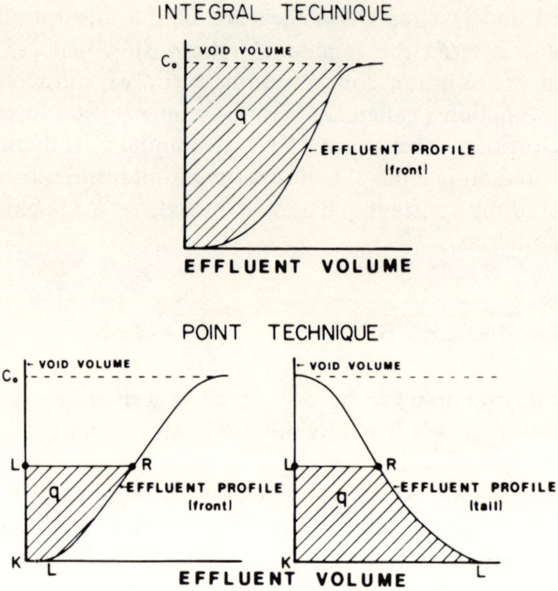

FIGURE 2. The integral and point techniques used in constructing adsorption isotherms.

by choosing a characteristic point, such as R in Figure 2, and drawing a line LR parallel to the baseline. The shaded area is equal to q at $c = R$, and thus plots of c_R versus q_R define the isotherm. This technique has been used in both Langmuir[33] and Freundlich isotherm[34] cases, and assumed both rapid equilibrium and no diffusional contribution. The latter effect can be evaluated by following the elution profile of a similar, non-sorbing compound and subtracting its contribution to the adsorbate's elution profile.[35,36]

Dynamic measurements, such as mass-transfer kinetics and diffusion, are made accessible through frontal analysis by the following experiment,[37] illustrated in Figure 3. If the leading edge of a front is marked with a radiolabeled component, the observed sorption profile remains the same as if unlabeled. The radiolabeled component produces a normal Gaussian peak shape, being eluted by the later arriving molecules, in effect, the same way as in a normal peak elution experiment. Thus, the frontal profile is simply an integral of many individual elution-peak profiles. This observation allows application of techniques used to analyze peak shapes to frontal analysis. This includes both plate theory and the more modern statistical moment analysis.

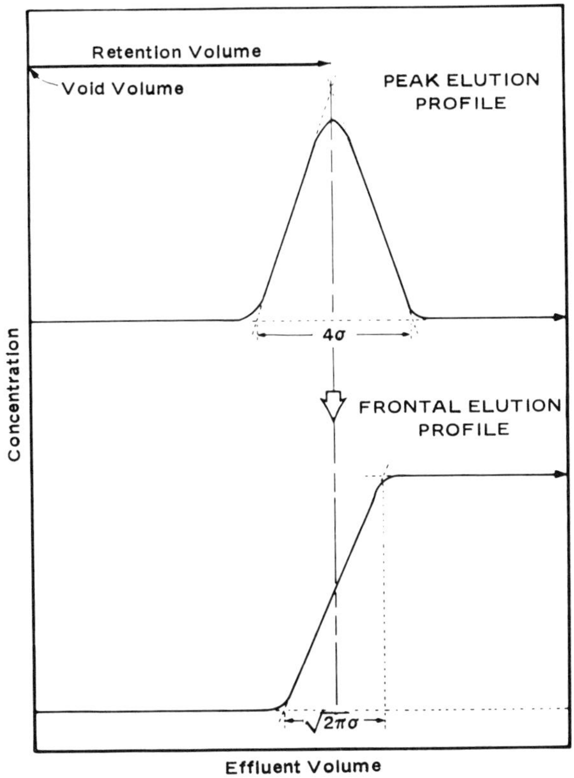

FIGURE 3. Integral relationship between peak and frontal elution profiles, with σ being the standard deviation of the zone.

For example, HETP, or height of an equivalent theoretical plate, which is a measure of column efficiency[(15)] is evaluated by the relation,

$$\mathrm{HETP} = \frac{\sigma^2}{L} \tag{5}$$

where σ^2 is the variance of the zone in units of time and L is the column length. This can be related to the diffusional properties of the solute, noting that

$$\sigma^2 = 2Dt_p \tag{6}$$

where t_p is the time to diffuse into a Gaussian profile, and D is the solute diffusion coefficient, which results in

$$\mathrm{HETP} = \frac{2D}{v} \tag{7}$$

Examples of this type of analysis applied to frontal profiles include determination of axial and inter-particle diffusion by their additive contribution to plate heights[23] and evaluation of the number of theoretical plates and of partition coefficients.[38]

4. ADSORPTION SYSTEM

The general features of the adsorption apparatus that may be used in this type of study are illustrated in Figure 4.[39] The major components include a jacketed adsorption column, a spectrophotometer (generally set at 280 nm to detect changes in protein concentration), a peristallic pump to maintain constant flow rate, and a fraction collector. Protein solution is pumped from a flask immersed in the regulated temperature bath through the column, which exits to a flow cell in the spectrophotometer. Sub-

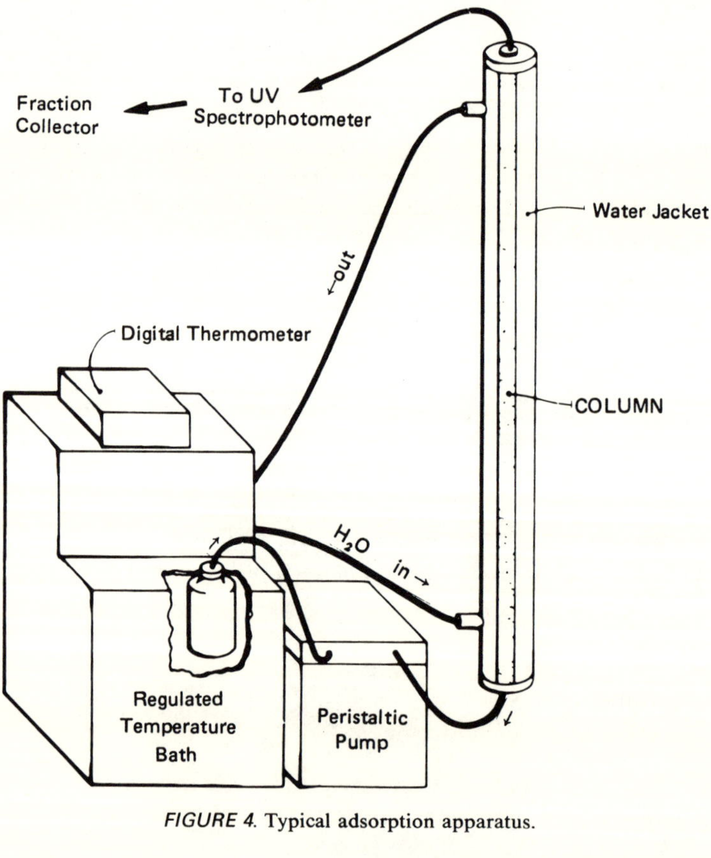

FIGURE 4. Typical adsorption apparatus.

sequently, the effluent may be analyzed for conformation by circular dichroism spectroscopy and other methods, and for aggregation state or molecular weight by high pressure liquid chromatography (HPLC). Dead volume within the system should be kept to a minimum by use of small-diameter connecting lines.

The following examples of the application of chromatographic analysis are taken from the authors' work and are quoted as specific examples. It seems reasonable to suppose that the same method is applicable to other proteins and protein mixtures and other surfaces.

In the current example the adsorption columns were glass barrel Econocolumns (0.7 × 50 cm) utilizing porous (70 μm) polyethylene bed supports and end caps (Bio-rad Laboratories) and a high surface area packing of 0.14-mm glass beads (soda-lime, true glass spheres, Cataphote Ferro Corp.) The beads were extensively cleaned and coated with silane agents, including *N*-(trimethoxysilylpropyl)ethylenediamine triacetic acid trisodium salt (carboxyl coating), aminopropyltrimethoxylsilane (amine coating), and dimethyldimethoxysilane (methyl coating). After copious rinsing, the beads were packed into the columns and equilibrated with buffer solution.

In general, the columns contained a surface area of 0.24 m^2 and a void volume of 7.6 ml. Flow rates of 0.09 to 0.7 ml/min were used in this study, which corresponded to column flow velocities of 0.01- 0.09 cm/s. On the average, Re values ranged from 10^{-6} to 10^{-4}.

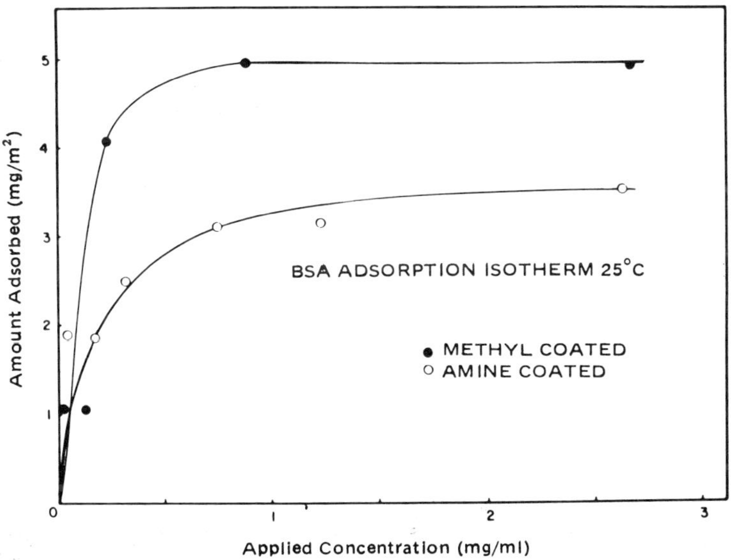

FIGURE 5. Adsorption isotherms of BSA on both amine- and methyl-coated surfaces [0.05*M* Tris/HCl, pH 7.4, 0 .02% NaN_3, .15 ml/min (ave), 25°C].

The proteins investigated were plasma fibronectin (PF), highly purified in our laboratory,[39] and bovine serum albumin (BSA), monomer standard powder, and γ-globulin (98%), both purchased from Miles Biochemicals. Purity and aggregation were analyzed by HPLC, using a Waters Associates HPLC with an I-250 protein column, capable of resolving proteins in the molecular weight range of 10,000–500,000.

Several types of experiments may be performed. The first technique, single-step adsorption, was used to develop dynamic adsorption isotherms. In the adsorption mode, a protein concentration c_0 was applied to the column until the buffer solution was displaced and the effluent concentration reached c_0. The amount of protein adsorbed at each c_0 value was evaluated by the integral technique. Examples of isotherms obtained in this manner are given in Figures 5 through 7.

Figure 5 illustrates the adsorption behavior of BSA on the methyl and amine surfaces. In both cases, a plateau surface concentration of protein was achieved, being slightly higher for the methyl surface (5.0 mg/m^2) than the amine surface (3.5 mg/m^2). For the γ-globulin (Figure 6), however, much greater uptake of protein was observed on the amine surface compared with BSA. Within the concentration range used in this experiment, no leveling in uptake was achieved, and the surface concentration of protein exceeded 33 mg/m^2. Note that the first four points of the isotherm were

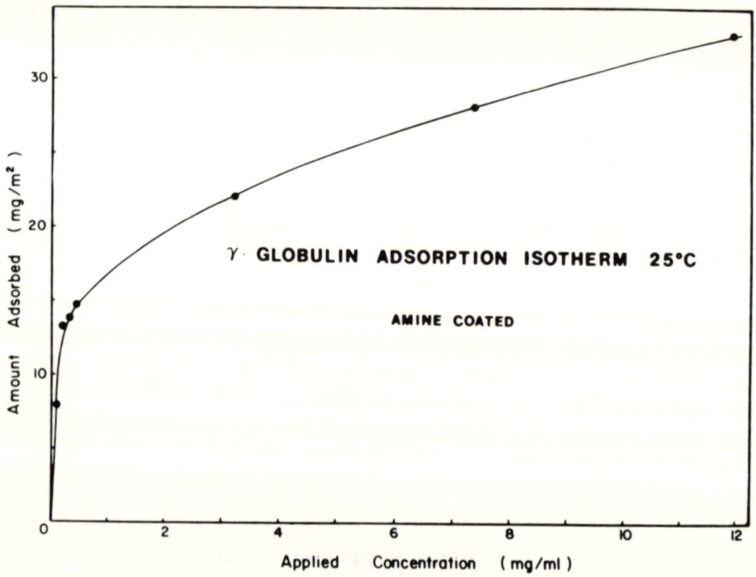

FIGURE 6. Adsorption isotherm of γ-globulin on amine-coated surface (0 .05 *M* Tris/HCl, pH 7.4, 0 .02% NaN_3, 0.08 ml/min, 25°C).

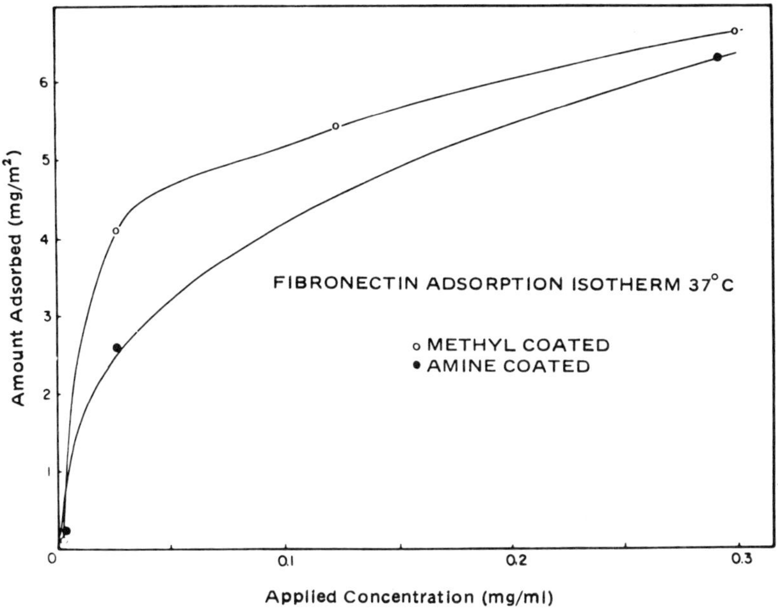

FIGURE 7. Adsorption isotherms of PF on both amine- and methyl-coated surfaces (0.05 *M*Tris/HCl, pH 7.4, 0 .02% NaN_3, 0 .08 ml/min, 37°C).

derived by successive additions of increasing concentrations of γ-globulin solution to the same column after the effluent concentration reached a constant value.

The isotherms for PF are illustrated in Figure 7. PF exhibited similar uptake on both surfaces, and, in contrast to BSA, showed much greater surface concentration at low c_0 values. A surface concentration of 7.0 mg/m^2 was approached at higher concentrations. Both lower flow rates and lower concentrations were used in order to inhibit the shear- and concentration-induced aggregation we observed in handling this protein.

In order to calculate the amount of reversibly adsorbed protein, single-step adsorption experiments were combined with single-step desorption experiments. After the surface concentration of protein reached its equilibrium value, as was observed when the effluent concentration reached the c_0 value, pure buffer was applied to the column to displace the protein solution. The amount adsorbed was evaluated as in the previous section, and the amount desorbed evaluated as the amount of protein in the effluent profile eluted after the void volume. The ratio of these two values is plotted in Figures 8 and 9 for each run of BSA on the methyl and amine surfaces as a function of applied concentration and flow rate. Both surfaces exhibited a general increase in the percentage of BSA desorbed with increase in flow

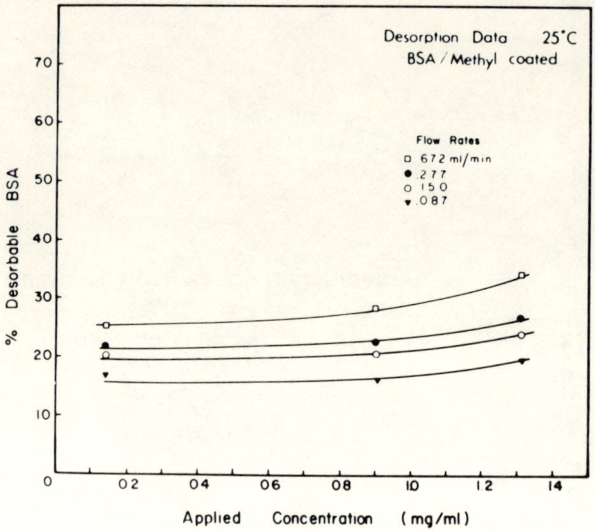

FIGURE 8. Desorption plot of BSA on methyl-coated surface showing the percentage of desorbable protein as a function of applied concentration and flow rate (0.05 *M* Tris/HCl, pH 7.4, 0.02% NaN_3, 25°C).

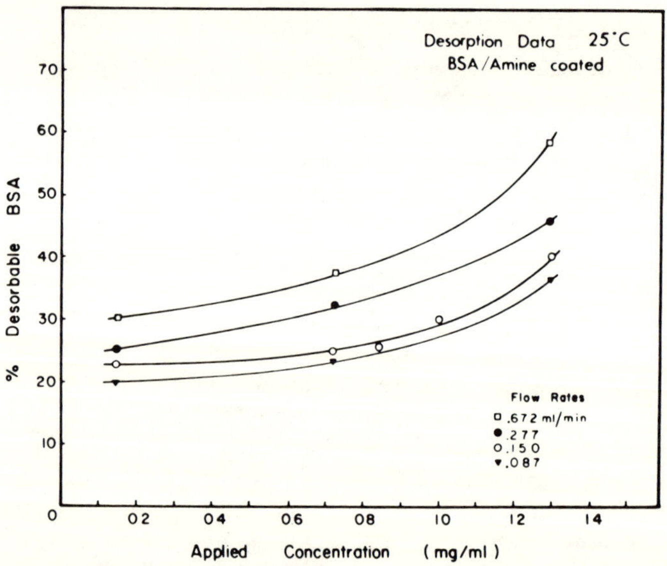

FIGURE 9. Desorption plot of BSA on amine-coated surface illustrating the percentage of desorbable protein as a function of applied concentration and flow rate (0.05 *M* Tris/HCl, pH 7.4, 0.02% NaN_3, 25°C). Reproduced by permission of the American Chemical Society from Reference (44).

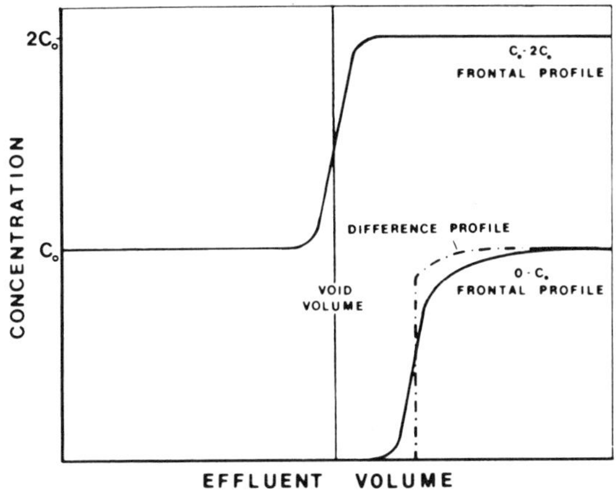

FIGURE 10. Two-step adsorption experiment with c_0 large enough to form monolayer coverage. Then, $2c_0$ is applied and the resultant frontal profiles are subtracted as shown (- · -).

and concentration; however, the effect was more pronounced on the amine surface. The values of percent desorbable BSA ranged from 17 to 38% for the methyl surface and 20 to 58% for the amine surface.

Another type of experiment may be performed in order to separate the adsorption process from diffusional effects. This involves a two-step adsorption technique illustrated in Figure 10, in which a solution of c_0 is applied first to a column, and then after equilibrium is reached, a second solution, $2c_0$, is applied. The first concentration of protein is large enough to produce monolayer coverage of the surface (typically 1 mg/ml for BSA), in order

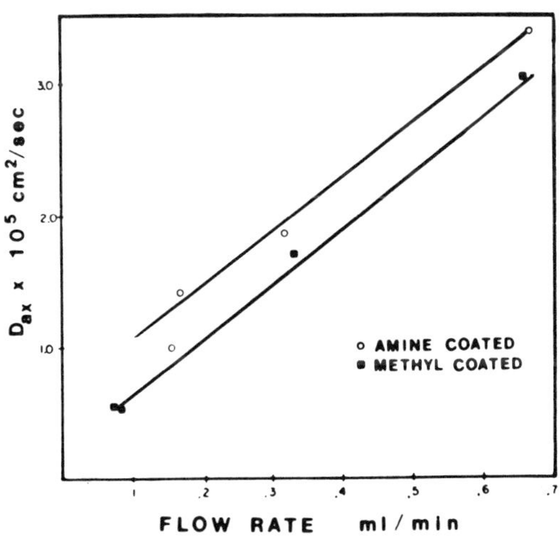

FIGURE 11. Plot of D_{ax} as a function of flow rate for BSA on amine- and methyl-coated surfaces (0.05M Tris/HCl, pH 7.4, 0.02% NaN_3, 25°C).

that the second profile exhibits no additional protein uptake. In effect, the obtained $c_0 \rightarrow 2c_0$ frontal adsorption profile shows broadening due to only molecular diffusion and flow phenomena, and not additional adsorption. A value for D_{ax} can then be evaluated by using Eqs. (5) and (7) and determining the standard deviation of the profile as in Figure 3.

In Figure 11, D_{ax} is plotted as a function of flow rate for both the methyl and amine surfaces. If pure molecular diffusion were the primary factor responsible for the broadening effect, one would expect broadening to increase with decreased flow rate.[40] The upward trend in these data indicates that the diffusional broadening is primarily due to flow phenomena and not pure diffusion within the range of flow rates used in this study. The effect appears to be equivalent on both surfaces as expected, since both surfaces are coated with a similar protein layer.

It was hoped that by subtracting the $c_0 \rightarrow 2c_0$ front from the $0 \rightarrow c_0$ front, the adsorption kinetics could be analyzed, since this difference profile represents the rate of uptake of material in the column. From the observed zero initial slope illustrated in Figure 11, it appears that the adsorption

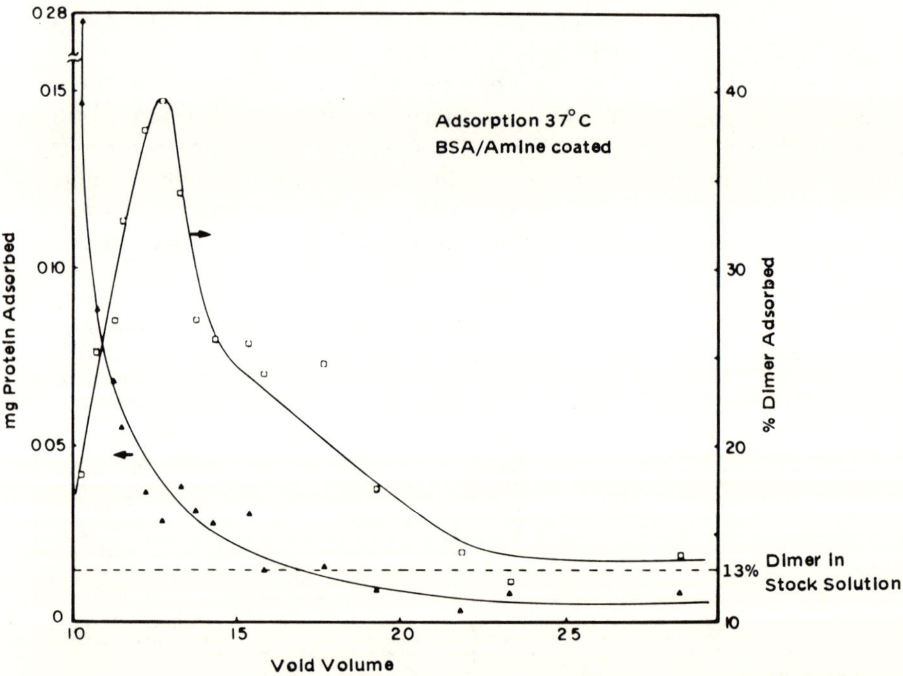

FIGURE 12. Instantaneous adsorption of BSA on an amine-coated surface. The stock solution of BSA contained 13% dimer (0.05 *M* Tris/HCl, pH 7.4, 0.02% NaN_3, .15 ml/min, 37°C, c_0 = 1.0 mg/ml). Key: ▲, the amount of BSA adsorbed from a single effluent volume; □, the percentage of dimer adsorbed in the same effluent volume.

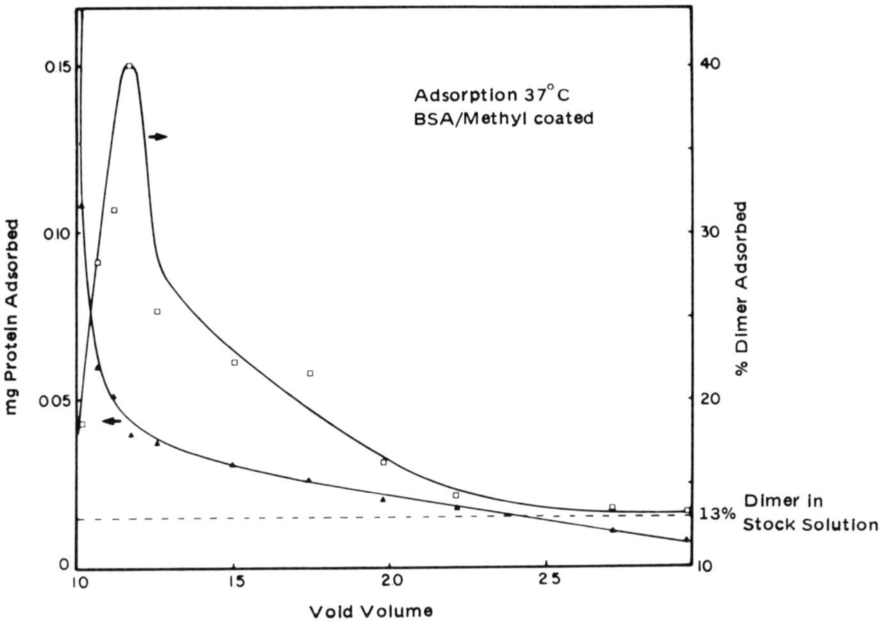

FIGURE 13. Instantaneous adsorption of BSA on a methyl-coated surface. The stock solution of BSA contained 13% dimer (0.05 *M* Tris/HCl, pH 7.4, 0.2% NaN_3, .15 ml/min, 37°C, c_0 = 1.0 mg/ml). Key: ▲, the amount of BSA adsorbed from a single effluent volume; □, the percentage of dimer adsorbed in the same effluent volume. Reproduced by permission of the American Chemical Society from Reference (44).

process is virtually instantaneous with an inflection point due to the slowing of site filling on the surface. These types of fronts are characteristic of Langmuir-type adsorption processes as discussed earlier. However, HPLC analysis of these fronts revealed a preferential adsorption for BSA dimer over BSA monomer.

Figures 12 and 13 illustrate the dynamic adsorption behavior of BSA containing 13% dimer on both the methyl and amine surfaces. Both exhibit similar features; the first few ml of effluent after the void volume show no preference for dimer, but as adsorption continues, dimer is selectively removed from solution. This preference reaches a peak at about 1.2 void volumes, and drops slowly to the stock solution value at 2.5 void volumes. If the total percent area is integrated to obtain total dimer adsorbed, it was found that both surfaces adsorb about 20% of the total dimer. The carboxyl surface exhibits a much greater preference; 47% of the surface contains dimer (Figure 14).

Surface exchange of protein was also analyzed by HPLC analysis. Experimentally, a column was coated with one protein until equilibrium

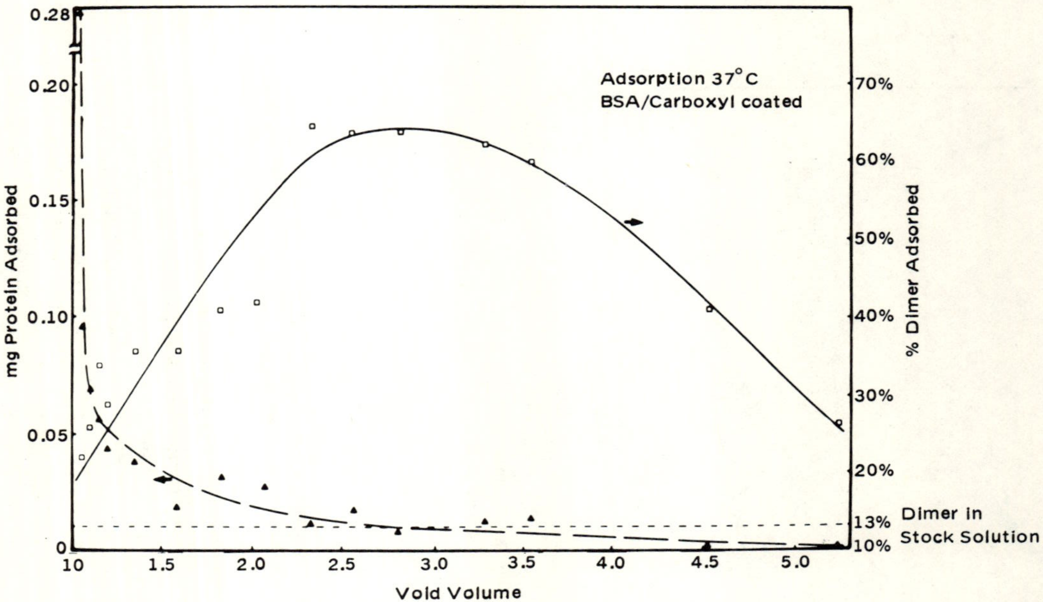

FIGURE 14. Instantaneous adsorption of BSA on a carboxyl-coated surface. The stock solution of BSA contained 13% dimer (0.05 *M* Tris/HCl, pH 7.4, 0.02% NaN_3, 0.15 ml/min, 37°C, $c_0 = 1.0$ mg/ml). Key: ▲, the amount of BSA adsorbed from a single effluent volume; □, the percentage of dimer adsorbed in the same effluent volume.

was reached. Then, without desorbing, a second protein was applied to the column until equilibrium was reached ($c = c_0$). Fractions were collected during the experiment and the concentration of each component in the effluent was determined by HPLC. Under conditions of no exchange of the two protein species, it would be expected that at the void volume, the concentration of the first protein would drop to zero, with a concurrent rise in concentration of the second protein.

Illustrated in Figures 15 and 16 are the data for the amine and methyl columns first coated with BSA, then exchanged with PF. On the amine surface, both BSA monomer and dimer begin to approach zero concentration while the PF effluent concentration slowly reaches its applied value after 4 void volumes. At about 3 void volumes, a peak is observed in the BSA monomer concentration in the effluent. From this result, it appears that PF is adsorbed by effectively displacing BSA monomer from the surface. On the methyl surface, the same trends in PF and BSA dimer concentration are noted. However, BSA monomer slowly drops to zero concentration in the effluent, indicating, as before, that monomer is predominantly displaced, but possibly at shorter times.

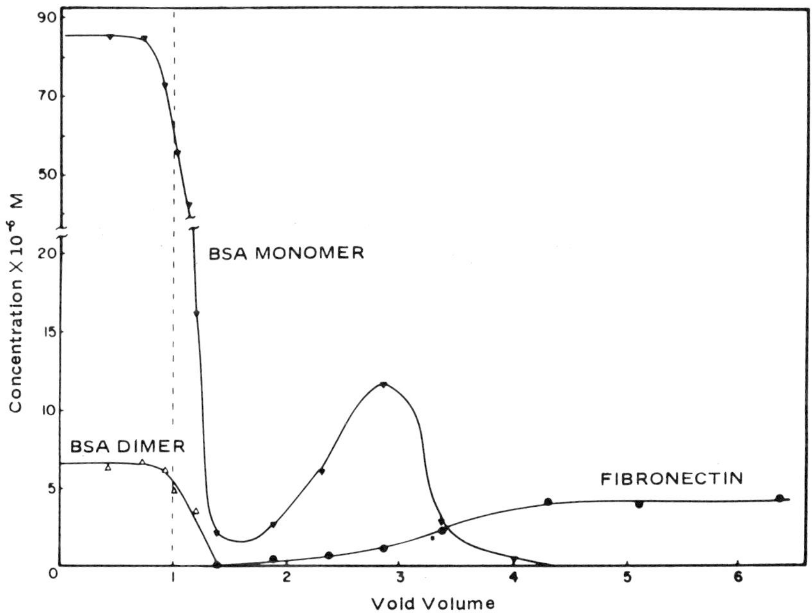

FIGURE 15. Exchange of BSA by fibronectin on amine-coated surface. Profiles illustrate the molar concentrations of each protein in the column effluent. The dashed line indicates the column void volume (0.05 *M* Tris/HCl, 0.02% NaN_3, pH 7.4, 0.15 ml/min, 37°C). Key: ●, fibronectin; ▼, BSA monomer; and △, BSA dimer. Reproduced by permission of the American Chemical Society from Reference (44).

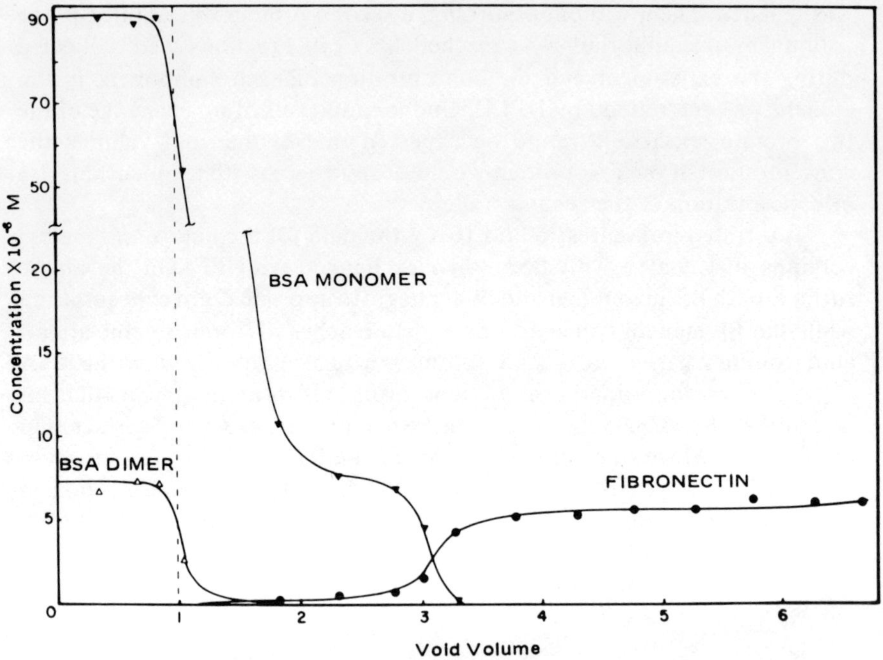

FIGURE 16. Exchange of BSA by fibronectin on methyl-coated surface. Profiles illustrate the molar concentrations of each protein in the column effluent (0.05M Tris/HCl, 0.02% NaN_3, pH 7.4, 0.15 ml/min, 37°C). Key: ●, fibronectin; ▼, BSA monomer; and △, BSA dimer.

5. DISCUSSION

In the previous section, it has been shown that dynamic isotherms can be constructed using frontal analysis. The utility of this technique can be checked by comparing quantitative results to static values. In the case of BSA, a greater total amount of protein was adsorbed on the hydrophobic methyl surface (5.0 mg/m^2) than the more ionic amine surface (3.5 mg/m^2). These values are comparable to previous static studies[41] by our group using the same glass bead technique with similar coatings, such as an alkane coating (4.06 mg/m^2) and a Lys^{20} : Leu^{80} copolymer coating (2.9 mg/m^2). These values are all indicative of a side-on (i.e., flat) orientation for BSA molecules (which are highly asymmetric) in agreement with other static studies.[42,43]

γ-Globulin has been shown to have a great affinity for many surfaces with an indicated preferential end-on orientation.[41,43] The data presented here have shown a great uptake of γ-globulin on the amine surface, and also the inability to reach a plateau concentration. This latter effect may be indicative of a continuous exchange of lower molecular weight protein for higher aggregates which were present in this material. Few comparative data are available for PF; however, in a previous report, we have shown that the molecular dimensions of PF can be estimated since an asymptotic surface concentration of 7.0 mg/m^2 is approached on both methyl and amine surfaces.[44] The dimensions calculated were approximately $130 \times 80 \times 100$ Å.

While the data are not shown, preliminary results with BSA have indicated that a lesser amount of protein is adsorbed at higher flow rates, which may be indicative of the existence of a more loosely bound protein layer, as observed in other flow studies.[8] In addition, increased ionic strength and temperature decreased the total amount of BSA adsorbed, which indicates that this technique is sensitive to slight changes in the adsorption conditions, not only in the protein-surface interactions, but also in the hydration layers.

Desorption experiments have indicated that adsorption of BSA on both surfaces was partially reversible (Figures 8 and 9). Reversibility was greater on the more ionic amine surface (58% desorbable protein) than on the methyl surface (38% desorbable protein). This result is in agreement with previous studies which have suggested that ionic surfaces exhibit greater adsorption reversibility than hydrophobic surfaces.[45,46] It is also noted that both flow rate and applied concentration are inversely proportional to the adsorption equilibrium time. In both figures, the data point denoting the lowest concentration and lowest flow rate required the longest amount of time to reach equilibrium (13.6 h) and exhibited the least amount of desorbable protein. In contrast, the point indicating the highest flow rate and greatest applied concentration requires the least amount of time to reach equilibrium (15 min) and exhibited the greatest amount of desorbable protein. It is apparent that the longer protein molecules are in contact with a surface, the less they are able to desorb. This is a direct result of the overall adsorption process. Reversible adsorption, such as that observed in a short time framework, is typical of "hard sphere" behavior where both arriving and departing molecular species have the same tertiary or three-dimensional structure. On the other hand, the longer-term, essentially irreversible adsorption appears to involve a surface denaturation or loss of secondary and tertiary structure of the adsorbed protein molecules. The unfolding of surface-adsorbed proteins apparently increases the hydrophobic bonding with the substrate and suggests a strong entropic contribution to the long-term adsorption process. Although it is difficult to collect sufficient long-term adsorbed protein for conformational analysis, circular dichroism studies have indeed provided evidence that unfolding is an integral part of the long-term behavior of adsorbed proteins.[41] Our present results also support these findings; however, in this case, flow rate also contributes to the equilibrium time, and therefore the shearing process must aid in the ability of the protein to desorb. This additional contribution is observed in the increase in the amount of protein desorbable with flow rate.

Since significant amounts (13%) of BSA dimer were found in the monomer standard BSA used in this study, it was of interest to determine the effect of these higher molecular weight species on the adsorption process. Overall, the amine and methyl surfaces exhibited about a 7% greater adsorption of dimer over the stock solution value. From the rate data in Figures 12 and 13, it appears that while the dimer may be slightly slower reaching the surface than the monomer, once it binds to the surface, it desorbs at a much slower rate than the monomer. This type of "competitive desorption" has also been used to explain the ability of fibrinogen to adsorb preferentially in the presence of albumin.[47] In effect, greater amounts of dimer are being removed from solution, and a peak surface concentration of dimer is reached at about 1.2 void volumes. Since the amount of dimer removed from solution returns to the original stock solution value, it appears that a surface equilibrium concentration of dimer is reached. Similar findings have been observed in the static adsorption of BSA monomer, dimer, and higher aggregates on polyethylene[48] with the adsorption of monomer decreased as aggregates bound to the surface.

While the carboxyl surface adsorbed more total BSA (2.4 mg/m^2), it exhibited a much greater preference for dimer compared with the above surfaces. It has been shown in this work (Figures 15 and 16) that on both the methyl and amine surfaces PF can displace BSA, and is more effective at displacing monomer than dimer.

The ability of PF to displace albumin monomer has important physiological implications since PF attachment to a substrate is important in cellular adhesion.[49] The ability of PF to remove other pre-adsorbed proteins from a surface may be indicative of greater potential cellular attachment, which can subsequently control the thrombogenic character of a surface.

6. SUMMARY AND CONCLUSIONS

In this review, we have shown that chromotographic analysis, and in particular frontal analysis, can be used to construct dynamic adsorption isotherms. These are similar to static isotherms which exhibit monolayer coverage, and are dependent upon both protein and surface chemistry, and are influenced by flow rate, temperature, and ionic strength. In the case of BSA, the reversibility of the adsorption process may be monitored by following both the adsorption fronts and desorption tails. It is found that the adsorption process is at least partially reversible, and is affected by residence time in the column.

When frontal analysis is coupled with HPLC analysis, the dynamics of both competitive adsorption/desorption processes and protein-protein exchange processes can be followed. In the former case, it has been observed that on all surfaces presented as examples, BSA dimer was preferentially adsorbed. In the latter case, it has been shown that PF primarily displaced BSA monomer and not the dimer, which in turn may be significant to the thrombogenic character of a surface.

REFERENCES

1. R. G. Lee and S. W. Kim, Adsorption of proteins onto hydrophobic polymer surfaces—adsorption isotherms and kinetics, *J. Biomed. Materials Res.* **8**, 251-259 (1974).
2. R. W. Watkins and C. R. Robertson, Total internal reflection technique for examination of protein adsorption, *J. Biomed. Materials Res.* **11**, 915-938 (1977).
3. J. L. Brash and G. M. Samak, Dynamics of interactions between human albumin and polyethylene surface, *J. Colloid Interface Sci.* **65**, 495-504.
4. R. M. Gendreau and R. J. Jakobsen, Blood-surface interactions—Fourier transform studies of protein-surface adsorption from flowing blood plasma and serum, *J. Biomed. Materials Res.* **13**, 893-906 (1979).
5. R. L. Beissinger and E. F. Leonard, Sorption kinetics of binary protein solutions—General approach to multicomponent systems, *J. Colloid Interface Sci.* **85**, 521-533 (1982).
6. R. C. Eberhart, M. E. Lynch, F. H. Bilge, J. F. Wissinger, M. S. Munro, S. R. Ellsworth, and A. J. Quattrone, Protein adsorption on polymers, visualization study of fluid shear and roughness effects, and methods to enhance albumin binding, *Adv. Chem. Ser.* **199**, 293-315 (1982).
7. B. M. Chan and J. L. Brash, Adsorption of fibrinogen on glass—Reversibility aspects, *J. Colloid Interface Sci.* **82**, 217-255 (1981).
8. R. A. VanWagenen, S. Rockhold, and J. D. Andrade, Probing protein adsorption: Total internal reflection intrinsic fluorescence, *Adv. Chem. Ser.* **199**, 351-370 (1982).
9. R. M. Gendreau, R. I. Leinger, S. Winters, and R. J. Jakobsen, Fourier transform infrared spectroscopy for protein-surface studies, *Adv. Chem. Ser.* **199**, 371-394 (1982).
10. H. V. Roohk, J. Pick, R. Hill, E. Hung, and R. H. Bartlett, Kinetics of fibrinogen and platelet adherence to biomaterials, *Trans. Am. Soc. Art. Int. Org.* **22**, 1-8 (1976).
11. R. Kobayashi, P. Chappelear, and H. Deans, Physico-chemical measurements by gas chromatography, *Ind. Eng. Chem.* **59**, 63-67 (1967).
12. D. C. Locke, Physicochemical measurements using chromatography, *Adv. Chromatogr.* **14**, 87-198 (1976).
13. R. M. Gendreau, Breaking the one second barrier—Fast kinetics of protein adsorption by FT-IR, *Appl. Spectrosc.* **36**, 47-49 (1982).
14. S. Whitaker, *Introduction to Fluid Dynamics*, Prentice Hall, Englewood Cliffs, New Jersey (1968).
15. J. C. Giddings, *Dynamics of Chromatography*, Vol. 1, Dekker, New York (1965).
16. J. R. Cluff and S. J. Hawkes, Zone dispersions in packings of impermeable spheres, *J. Chromatogr. Sci.* **50**, 248-255 (1976).
17. J. Lighthill, *Mathematical Biofluiddynamics*, Soc. for Ind. Appl. Math., Philadelphia (1975).
18. J. C. Sernberg and R. E. Poulson, Particle-to-column diameter ratio effects on band spreading of a fixed bed, *Anal. Chem.* **36**, 1492-1502 (1964).
19. J. B. Rosen, Kinetics of a fixed bed system for solid diffusion into spherical particles, *J. Chem. Phys.* **20**, 387-394 (1952).

20. E. E. Graham and C. F. Fook, Rate of protein adsorption and desorption on cellulosic ion-exchangers, *AIChE J.* **28**, 245–250 (1982).
21. M. Susuki and J. M. Smith, Kinetic studies by chromatography, *Chem. Eng. Sci.* **26**, 221–235 (1971).
22. J. F. Parcher, Applications and theory of finite concentration frontal chromatography, *Adv. Chromatogr.* **16**, 152–174 (1978).
23. J. J. Van Deemter, F. J. Ziuderweg, and A. Klinkenberg, Longitudinal diffusion and resistance to mass transfer as causes of nonideality in chromatography, *Chem. Eng. Sci.* **5**, 271–289 (1956).
24. K. Wiedemann, A. Roethke, K. H. Radeke, and D. Gelbin, Modelling of adsorption-desorption breakthrough curves using statistical moments, *Chem. Eng. J.* **16**, 19–26 (1978).
25. D. R. Garg and D. M. Ruthven, Performance of molecular-sieve adsorption columns-system with micropore diffusion control, *Chem. Eng. Sci.* **29**, 571–581 (1974).
26. K. H. Radeke, H. J. Ortlieb, and D. Gelbin, Evaluating breakthrough curves with the method of moments for systems obeying the Langmuir isotherm, *Chem. Eng. Sci.* **36**, 11–17 (1981).
27. S. Katoh, T. Kambayashi, R. Deguchi, and R. Yoshida, Performance of affinity chromatography columns, *Biotech. Bioeng.* **20**, 267–280 (1978).
28. K. Miura, H. Kurahashi, Y. Inokuchi, and K. Hasimoto, Method for calculating breakthrough curves of bicomponent fixed-bed adsorption under constant pattern and linear driving force, *J. Chem. Eng. Japan* **12**, 281–288 (1979).
29. C. C. Shepard and A. Tiseilius, Chromatography of proteins: The effect of salt concentration and pH on the adsorption of proteins on silica gel, *Disc. Faraday Soc.* **7**, 275–285 (1949).
30. H. W. Habgood, Chromatography and the solid-gas interface in: *The Solid-Gas Interface*, Vol. 2 (E. Alison Flood, ed.), Dekker, New York (1967).
31. H. L. Wang, J. L. Duda, and C. J. Radke, Solution adsorption from liquid-chromatography, *J. Colloid Interface Sci.* **66**, 153–165 (1978).
32. E. Gluekauf, Adsorption isotherms from chromatographic measurements, *Nature* **160**, 301 (1947).
33. P. E. Eberly, Jr., Measurement of adsorption isotherms and surface areas by continuous flow method, *J. Phys. Chem.* **65**, 1261–1265 (1967).
34. P. Urone and J. F. Parcher, Support effects on retention volumes in gas chromatography, *Anal. Chem.* **38**, 270–274 (1966).
35. D. R. Owens, A. G. Hamlin, and T. R. Phillips, Measurement of adsorption isotherms of reactive gases by gas chromatographic techniques, *Nature* **201**, 901–902 (1964).
36. S. I. Zverev, O. G. Larinov, and K. V. Chmutov, Factors of washing-out in liquid column chromatography, *Russ. J. Phys. Chem.* **48**, 2126–2127 (1974).
37. C. N. Reilley, G. P. Hilderbrand, and J. W. Ashley, Gas chromatographic response as a function of sample input profile, *Anal. Chem.* **34**, 1198–1202 (1962).
38. H. D. Spitz, Frontal and Elution Chromatography of Nonelectrolytes, Ph.D. Thesis, Rutgers University, New Brunswick, New Jersey (1966).
39. B. M. Koltisko, Chromatographic Analysis of Protein Adsorption, M.S. Thesis, Case Western Reserve University, Cleveland, Ohio (1981).
40. V. R. Maynard and E. Grushka, Measurement of diffusion coefficients by gas chromatography broadening techniques, *Adv. Chromatogr.* **12**, 99–140 (1975).
41. M. E. Soderquist and A. G. Walton, Structural changes in proteins adsorbed on polymer surfaces, *J. Colloid Interface Sci.* **75**, 386–397 (1980).
42. W. Norde, Ph.D. Thesis, Agricultural University, Wageningen, The Netherlands (1976).
43. B. D. Fair and A. M. Jamieson, Studies of protein adsorption on polystyrene latex surfaces, *J. Colloid Interface Sci.* **77**, 525–534 (1980).
44. A. G. Walton and B. Koltisko, Protein structure and the kinetics of interaction with surfaces, *Adv. Chem. Ser.* **199**, 245–260 (1982).

45. W. J. Willman, Jr. and I. F. Miller, Adsorption of serum-proteins on polymer membrane surface, *J. Colloid Interface Sci.* **44**, 221–241 (1973).
46. H. Y. K. Chuang, W. King, and R. Mason, Interaction of plasma-proteins with artificial surfaces—Protein adsorption isotherms, *J. Lab. Clin. Med.* **92**, 482–496 (1978).
47. R. D. Bagnall, J. A. D. Annis, and S. J. Sherliker, Adsorption of plasma-proteins on hydrophobic surfaces—4. Contact-angle studies on implanted polymers, *J. Biomed. Materials Res.* **14**, 1–10 (1980).
48. L. E. Smith, R. E. Dehl, W. H. Grant, R. R. Stromberg, and B. W. Morrissey, Interaction of Blood Proteins with Solid Surfaces, Nat. Bureau Standards Internal Dept., p. 1128 (1976).
49. L. A. Culp, B. A. Murray, and B. J. Rollins, Fibronectin and proteoglycans as determinants of cell-substratum adhesion, *J. Supramol. Struct.* **11**, 401–427 (1979).

7

Proton Titration and Electrokinetic Studies of Adsorbed Protein Layers

W. Norde and J. Lyklema

1. INTRODUCTION

In Volume 1, Chapter 8 of this series, henceforth called (1), one of us has discussed in some detail the characterization of the electrical state of biomedical polymer surfaces by a number of methods, in particular, proton titration and electrokinetics. At issue in this chapter is the application of these techniques to the more complex situation of a surface covered by a proteinaceous adsorbate.

Proton titration and electrokinetics are complementary electrochemical techniques. Titration counts all the groups that can take up or lose a proton in the pH range under study. It yields the *total proton charge* Q_H (with respect to a reference state that is the object of further study).* If applied to proteins adsorbed on a substrate that, in the given pH range, also contains groups that are dissociated or associated, Q_H (total) = Q_H (protein) + Q_H (sorbent).

Electrokinetic methods yield the *electrokinetic* or *zeta potential* ζ, that is the potential at the slipping plane [for a discussion, see (1)]. Using Gouy-Chapman theory [see below, Eq. (15)], ζ can be converted into the *electrokinetic charge density* σ_{ek}, which in turn gives the total electrokinetic charge $Q_{ek} = \sigma_{ek} \times$ total area of the particle. This applies both to free and adsorbed proteins, where in the latter case Q_{ek} includes contributions of the sorbent and of the counterions adsorbed within the slipping plane.

* $Q_H = Z_H F$ where Z_H is the number of titrated groups and F the Faraday constant. Likewise, one can write $Q_{ek} = Z_{ek} F$.

W. Norde and J. Lyklema • Laboratory for Physical and Colloid Chemistry, Agricultural University, De Dreijen 6, 6703 BC Wageningen, The Netherlands.

The difference between Q_H and Q_{ek} is therefore twofold: (i) Q_H counts only proton charges, whereas Q_{ek} measures all types of charge and (ii) Q_H refers to protein + substrate only, whereas Q_{ek} includes also the bound part of the countercharge ("bound" in the sense as defined above).

Because of these differences, it is recommendable to determine both Q_H and Q_{ek}, in order to achieve an as complete as possible electrical characterization.

The two techniques have their specific advantages and disadvantages. Experimentally, proton titrations can be conducted only with sufficient amounts of adsorbed proteins. The only viable way to achieve this is to disperse the polymeric substrate finely in the aqueous solution and adsorb the protein on the particles obtained in this way. Alternatively, the study can be made with polymer colloids (latices) consisting of the same material as the polymer under study. Automatically the question arises of to which extent the surface properties of the dispersed polymer or the latex are identical to those of the original prosthetic material under study and, therefore, to which extent are the measurements of proteinaceous adsorbates on dispersed substrates representative of *in situ* adsorbates on biomedical polymers. If electrokinetic charges are determined by electrophoresis the same difficulty arises, but *in situ* measurements *can* be done electro-osmotically or invoking streaming potentials [see (1), Section 5].

Potentiometry has a zero point problem that is not encountered in electrokinetics. By titrating over a certain pH range, say from pH 4 to 10 one measures $Q_H(\text{pH}) - Q_H(\text{pH} = 4)$ but there is no independent way to establish whether or not $Z_H(\text{pH} = 4) = 0$ or, perhaps, if there is another pH at which $Q_H = 0$. Specific features of the titration curves can sometimes be used to locate this *point of zero charge* (p.z.c).

Another possibility is to determine the *isoionic point* (i.i.p.), that is, the pH at which all ions are replaced by H^+ and OH^- ions. The exchange can be achieved by treating the system with cation and anion exchangers. From the i.i.p. the excess (in equivalents/liter) of H^+ or OH^- in the solution can be computed, which equals minus the excess charge on the protein because the whole is electroneutral. This procedure can also be applied to adsorbed proteins provided the protein is not desorbed from the substrate to become attached to the ion exchanger. Starting at the i.i.p., Q_H under non-isoionic conditions can be found by following pH changes in solution upon addition of electrolytes.

The zero point of σ_{ek} is the *isoelectric point* (i.e.p.) which is easily established as the condition where the measured electrokinetic effect is zero. As explained in (1), p.z.c. and i.e.p. are different quantities and identical only in the absence of specific counterion adsorption. The titration charge per unit area is the *surface charge* σ_0.

The main problems involved in the determination of σ_{ek} or ζ are those inherent in electrokinetic techniques, such as the proper accounting for retardation in electrophoresis and surface conductivity in streaming-potential and electro-osmosis measurements.

In principle, proton titration is a more sensitive technique than those based on electrokinetic phenomena, but the greater sensitivity cannot always be fully exploited. Basically, the various dissociable groups (carboxyls, imidazoles, amino groups, etc.) are titrated at a pH, characteristic for that group, with 50% of the groups titrated at $\mathrm{pH} = \mathrm{p}K_a = -\log K_a$ where K_a is the dissociation constant (or binding constant, depending on the definition). If, due to adsorption, the environment of a certain type of group is changed, this is reflected in the titration behavior. However, in practice there are so many groups of differing pK_a that usually overlap of the individual titration ranges takes place, thwarting simple unraveling.

Electrokinetic characterization is by its very nature less sensitive, because it measures σ_0 plus the specifically adsorbed countercharge σ_s. As charge compensation due to the latter increases with the former, there is a tendency for σ_{ek} to become independent of σ_0 if σ_0 is high. Indeed, σ_{ek} does not usually exceed values of a few $\mu\mathrm{C\ cm^{-2}}$, although σ_0 for biomedical polymers, both in the absence and presence of an adsorbed protein, can attain values of tens of $\mu\mathrm{C\ cm^{-2}}$.

The two techniques under consideration are of greater consequence if they are not applied to the system as such, but are used to monitor the protein adsorption *process*. For instance, if σ_{ek} of the substrate and protein together are measured before and after adsorption, it is possible to deduce if co-adsorption of low-molecular weight ions has taken place. Likewise, proton titration curves of bound proteins may be compared with the corresponding ones for the same amount of free (i.e., dissolved) proteins, the difference between the two reflecting changes in pK_a's upon adsorption.

In this contribution we intend to give a few illustrations of this procedure.

2. PROBING PROTEIN ADSORPTION THROUGH PROTON TITRATION

2.1. TITRATION OF ADSORBED POLYELECTROLYTES

As titration curves of adsorbed proteins are difficult to unravel because of the often occurring overlap of the ranges where the various groups are (de-)protonated, we shall start the discussion with an example where this difficulty does not occur, viz. the titration of adsorbed polymethacrylates

(PMA). As compared with proteins, this model system is much simpler in that it contains only carboxylic groups. On the other hand, it has the interesting feature that it shows a conformational transition as a function of pH: at low pH the molecule is in a very condensed form, the so-called hypercoiled or α-form, whereas at high pH it is in the expanded or β-form. The specific characteristic by which PMA stands out from most other anionic polyelectrolytes is the relatively strong compaction at low pH. This compact conformation is likely due to hydrophobic bonding between the methyl groups in the main chain. Polyacrylates (PAA), in which these methyl groups are absent, do not show the hypercoiled conformation. It is important for our purpose that the configurational transition of PMA not only occurs in the free polyelectrolyte, but also in the adsorbed state as van Vliet has shown.[2,3] In this particular example, adsorption of partially esterified poly(methacrylic acid), PMA-pe, took place on paraffin oil droplets, on which sufficiently thick layers of PMA could be obtained by emulsification. It is not yet established if the same phenomenon applies also to PMA-pe adsorbed in thick layers on latices; but at any rate for a number of other polymeric adsorbates there are great similarities between adsorbates on polystyrene latex and on paraffin oil droplets. For instance, such a similarity has been observed for the influence of solvent quality on the properties of long-chain nonionic surfactants of the polyoxyethylene type.[4] Similarities in adsorption behavior were also established for PMA-pe.[5] [The partial esterification (30% by methanol) is merely a trick to ensure better adsorption: due to the ester groups, the molecules are so strongly anchored to the interface that they do not desorb, even if at high pH the adsorbed loops tend to expand due to the charging.] Probably the capabilities of macromolecular adsorbates to undergo conformational alterations are the more restricted, the stronger the segments are adsorbed on the surface.

Figure 1 is a typical example of a titration plot. In a system like this, where only one class of titratable groups is present, it is customary and rewarding to plot the data as pH vs. $\log[(1-\alpha)/\alpha]$ where α is the degree of dissociation. One of the difficulties is to obtain absolute values of α, i.e., it is not always easy to ascertain that at sufficiently low pH really all groups are protonated. For a discussion, see References (2) and (3).

A plot like the one in Figure 1 is called a *Henderson–Hasselbalch* (HH) plot, after Reference (6).

For a monobasic weak acid in aqueous solution with dissociation contant K_{am}

$$\text{pH} = \text{p}K_{am} - \log[(1-\alpha)/\alpha] \tag{1}$$

with

$$\text{p}K_{am} = 0.434\,\Delta_{\text{diss}}G^0/RT \tag{2}$$

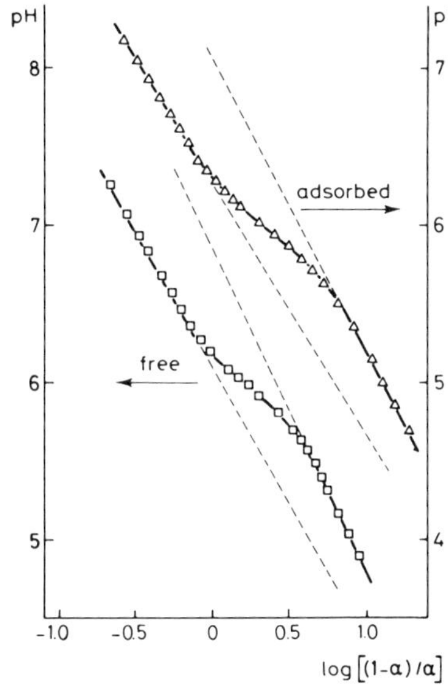

FIGURE 1. Henderson–Hasselbalch plot of partially esterified poly(methacrylic acid) adsorbed on paraffin oil droplets. Electrolyte: 0.05 *M* NaCl. Adsorbed amount *ca.* 1.3 mg m^{-2}.

where $\Delta_{\text{diss}}G^0$ is the standard free energy of dissociation. For such monomeric acids, a HH plot is a straight line with unit slope, passing through the point pH $= pK_{am}$ at $\alpha = 0.5$.

For a polyelectrolyte, dissociation of each group is affected by the electric field of other neighboring groups. Generally, the free energy of dissociation now consists of an electrical term $\Delta_{\text{el}}G(\alpha)$ in addition to the intrinsic "chemical" term of Eq. (2). Hence Eq. (1) must be modified to read

$$\text{pH} = pK_{am} - \log[(1-\alpha)/\alpha] + [0.434\Delta_{\text{el}}G(\alpha)/RT] \qquad (3)$$

In this equation, $\Delta_{\text{el}}G$ is a compounded term, including all electrical effects that make pK_a of the polyelectrolyte different from that of the monomer pK_{am}. For this reason $\Delta_{\text{el}}G$ is a function of α but also of the ionic strength ω, because electrolytes screen the electric interaction. If there are chemical polyelectrolyte effects in addition to the electrical ones (for instance, if the electron density near a carboxyl group is altered because of polymerization) pK_a of the polyelectrolyte is different from pK_{am}.

As an alternative to Eq. (3), it has been *empirically* established that for many polyelectrolytes the titration behavior is also well represented by [7,8]

$$\text{pH} = A - n\log[(1-\alpha)/\alpha] \qquad (4)$$

Here, A and n are constants. A is sometimes written as pK_{app} or pK_{av} but the physical meaning of this apparent or averaged dissociation constant is not easily established, because the corrections for electrostatical interactions are partly, and only in an empirical fashion, accounted for by the parameter n. The quantity n is a measure of the polyelectrolyte effect: $n > 1$ and increases if the electrical influence of groups on each other increases. Generally, n depends on the nature and concentration of the polyacid and on ω.[9] It follows from Eq. (4) that for many polyelectrolytes the HH plot is a straight line with a slope differing from unity.

Figure 1 illustrates these features. There are two straight parts, representing the α and β conformations, separated by a transition range. The slope n is steeper for the more compact conformation because then the polyelectrolyte effect is more pronounced.

The electrical free energy is an important variable. In Eq. (3) $\Delta_{el}G$ is the electrical reversible isothermal work to remove a proton from the surface and bring it to infinity:

$$\Delta_{el}G = \frac{\delta G_{el}(Z_H)}{\delta Z_H} = \frac{\delta G_{el}(\alpha)}{Z_{H,max}\mathrm{d}\alpha} \tag{5}$$

where we have used $\alpha = Z_H/Z_{H,max}$, $Z_{H,max}$ being the total number of ionizable groups. Hence, from Eq. (3) by integration

$$G_{el}(\alpha) = 2.303 Z_{H,max}RT \int_0^\alpha \mathrm{pH} - \mathrm{p}K_a + \log\left(\frac{1-\alpha}{\alpha}\right) \mathrm{d}\alpha \tag{6}$$

implying that the electric free energy of a polyacid is obtainable as the area under the graph of $\{\mathrm{pH} - \mathrm{p}K_a + \log[(1-\alpha)/\alpha]\}$ versus α. Information on $G_{el}(\alpha)$ is useful for the verification of polyelectrolyte theories because, on the basis of assumptions on expansion, ion binding, flexibility, etc., $G_{el}(\alpha)$ can be calculated, using some appropriate charging mechanism.

A variant of the above is the possibility to calculate the nonelectrostatic free energy of the conformational transition. This quantity is found at[10]

$$\Delta_{trans}G^0 = 2.303 Z_{H,max}RT \int_0^1 [\mathrm{p}K(\alpha\text{-conformation}) - \mathrm{p}K(\beta\text{-conformation})]\, \mathrm{d}\alpha \tag{7}$$

where the required $\mathrm{p}K(\alpha)$ plots (which cannot be directly measured over the entire α-range because the two conformations persist only over a restricted range of α's) can be obtained by extrapolation of HH plots (Figure 1). For the partially esterified poly(methacrylic acid) of Figure 1, this free energy corresponds to about 0.3–0.5 RT per mole of monomer, depending on the nature of the cation in the system.[3,11]

2.2. TITRATION OF ADSORBED PROTEINS

As compared with monofunctional polyelectrolytes the main complication with proteins is that several classes of groups ($i, j, k, \ldots$) are simultaneously titrated. If the pK_a's of these casses are sufficiently different, titration curves can be decomposed into constituent parts, each applying to one class of functional groups only. For such parts the picture of Section 2.1 is directly applicable. If this is not the case, such a detailed unraveling is not possible without additional assumptions or information. An additional problem is that due to adsorption some groups can shift their positions inside the molecule, leading to different pK_a values. Moreover, some groups can bind so strongly to sites on the substrate surface that they are no longer "seen" in proton titrations.

General experience has shown that Eqs. (3) and (4) apply well to monofunctional polyelectrolytes. This means that the electrostatic influence of the neighboring groups can satisfactorily be formulated either by adding an $\Delta_{el}G$ term or by the introduction of the prelogarithmic factor n. As it is immaterial whether the electric field at the site of dissociation of a proton originates from similar or dissimilar groups, it is likely that Eqs. (3) and (4) apply also to the various classes of groups in protein molecules:

$$\begin{aligned} \text{pH} &= \text{p}K_{aj} - \log[(1-\alpha_j)/\alpha_j) + [0.434\Delta_{el}G(\alpha_j)/RT] \\ &= \text{p}K_{ak} - \log[(1-\alpha_k)\alpha_k] + [0.434\Delta_{el}G(\alpha_k)/\text{RT}] \\ &= \cdots \text{etc.} \end{aligned} \tag{8}$$

Alternatively, the empirical HH equation [Eq. (4)] can be generalized to

$$\text{pH} = A_j - \log[(1-\alpha_j)/\alpha_j] = A_k - n_k \log[(1-\alpha_k)/\alpha_k] = \cdots \tag{9}$$

Thus, these equations establish a formal relationship between the α's of the various classes, at any pH.

Equation (9) is empirical and therefore less amenable to further analysis than Eq. (8). If decomposition is possible, values of n_j can be obtained, giving an impression of the polyelectrolyte character for group j. We may add that inside an adsorbed protein the electric field can be heterogeneous, so that for a given class j the quantities α_j and n_j can assume different values. This can, for instance, be the case for carboxyl groups inside a protein, adsorbing on a highly negatively charged substrate: groups close to the surface have a lower α and higher n than those farther out.

In Eq. (8), pK_{aj} is the real dissociation constant of group j inside the adsorbed protein. As the electrostatic germ has already been set apart, pK_{aj} is identical to pK_{amj} except for the fact that the medium inside the protein differs from an aqueous solution. Therefore, we write

$$\mathrm{p}K_{aj} = \mathrm{p}K_{amj} + [0.434\Delta_{\mathrm{med}}G_j^0/RT] \tag{10}$$

where $\Delta_{\mathrm{med}}G_j^0$ is the nonelectrical free energy change involved in transporting a proton from the interior of the protein to a dilute aqueous solution. In synthetic polyelectrolytes, due to their usually substantial expansion in aqueous solutions, all groups behave as if the medium is water; therefore the term $\Delta_{\mathrm{med}}G_j^0$ is often neglected.

If Eq. (8) can be applied, $G_{\mathrm{el}}(\alpha_j)$ can be obtained for any class j. The total electrical free energy is found from the generalization of Eq. (6):

$$G_{\mathrm{el}}(\mathrm{pH}) = 2.303\ RT \sum_j Z_{j,\max} \int_0^{\alpha_j(\mathrm{pH})} \left\{\mathrm{pH} - \mathrm{p}K_{aj} + \log\left(\frac{1-\alpha_j}{\alpha_j}\right)\right\} \mathrm{d}\alpha_j \tag{11}$$

A similar analysis with a more statistical character can also be given. See for instance the review by Linderstrøm-Lang and Nielsen.[12] As far as the present authors are aware, such an approach has not yet been applied to adsorbed proteins.

The *total* free energy of an adsorbed or dissolved protein as a function of pH can be directly obtained from proton titration curves, following an argument similar to that given by Pfeil and Privalov.[13] Generally, this free energy is a function of composition, P, T, pH, and ionic strength. We write

$$\mathrm{d}G = \left(\frac{\partial G}{\partial Z_{\mathrm{H}}}\right)_{\mathrm{P,T}} \mathrm{d}Z_{\mathrm{H}} = \mu_{\mathrm{H}}(\mathrm{protein})\,\mathrm{d}Z_{\mathrm{H}} = \mu_{\mathrm{H}}(\mathrm{solution})\,\mathrm{d}Z_{\mathrm{H}}$$

$$= [\mu_{\mathrm{H}}^0 - (2.303RT)\mathrm{pH}]\,\mathrm{d}Z_{\mathrm{H}}$$

$$G(Z_{\mathrm{H}}) = G(Z_{\mathrm{H}}^0) - 2.303RT\int_{Z_{\mathrm{H}}^0}^{Z_{\mathrm{H}}} \mathrm{pH}\,\mathrm{d}Z_{\mathrm{H}} \tag{12}$$

where $G(Z_{\mathrm{H}}^0)$ is the free energy in some reference state in which the charge on the protein is FZ_{H}^0.

In Table 1 some data for dissolved proteins are tabulated giving the numbers and pK_a's of various classes of titratable groups in some proteins. These data are taken from work by Tanford *et al.*[14,15,16] and Bucci *et al.*[17] They may serve as the basis for comparison with the corresponding information for adsorbed proteins. The variation in pK_{aj} for the various proteins results from different environments in the molecules. This medium effect is usually small; as a rule it does not surpass RT per mole.

The interpretation of G_{el} in terms of electric charges and potentials is not without pitfalls. The electrical *energy* of a unit charge e at a point x

TABLE 1

Titratable Classes of Groups in Dissolved Proteins

	BPA[14]		RNASE[15]		Insulin[16]		Human hemoglobin[17]	
Group j	$Z_{j,\max}$	pK_{aj}	$Z_{j,\max}$	pK_{aj}	$Z_{j,\max}$	pK_{aj}	$Z_{j,\max}$	pK_{aj}
α-Carboxyl	1	3.8	1	3.75	4	3.6	1	3.6
β, γ-Carboxyl	105	4.0	10	4.0–4.7	8.5	4.7	12	4.5–4.7
Imidazole	17	6.9	4	6.5	4	6.4	10	7
α-Amino	1	7.8	1	7.8	4	7.4	1	7.8
ε-Amino	60	9.8	10	10.2	2	9.6	11	10.2–10.6
Phen-OH	20	10.4	3	9.95[a]	8		3	10.4
Guanidyl	23	>12	4	$\geqslant 12$	2	11.9	3	12

[a] Plus three groups inaccessible to titration.

where the potential is $\phi(x)$ amounts to $e\phi(x)$: it is the electrical work to bring the unit charge from infinity (where $\phi = 0$ by definition) to x. The electrical *free energy* contains also an entropic term; it is the reversible isothermal work. Of particular relevance here is the possibility of co-uptake of low-molecular weight ions during the charging, which has to be accounted for in G_{el}. For one unit charge at one given point x, there is no difference between the energy and the free energy because there is no entropic contribution. However, for a large assembly of charges (as in a polyelectrolyte with its counterions or in a protein) the incorporation of each additional charge leads to a change in the configuration, and hence involves an entropic contribution. In those cases, G_{el} can be obtained by a charging process, where the protein or the proteinaceous adsorbate is brought from the uncharged reference state into the final charged state by infinitesimal increments of charge.

For an electrical double layer, characterized by an interfacial potential difference ϕ_0 and a surface charge density σ_0

$$G_{el} = \int_0^{\sigma_0} \phi_0' \, d\sigma_0' \tag{13}$$

where ϕ_0' and σ_0' are the variable potential and charge during the charging process. The integral can be worked out if the relation $\phi_0'(\sigma_0')$ is known. For instance, in the (oversimplified) case that the double layer has a constant integral capacitance $K = \sigma_0'/\phi_0' = \sigma_0/\phi_0$, $G_{el} = K\sigma_0^2/2$ and G_{el} is identical to the electrical enthalpy H_{el}. However, the relationship between ϕ_0 and σ_0 is usually more complex. For proteinaceous adsorbates, the following equation, derived by Verwey and Overbeek,[18] is more appropriate.

$$G_{\text{el}} = \int_0^1 \frac{d\lambda}{\lambda} \int_0^\infty \phi'(x, \lambda)\rho'(x, \lambda)\, dx \tag{14}$$

Here the dependence of ϕ' on the distance x to the surface is explicit, as is the corresponding dependence of the space charge density ρ'. Equation (14) contains a charging parameter λ, running from zero to unity; the charge is considered to be brought onto the adsorbate in infinitesimal steps $ed\lambda$; as in Eq. (13), ϕ' and ρ' represent the values of potential and space charge density during the charging process. Clearly, Eq. (14) can be evaluated if some model is available for the distribution of charge and potential in the macromolecular adsorbate. Hence, proton titration and model studies can be combined to improve our insight into the electrical properties of adsorbed proteins.

2.3. CASE STUDIES: ADSORPTION OF HPA AND RNase

The potentiometric investigation of human plasma albumin (HPA) and bovine pancreas ribonuclease (RNase) adsorbates on polystyrene latices has been described by us in Reference (19). Polystyrene latices with sulfate groups on the surface are suitable substrates for such studies since (a) they provide sufficiently large areas to make Z_H analytically measurable, (b) the surface sulfate groups are not titrated in the same pH range as the protein is titrated, and (c) they resemble biomedical polymers. For the two proteins mentioned, the adsorption is semireversible with respect to pH in the following sense: if the pH is changed in such a way that $\Gamma(\text{new pH}) > \Gamma(\text{old pH})$, additional adsorption takes place till a final value is attained that would also have been obtained if protein and latex had been mixed directly at the new pH. However, if $\Gamma(\text{new pH}) < \Gamma(\text{old pH})$, no desorption takes place, i.e., a nonequilibrium situation is frozen in. Hence, if the latex with adsorbed protein is centrifuged off and redispersed in a solution containing no dissolved protein, back and forward pH titrations can be done without desorption or re-adsorption of protein. (For systems in which adsorption–desorption processes take place as a function of pH, proton titrations become very difficult to interpret.) We shall now briefly review some of the findings for the present system.

Experimental details of this work have been described in Reference (19). Particular care was taken to exclude contamination with CO_2 at higher pH. An automatic titration device was used, as described by Van Os *et al.*[20] So much protein was added to the latex that the plateau value of the adsorption isotherm was attained. The protein remaining in solution was discarded by centrifugation. In Reference (19) the procedures used to establish the zero point for Q_H are discussed. Figures 2 and 3 summarize the most important results.

The first striking feature is that for HPA the Q_H values differ more from those of the dissolved protein than for RNase. Qualitatively, this distinction was related to the fact that upon adsorption RNase molecules retain their bulk conformation to a larger extent than HPA molecules do, an observation for which independent evidence was gathered. In connection with this conclusion, there is not much difference among the curves for RNase adsorbed at various pH values, whereas for HPA the curve systematically differs more from that of the dissolved molecule, the more the pH of adsorption differs from the i.e.p. In both cases, there is some influence of the latex surface charge σ_0 on the curves, but again this trend tends to be more pronounced for HPA than RNase. This is apparently a consequence of the polyelectrolyte effect.[3,4] The latex surface being negative (in this case due to covalently bound sulfate groups), dissociation of, say, carboxyl groups is reduced, so that at a given pH Q_H is increased. For RNase, which undergoes less conformational alterations, the polyelectrolyte effect is relatively more strongly dominated by the groups of the protein itself, so that the latex surface charge has not so much effect. In view of this, the trends of Figures 2 and 3 may also be taken as experimental evidence for stronger structural rearrangements in HPA than in RNase.

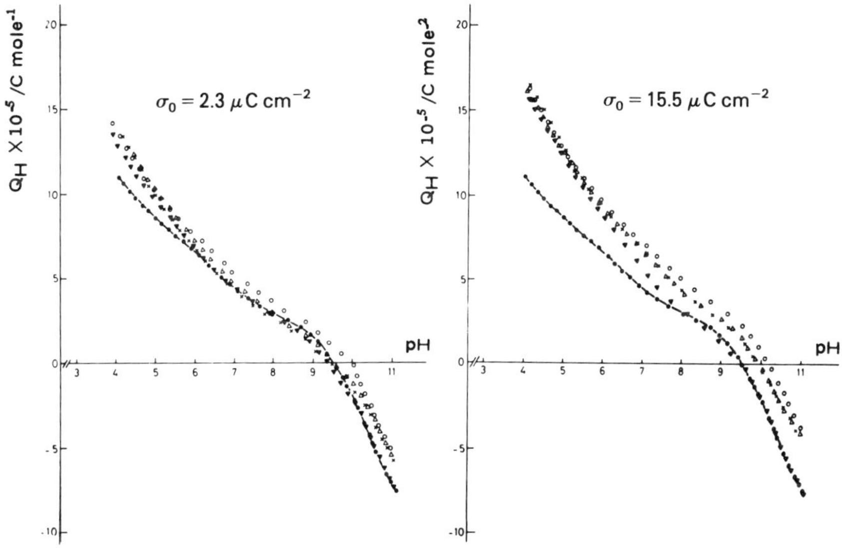

FIGURE 2. Potentiometric proton titration of RNase adsorbed on negatively charged polystyrene latex, of which the surface charge is given. Electrolyte, 0.05 M KNO_3, T = 25°C. —·— Dissolved RNASE. pH of adsorption: ○ 4; △ 7.0; × 9.3; ▼ 11.0.

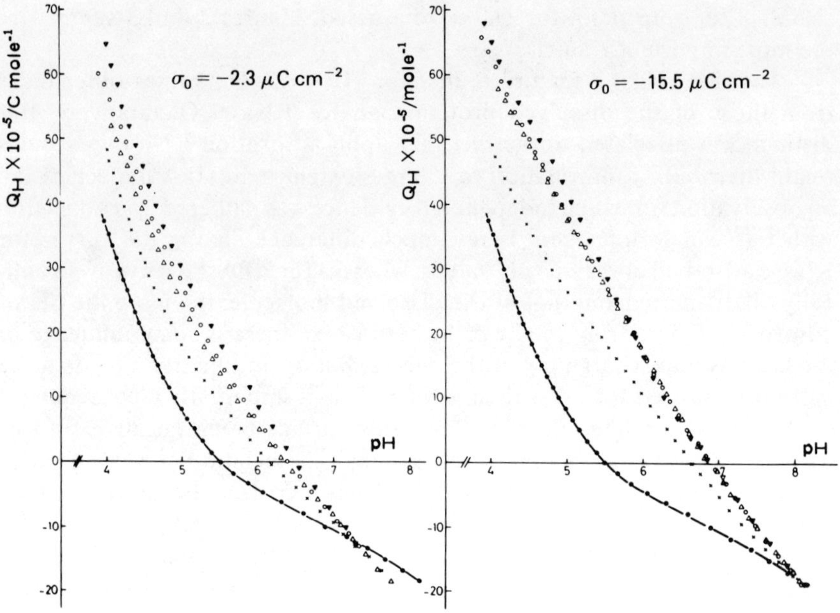

FIGURE 3. Same treatment as Figure 2 for HPA —·—. Dissolved HPA. pH of adsorption: ○ 3.6; △ 4.0; × 4.6; ▼ 7.3 or 8.0.

For adsorbed HPA and RNase in the low-pH part of Figures 2 and 3 almost exclusively carboxyl groups are titrated; hence in this range it is possible to check the applicability of the empirical HH equation [Eq. (4)]. Such plots are indeed linear, as illustrated by Figures 4 and 5. Moreover they show the expected trends: for RNase in bulk the slope (n = 1.4) is the same as in the adsorbed state, suggesting that the screening of the carboxyls is little or not at all influenced by the surface charge residing on the latex particles. On the other hand, for HPA the polyelectrolyte effect increases considerably upon adsorption: for dissolved HPA n = 1.6, on low-charge latex it amounts to 1.9, and for the highly charged substrate it rises to 2.4–2.6. Another trend is the systematic upward shift of the lines with increasing σ_0 of the latex and, for HPA, with increasing distance between the pH of adsorption and the i.e.p. In terms of Eq. (4) this may be interpreted as an increase of A but the physical meaning of this quantity is obscure.

As an alternative approach, the data could be analyzed in terms of Eq. (3) with $G_{el}(\alpha)$ interpreted according to Eq. (6).

The potentiometric data of Figures 2 and 3 were also used by the authors[21] as one of the building blocks of a rather comprehensive thermodynamic analysis of the protein adsorption. This process is very complex and involves various contributions, such as free energy changes due to

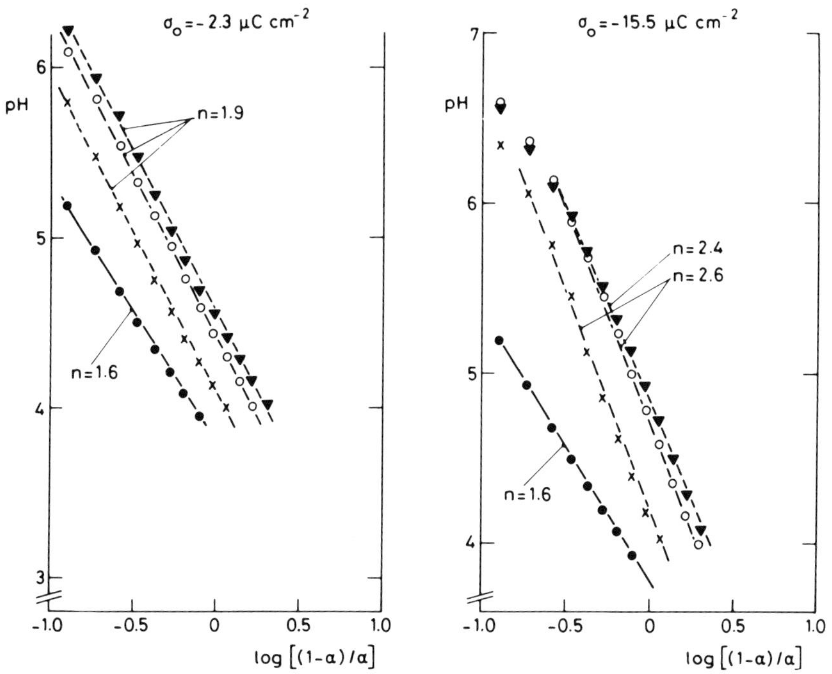

FIGURE 4. Henderson–Hasselbalch plot for the carboxyl groups in HPA adsorbed on polystyrene latex. Left: $\sigma_0 = -2.3$ C cm^{-2}, right: $\sigma_0 = -15.5$ μC cm^{-2}. Electrolyte, 0.05 M KNO_3. $T = 25$°C. —·— HPA in solution. pH of adsorption: ○ 3.6; × 4.6; ▼ 7.3 or 8.0.

rearrangements inside the molecule, (de-)hydration of protein and latex, and co-adsorption of low molecular weight ions besides protons, which are difficult to assess. The procedure followed was that overall adsorption enthalpies $\Delta_{ads}H$ were measured under various conditions and the measured values were treated as the sum of some six constituent parts, five of which were considered "known," (in the sense that from additional evidence and model considerations reasonable estimates can be made) so that the sixth, the contribution of reconformations, could be solved for. In Chapter 8 of this volume, one of us shall describe the microcalorimetrical assessment of $\Delta_{ads}H$ and the ensuing thermodynamic analysis in more detail. Amounts of low-M.W. ions incorporated can be obtained from electrokinetics (Section 3). At issue is now what can be inferred from proton titration.

First, potentiometry can be used to obtain the free energy of adsorption, $\Delta_{ads}G$. In principle, this quantity is found by using Eq. (12) three times: once for the latex with its envelope of adsorbed protein, once for the bare latex, and once for the dissolved protein. Subtraction gives $\Delta_{ads}G$ but a number of factors should be taken into account, among them the changes in pH and ω upon adsorption.[21]

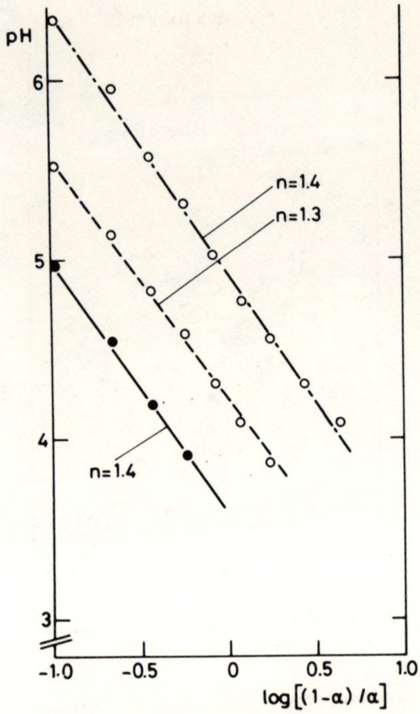

FIGURE 5. Henderson–Hasselbalch plot for the carboxyl groups in RNase adsorbed on polystyrene latex. Electrolyte, 0.05 M KNO_3. $T = 25°C$, pH of adsorption = 4. Surface charge of the polystyrene latex: $-2.3\ \mu C\ cm^{-2}$ (∘ - - - ∘) and $-15.5\ \mu C\ cm^{-2}$ (∘ - · - ∘). — · — RNase in solution.

Second, invoking model considerations, the electrical part $\Delta_{el}G$ is obtainable, in principle, by using Eq. (14) three times in the same way as above for the total free energy. In the model adopted for the computation of G_{el}, it was assumed that the adsorbed protein layer could be split up into three sublayers, with homogeneous properties within each of them. For the dissolved protein, a two-shell picture was similarly assumed, whereas for the naked PS surface simply Eq. (13) can be used, the required final surface charge being determined by conductometric and potentiometric titration [(1), Figure 14]. The sought relaton between ϕ' and ρ' at any x was deduced from Poisson's formula [(1), Eq. (9)] by integration.

One of the advantages of the availability of independent information on low-M.W. ion incorporation for the present system is that the total electrochemical free energy of ion incorporation can be split with some confidence into a purely electrical and a chemical part. The idea is that $\Delta_{med}G_i^0$ for ions of type i can be written as the product of a standard molar free energy of transfer g_i^0 and the number ΔN_i of ions i transported. Obviously, the linear relationship between $\Delta_{med}G_i^0$ and ΔN_i is a consequence of the fact that for the short-range nonelectrostatic contributons no polyelectrolyte effect need be accounted for. In the analysis referred to, g_i^0 acts as a parameter, but information on it is obtainable from literature tabulations on the free energy of transfer of ions from aqueous to various nonaqueous solvents. The quantity $\Delta_{\text{med}}G_i^0$, obtained in this way, can therefore be sequestered from the electrical contributions.

For a further discussion see Chapter 8. The above elaborations have indicated how potentiometry, in combination with other experiments and some theoretical considerations, can contribute to the understanding of protein adsorption. It goes without saying that this example by no means exhausts the potentialities of such titrations.

3. PROBING PROTEIN ADSORPTION ELECTROKINETICALLY

3.1. ELECTROKINETICS OF ADSORBED PROTEINS

Although the literature contains some examples of electrophoresis measurements on adsorbed proteins [see, e.g., References (11)-(28)] their number is by no means abundant and the experiments have not always been carried out systematically. In particular, measurements on materials that sufficiently resemble biomedical polymers are too scarce to be useful for the present purpose.

Especially in the older literature "the" zeta potential of dissolved proteins was sometimes obtained by adsorbing these proteins on electrophoretically inert particles, measuring the mobility of the covered particles, and assuming that ζ(dissolved protein) $\sim$ ζ(adsorbed protein). The advent and development of moving boundary electrophoresis (Tiselius) made this procedure obsolete. Nowadays we are in a position to measure independently both the mobilities of dissolved proteins and of protein-covered particles (or, for that matter, one can measure streaming potentials, electro-osmotic flow, etc. of flat, protein-covered surfaces; see (1), Section 6). The advantage is that one can now investigate if the electrokinetic charges σ_{ek} of proteins *are* identical before and after adsorption and in this way obtain valuable information on the characteristics of protein adsorption. In particular, changes of σ_{ek} due to adsorption reflect co-uptake or expulsion of low-M.W. ions upon attachment of the protein to the substrate. An example of such an analysis will be given in Section 3.2.

Although electrophoresis is a very popular technique in biological and medical science, the quantitative interpretation of the measured mobilities u is not without problems. The first and main difficulty is to obtain ζ-potentials. Only under a number of conditions is the relation between u and ζ straightforward and simple, viz. for spherical particles in solutions of (1-1) electrolytes and having either a low or a high value of the ratio between particle radius and double layer thickness κa. At intermediate κa numerical solutions are available.[29,30] These limitations are due to the difficulties encountered in computing the relaxation corrections that have to be applied. In solutions of asymmetrical electrolytes and for nonspherical geometry, only approximate solutions are available, usually applying to low ζ only. For nonspherical particles toppling can occur and for that case the mathematical formulation of the hydrodynamics is only in a rudimentary state. For insulating particles and $\kappa a \gg 1$, Helmholtz and Smoluchowski's equation [(1), Eq. (34)] is generally valid.[31] For cylinders, ellipsoids, etc. of partially or completely conducting particles calculations are available without accounting for the relaxation effect. Tables for computation have been given by Henry[32] and Abramson *et al.*[33]

Considering that electrophoresis measurements, due to various perturbing factors, are routinely subject to an error of 5–10%, the systematic error that is made in converting mobilities into ζ-potentials compounds the uncertainty. One should therefore not overestimate the absolute accuracy of the ζ-potentials obtained, although relative measurements can be informative if there are reasons to presume invariance of the systematic errors.

The problem of accounting for relaxation phenomena is absent in measurements of streaming potentials and electro-osmosis, but in this case surface conduction may become a serious problem. Surface conduction is the flow of electricity not through the aqueous medium but through the surface of the protein (or the surface of some prosthetic material with protein adsorbed onto it). For a discussion of this complication and a description of the conditions under which significant measurements remain possible, see Overbeek.[34]

A second, general limitation of electrokinetic measurements is that ζ-potentials and σ_{ek} are sensitive characteristics of the protein only if the surface charge σ_0 of the protein is low. In (1) it has been discussed that at increasing σ_0, σ_{ek} becmes independent of σ_0, so that then virtually only the sign of σ_{ek} relays any information.

Conversion of ζ-potentials into σ_{ek} is no problem because Gouy-Chapman theory applies outside the slipping plane so that Eq. (19) of (1) may be used in this modification:

$$\alpha_{ek} = (8\varepsilon\varepsilon_0 cRT)^{1/2} \sinh(ze\zeta/2kT) \tag{15}$$

On the other hand, locating the slipping plane for a protein (which may have extending parts of the polypeptide chain) is elusive, the more so as the slipping plane is in reality a slipping layer of finite thickness. By virtue of the underlying slipping-layer model, σ_{ek} is a measure of the protein, inclusive of its bound countercharges, where "bound" must be interpreted in the hydrodynamic sense. For nonporous solid surfaces it has been established that these ions may be identified with the ions bound in the Stern layer (i.e., that ζ may be put equal to the potential of the diffuse part of the double layer, ϕ_d) but for proteins and "hairy" surfaces it is questionable if this identification is allowed. For a further discussion, see (1), Section 5.

3.2. CASE STUDIES

Accepting the limitations and potentialities of electrokinetic measurements, we shall now briefly describe some measurements with HPA and RNase, emphasizing changes occurring upon adsorption.

Phenomenologically, Figures 6 and 7 are different. With HPA, the $u/E(\mathrm{pH})$ curve of adsorbed protein is rather similar to that of free protein, especially at not too high pH, whereas for RNase the curve shifts drastically. For adsorbed RNase, the curves behave more or less as if they are additively built up of contributions of the protein and of the substrate. With HPA, on the other hand, the effect of the latex is screened.

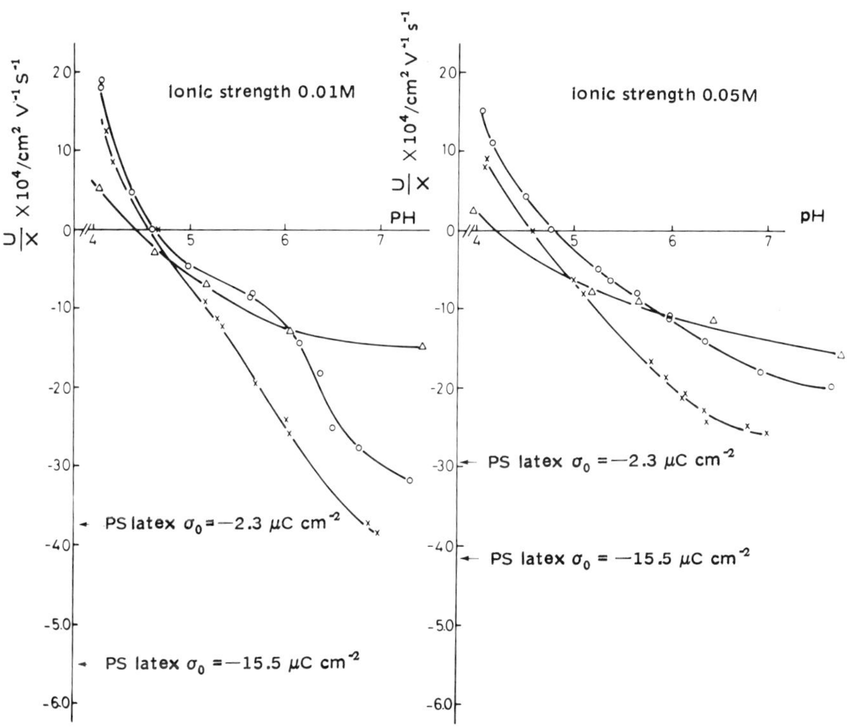

FIGURE 6. Electrophoretic mobilities of dissolved HPA (△) compared with the same for HPA adsorbed on polystyrene latex particles with latex surface charge $\sigma_0(\text{latex}) = -15.5\ \mu\text{C cm}^{-2}$ (×) or $-2.3\ \mu\text{C cm}^{-2}$(○). Buffer: veronal-acetate-KNO_3. The ionic strength is indicated. $T = 25°\text{C}$. The mobility of the bare latex is independent of pH and is indicated by an arrow.

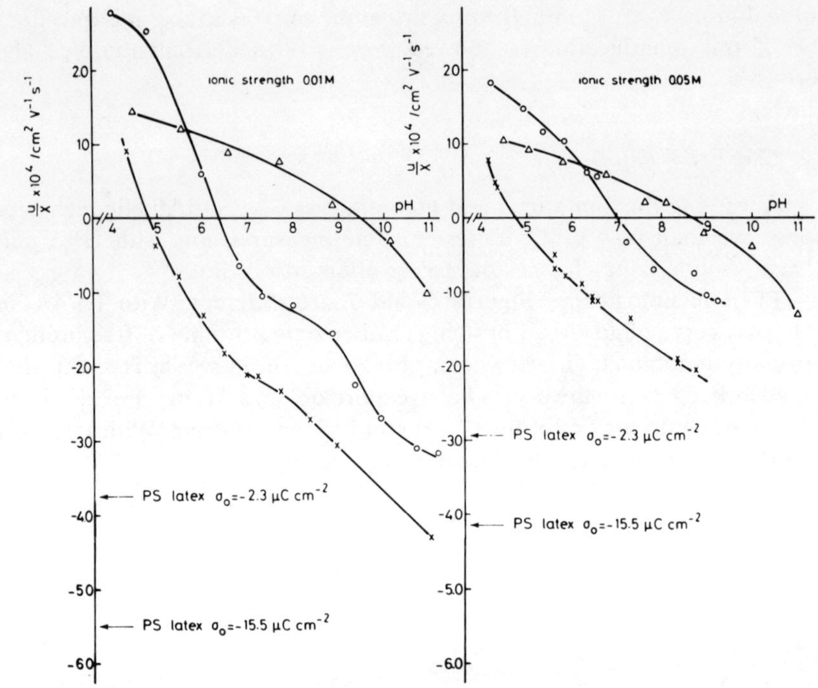

FIGURE 7. Electrophoretic mobilities of dissolved RNase (△) compared with the same for RNase adsorbed on polystyrene latex particles with latex surface charge σ_0(latex) = $-15.5\ \mu$C cm^{-2}(×) or $-2.3\ \mu$C cm^{-3}(○). Buffer: veronal-acetate-KNO_3. The ionic strength is indicated. $T = 25$°C. The mobility of the bare latex is independent of pH and is indicated by an arrow.

From this observation it can, in the first place, be deduced that the old method of measuring σ_{ek} (or ζ) of proteins by adsorbing them onto electrokinetically inactive carriers works reasonably well for HPA on PS latex but not at all for RNase on the same latex, not even with respect to the sign. In itself this is a conclusion of practical consequence, since the search towards inert carriers has drawn much attention in the past, whereas we know now that inertness is (also) a property of the protein. That it is really the combination of the substrate and the protein that counts is supported by our observation that HPA adsorbed on hematite (α-Fe_2O_3) has a u/E(pH) curve which differs substantially from that of the dissolved protein and, in fact, rather resembles the curve for the uncovered substrate particles (hematite is amphoteric; like HPA it has a positive σ_{ek} at low pH and a negative one at high pH).[35] From our experience, it can be concluded that the few instances where agreement between σ_{ek}(ads) and σ_{ek}(free) has been observed are fortuitous.

More relevant to the understanding of the protein adsorption processes is the question of why in some combinations screening is almost complete whereas in others it is all but absent. We have already argued that on PS latex the RNase molecule largely retains its bulk conformation, whereas HPA is subject to some reconformation, increasingly so, the further the pH of adsorption differs from the i.e.p. It is conceivable that a relatively "pliable" molecule, such as HPA, more easily adjusts the charges on its chain to screen the surface charge of the substrate than a more "rigid" molecule, such as RNase. However, this interpretation is too simple. For one thing, HPA appears to be a much poorer screener for hematite, but, in addition, from potentiometry there is no indication of accumulation of the positive groups of the HPA molecule near the negative PS surface; on the contrary, there is even a tendency for the negative (carboxyl) groups to enrich near the surface (Section 2 of this chapter). Excessively negative potentials do not develop because of co-adsorption of low-M.W. cations. Therefore, the screening phenomenon cannot be understood quantitatively unless this co-uptake of low-M.W. electrolyte is accounted for.

In our analysis[36] we assessed the incorporation of low-M.W. ions by comparing σ_{ek}(protein + latex) with σ_{ek}(protein) + σ_{ek}(latex), accounting for the adsorbed amounts. This procedure cannot be exact since for the geometry of the HPA molecule in solution (an ellipsoid of revolution) ζ cannot be rigorously derived from u/E. In this way we found the *total* ion uptake, i.e., the charge uptake due to (in this case) protons and other cations (K^+). The proton contributon can be obtained potentiometrically (see previous section) and subtracted so that $\Delta_{ads}\sigma_{ek}$(ions) is left. The pH dependence of this quantity is illustrated by Figure 8. For both proteins, uptake of anions takes place at low pH, whereas cations are co-adsorbed at high pH. Our procedure gives no clue as to the nature of these ions. The positive values of $\Delta_{ads}\sigma_{ek}$(ions) are likely attributable to K^+ ions, whereas the negative values are either due to adsorption of NO_3^- or to negative adsorption of K^+. Using labeled ions, it has been possible to measure the ion uptake independently, thus confirming the data obtained as described above.[37] The influence of σ_0 (latex) is to a large extent a consequence of its effect on Q_H (not shown in the figure). In particular, the incorporation of K^+ ions at high pH is worth mentioning. Above pH = 4.8 for HPA and above pH = 9.5 for RNase, the protein molecule is negatively charged and so is the latex. Nevertheless, spontaneous adsorption takes place and apparently the adsorbed K^+ ions are sandwiched between the negative latex surface and the protein. Electrical interactions are therefore not the driving forces of the adsorption process, but where they would oppose, their action is minimized by this co-adsorption. From the figures it is learned that at high pH, where, due to the negligible amount of protons in solution, $\Delta_{ads}\sigma_{ek}(K^+) \cong \Delta_{ads}\sigma_{ek}$(total), the value of this quantity is not sensitive to

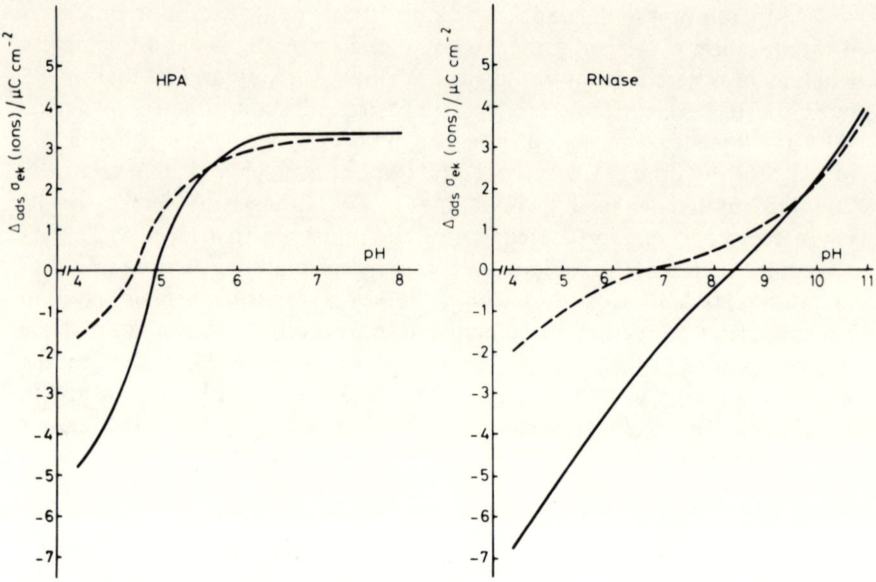

FIGURE 8. Uptake of charge due to the transfer to or from the solution of ions other than protons (mainly K^+). Electrolyte, 0.05 M KNO_3, $T = 25°C$. —— $\sigma_0 = -15.5\ \mu C\ cm^{-2}$; - - - - $\sigma_0 = -2.3\ \mu C\ cm^{-2}$.

σ_0(latex) and is solely determined by the nature of the protein. Note that the absolute amount of ions is substantial; for the low-charge latex it even exceeds σ_0(latex).

The analysis given shows that electrophoretic measurements, if performed under the appropriate conditions and with due consideration of the various variables, can contribute very much to our understanding of protein adsorption.

REFERENCES

1. J. Lyklema, this series, Volume 1, Chapter 8.
2. T. van Vliet, Thesis, Agricultural University, Wageningen, The Netherlands (1977).
3. T. van Vliet, Rheology of polyelectrolyte-stabilized emulsion-effects due to conformational properties of adsorbed polymethacrylate, *J. Colloid Interface Sci.* **63**, 97–105 (1978).
4. T. Van den Boomgaard, Thesis, Agricultural University, Wageningen, The Netherlands, 1985.
5. J. Th. C. Bohm, Thesis, Agricultural University, Wageningen, The Netherlands (1974).
6. a. L. J. Henderson, Equilibrium in solution of phosphates, *Am. J. Physiol.* **15**, 257–271 (1906). b. L. J. Henderson, Concerning the relationships between the strength of acids and their capacity to preserve neutrality, *Am. J. Physiol.* **21**, 173–179 (1908). c. L. J.

Henderson, Diagrammatic representation of equilibria between acids and bases in solution, *J. Am. Chem. Soc.* **76**, 954-960 (1980). d. L. J. Henderson, Equilibrium between oxygen and carbonic acid in the blood, *J. Biol. Chem.* **41**, 401-430 (1920). e. C. Bohr, K. Hasselbalch, and A. Kragh, *Skand. Arch. Physiol.* **16**, 402 (1904).

7. A. Katachalsky and P. Spitnik, Potentiometric titrations of polymethacrylic acid, *J. Polym. Sci.* **2**, 432-448 (1947).
8. S. Fischer and R. Kunin, Effect of cross-linking on the properties of carboxylic polymers, *J. Phys. Chem.* **60**, 1030-1032 (1956).
9. M. Mandel, Potentiometric titration of weak polyacids, *Eur. Polym. J.* **6**, 807-822 (1970).
10. J. C. Leyte and M. Mandel, Potentiometric behavior of polymethacrylic acid, *J. Polym. Sci. Part A*, **2**, 1879-1891 (1964).
11. J. Ralston, T. van Vliet, and J. Lyklema, Inference on ion binding to polyelectrolytes from emulsion rhelology, *J. Colloid Interface Sci.* **82**, 53-61 (1981).
12. K. Linderstrøm-Lang and S. O. Nielsen, Acid-base equilibria of proteins, in: *Electrophoresis* (M. Bier, ed.), Academic Press, New York (1959).
13. W. L. Pfeil and P. L. Privalov, Thermodynamic investigation of proteins—Standard functions for proteins with lysozyme as an example, *Biophys. Chem.* **4**, 23-32 (1976).
14. C. Tanford, S. A. Swanson, and W. S. Shore, Hydrogen ion equilibria of bovine serum albumin, *J. Am. Chem. Soc.* **77**, 6414-6421 (1955).
15. C. Tanford and J. D. Hauenstein, Hydrogen ion equilibria of ribonuclease, *J. Am. Chem. Soc.* **78**, 5287-5291 (1956).
16. C. Tanford and I. Epstein, Physical chemistry of insulin. 1. Hydrogen ion titration curve of zinc-free insulin, *J. Am. Chem. Soc.* **76**, 2163-2169 (1954).
17. E. Bucci, C. Fronticelli, and B. Ragatz, Proton binding behavior of human hemoglobin and its subunits in their native states, *J. Biol. Chem.* **243**, 241-249 (1968).
18. E. J. W. Verwey and J. Th. G. Overbeek, *Theory of the Stability of Lyophobic Colloids*, Elsevier, Amsterdam (1948).
19. W. Norde and J. Lyklema, Adsorption of human-plasma albumin and bovine pancreas ribonuclease at negatively charged polystyrene surfaces. 2. Hydrogen-ion titrations, *J. Colloid Interface Sci.* **66**, 266-276 (1978).
20. G. A. J. Van Os, S. H. de Bruin, and L. H. M. Janssen, Hydrogen ion titration curves of proteins as a source of structural and functional information, *J. Electroanal. Chem.* **37**, 303-311 (1972).
21. W. Norde and J. Lyklema, Thermodynamics of protein adsorption. Theory with special reference to the adsorption of human-plasma albumin and bovine pancreas ribonuclease on polystyrene surfaces, *J. Colloid Interface Sci.* **71**, 350-366 (1979).
22. D. K. Chattoraj and H. B. Bull, Electrophoresis of adsorbed protein, *J. Am. Chem. Soc.* **81**, 5128-5133 (1959).
23. L. B. Barnett and H. B. Bull, Electrophoresis of ribonuclease. Isoelectric points on various adsorbents, *J. Am. Chem. Soc.* **81**, 5133-5136 (1959).
24. T. H. Chiu, E. Nyilas, and D. M. Lederman, Thermodynamics of native protein foreign surface interactions—Calorimetric and microelectrophoretic study of human fibrinogen sorption onto glass and LTI-carbon, *Trans. Am. Soc. Art. Int. Org.* **22**, 498-513 (1976).
25. E. Nyilas, T. H. Chiu, and D. M. Lederman, Thermodynamics of native protein/foreign surface interactions. 2. Calorimetric and electrophoretic mobility studies on the human γ-globulin/glass system, in: *Colloid and Interface Science*, Vol. 5 (M. Kerker, ed.), pp. 77-89, Academic Press, New York (1976).
26. B. Dobias, Electrokinetic study of the interaction of proteins with lipid layers, *Progr. Colloid Polym. Sci.* **67**, 63-69 (1980).
27. P. Bagchi, S. M. Birnbaum, Effect of pH on the adsorption of immunoglobulin-G on anionic poly(vinytoluene) model latex particles, *J. Colloid Interface Sci.* **83**, 460-478 (1971).

28. D. R. Absolom, I. Michaeli, and C. J. van Oss, Electrophoresis of adsorbed albumin, *Electrophoresis* **2**, 273–278 (1981).
29. P. H. Wiersema, A. L. Loeb, and J. Th. G. Overbeek, Calculation of electrophoretic mobility of a spherical colloid particle, *J. Colloid Interface Sci.* **22**, 78–99 (1966).
30. J. Th. G. Overbeek and P. H. Wiersema, Interpretation of electrophoretic mobilities, in: *Electrophoresis*, Vol. 2 (M. Bier, ed.), Academic Press, New York (1967).
31. J. Th. G. Overbeek, Smoluchowski's equation for the electrophoresis of colloidal particles, *Philips Res. Repts.* **1**, 315–320 (1946).
32. D. C. Henry, Cataphoresis of suspended particles, *Proc. Roy. Soc., Ser. A* **133**, 106–140 (1931).
33. H. A. Abramson, L. S. Moyer, and M. H. Gorin, *Electrophoresis of Proteins*, New York (1942).
34. J. Th. G. Overbeek, in: Colloidal Science, Vol. 1 (H. R. Kruyt, ed.), Elsevier, Amsterdam (1948).
35. P. G. Koutsoukos, W. Norde, and J. Lyklema, Protein adsorption on hematite, *J. Colloid Interface Sci.* **95**, 385–397 (1983).
36. W. Norde and J. Lyklema, Adsorption of human-plasma albumin and bovine pancreas ribonuclease at negatively charged polystyrene surfaces. 4. The charge distribution in the adsorbed state. *J. Colloid Interface Sci.* **66**, 285–294 (1978).
37. P. van Dulm, W. Norde, and J. Lyklema, Ion participation in protein adsorption at solid surfaces, *J. Colloid Interface Sci.* **82**, 77–82 (1981).

8

Probing Protein Adsorption via Microcalorimetry

W. Norde

1. INTRODUCTION

It may be clearly seen from the previous chapters that protein adsorption is an intricate process. It is the outcome of the interplay of various forces mutually acting between the sorbent surface, the protein molecules, the solvent molecules, and any other species present. For instance, in Chapter 7 the involvement of small ions in the overall adsorption process has been emphasized.

The feasibility of protein adsorption (at constant pressure and temperature) is determined by the Gibbs energy of adsorption, $\Delta_{\text{ads}}G$. For any process to occur spontaneously ΔG must be negative, the affinity being greater the more negative the value of ΔG. Under equilibrium conditions, the calculation of ΔG involves the measurement of the concentrations of reactants and products. These concentrations are determined by the equilibrium constant K of the process, according to

$$K = \frac{\Pi c_p}{\Pi c_r} \tag{1}$$

where Πc_p and Πc_r refer to the products of the concentrations of products and reactants, respectively, each raised to the power corresponding to its coefficient in stoichiometric equation for the process.

The standard Gibbs energy change ΔG^0 is related to K through

$$\Delta G^0 = -RT \ln K \tag{2}$$

where R is the gas constant and T the temperature in Kelvin.

W. Norde • Laboratory for Physical and Colloid Chemistry, Agricultural University, De Dreijen 6, 6703 BC Wageningen, The Netherlands.

In the case of a reversible adsorption process the equilibrium state would be represented by the adsorption isotherm that shows the amount of protein adsorbed (per unit area of sorbent surface) as a function of the equilibrium concentration of the protein in solution. Then, after adopting some model assumptions, $\Delta_{ads}G^0$ can be evaluated. For this purpose the Langmuir theory for adsorption is often applied.

Since ΔG is composed of the changes in enthalpy ΔH and entropy ΔS, according to

$$\Delta G = \Delta H - T\Delta S \tag{3}$$

additional assessment of either ΔH or ΔS makes it possible to obtain the changes in all three thermodynamic state functions. Knowledge of the relative contributions of the enthalpy and the entropy to the change in the Gibbs energy is very useful for the elucidation of the adsorption mechanism.

For reversible processes ΔS and ΔH can be obtained from the temperature dependence of ΔG.

$$\partial \Delta G/\partial T = -\Delta S \tag{4}$$

and

$$\partial(\Delta G/T)/\partial T = -\frac{\Delta H}{T^2} \tag{5}$$

Some authors [e.g., References (1)–(3)] have treated protein adsorption as a reversible process. Accordingly, as discussed above, adsorption isotherms at different temperatures may serve as a basis for a complete thermodynamic analysis of the adsorption process. However, in most cases such a treatment is not justified. As a rule, protein adsorption is not completely reversible, i.e., exposure of the adsorbed molecule to the diluent does not lead to desorption, at least not within the time course of the experiment. This is due to the fact that adsorbed polymers (including proteins) are attached to the sorbent surface with many segments. As a consequence of this irreversibility it is not valid to apply Eqs. (2), (4), and (5) to the isotherms that represent the adsorption of whole protein molecules. For this reason a thermodynamic analysis of irreversible protein adsorption should be based on a direct determination of $\Delta_{ads}S$ and/or $\Delta_{ads}H$. In view of the many possible conformational and configurational changes that protein molecules and solvent molecules may undergo during adsorption, it is obvious that a direct statistical calculation of $\Delta_{ads}S$ is practically impossible. We are then left with $\Delta_{ads}H$. Under conditions of constant pressure, which is usually the case for adsorption from solution, $\Delta_{ads}H$ is equal to the heat of adsorption. In that case $\Delta_{\text{ads}}H$ can be obtained directly by (micro)calorimetry.

Anticipating the discussion in Section 4 it is noted here that $\Delta_{\text{ads}}H$ reflects the overall adsorption process. It contains contributions from several factors, i.e., redistribution of charged groups, changes in the state of hydration of both the protein and the sorbent surface, and rearrangements in the protein structure.

To study the adsorption mechanism it may be informative to measure $\Delta_{\text{ads}}H$ at different temperatures. The temperature coefficient (at constant pressure) of $\Delta_{\text{ads}}H$ equals the change upon adsorption of the apparent heat capacity $\Delta^{\phi}C_P$ of the system.

$$\Delta^{\phi}C_P = (\partial \Delta_{\text{ads}}H/\partial T)_P \tag{6}$$

$\Delta^{\phi}C_P$ can be interpreted in terms of changes in configurations of the components that are involved in the process.[5] In aqueous media changes in ${}^{\phi}C_P$ are usually to a large part due to changes in the water structure. For instance, dehydration of hydrophobic materials results in large negative values for $\Delta^{\phi}C_P$.[6] This suggests that hydrogen bonds between the water molecules surrounding nonpolar solutes are more labile than they are in bulk water. The transfer of ions from an aqueous to a nonaqueous medium contributes positively to $\Delta^{\phi}C_P$.[7] Loss of secondary structure (i.e., α-helix, β-sheet), which allows a greater rotational mobility along the polypeptide chain of the protein, would lead to a larger value for C_P.[8]

In view of the foregoing, the relevance of a calorimetric determination of ΔH for irreversible processes is evident. It is, therefore, surprising that calorimetric data on protein adsorption have been scarcely reported in literature. Before discussing experimental data, a brief outline of the principles of calorimetry will be given.

2. PRINCIPLES OF CALORIMETRY

Calorimetry is the measurement of heat changes. Based on their mode of operation calorimeters may be classified in two extreme types: (a) the adiabatic and (b) the conduction calorimeter. The ideal adiabatic calorimeter is completely insulated from its surroundings so that the heat evolved in the calorimeter is totally retained therein. Using this type of calorimeter one measures the temperature change of the process under investigation. The corresponding heat effect is proportional to the temperature change. The ideal conduction calorimeter, on the other hand, is perfectly connected thermally to a surrounding heat sink to which the heat developed in the reaction vessel is completely transferred. In this arrangement the heat exchange between the vessel and the heat sink takes place through a thermopile (multiple thermocouples). The voltage signal generated by the thermopile is proportional to the heat flux. As the ideal heat sink is of infinite capacity there is no change in temperature resulting from the process.

Between these two extremes lies another type of calorimeter, the so-called isothermally-jacketed calorimeter. In this apparatus the reaction vessel is imperfectly insulated from its surroundings, but the heat sink is maintained at constant temperature throughout. The heat developed causes a temporary increase in the temperature of the calorimeter followed by a gradual decrease towards the temperature of the heat sink. In this case the heat effect is calculated from the variation of the temperature with time, before, during, and after the process.

Until some ten years ago most calorimeters were of the isothermal-jacket type. However, recent technical advancements have improved the construction of adiabatic shields and of thermopiles so that the ideal conditions for both the adiabatic and the conduction calorimeter can practically be met.

In what way does *microcalorimetry* differ from ordinary or *macrocalorimetry*? There is no principal or well-defined distinction between these two. The prefix "micro" indicates that the calorimeter is very sensitive and therefore requires only small amounts of reactants. Sample volumes are in the range of a few cm^3 and the corresponding heat effects in the order of mJ's.

It is obvious that when small quantities of heat are measured the risk of significant systematic errors is much greater than in the case of large heat effects. As a consequence, in designing microcalorimeters much attention is paid to minimizing errors of mechanical origin, e.g., from agitation of the reactants or from processes such as evaporation, condensation, etc. Over the last two decades microcalorimeters have been drastically improved, allowing high-quality thermochemical studies. Microcalorimetry is of special importance in the life sciences, where often the relevant compounds are available only in small amounts.

Most microcalorimeters that are in use presently are commercial instruments. They are all of the conduction type. Well-known microcalorimeters are those based on the design of Calvet[9,10] and manufactured by Setaram, Lyon, France, the Benzinger microcalorimeter[11] marketed by Beckman Instruments, Palo Alto, California, and those designed by Wadsö,[12,13] made commercially by LKB Produkter AB, Bromma, Sweden. In the LKB instruments the twin principle has been adopted, i.e., two vessels are arranged as perfect twins. The detector units around these vessels are in opposition so that the recorded signal reflects the difference between the heat effects in each vessel. It may be expected that disturbances from the surroundings affect each vessel to the same extent so that they cancel in the differential signal. Because of this elimination of possible error sources,

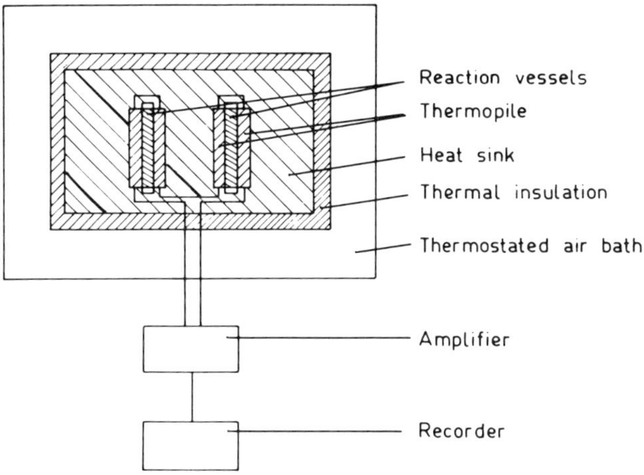

FIGURE 1. Arrangement in a twin calorimeter.

the twin arrangement is especially attractive in microcalorimetric systems. Figure 1 illustrates the principle of a twin calorimeter. In one of the vessels the reaction takes place whereas the other vessel serves as the reference. The two vessels make it possible to monitor simultaneously the reaction and the blank, for instance, in a dilution process.

Another important difference is that between a *batch* and a *flow* calorimeter. This distinction is based on the way in which the reactants are mixed. In a batch calorimeter the reactants are brought together instantaneously from two separate compartments. In a flow calorimeter the reactants are mixed in a continuous flow. The thermal disturbances related to the mixing process are smaller in a flow than in a batch calorimeter. However, the use of a flow calorimeter is limited to relatively fast processes. The reaction time must be shorter than the retention time for the reactants in the flow cell. Because of their different modes of operation the signals from the batch and the flow calorimeter are different. A typical voltage (V)-time (t) record obtained from a reaction in the batch calorimeter is shown in Figure 2a. The heat effect Q is proportional to the voltage-time integral, i.e., to the area under the curve shown in the figure.

$$Q = \text{const.} \times \int V \, dt \qquad (7)$$

The response obtained from a flow calorimeter is shown in Figure 2b. When the two liquids are pumped through the mixing vessel the baseline undergoes a deflection and a new steady state is reached. The heat flux W to or from

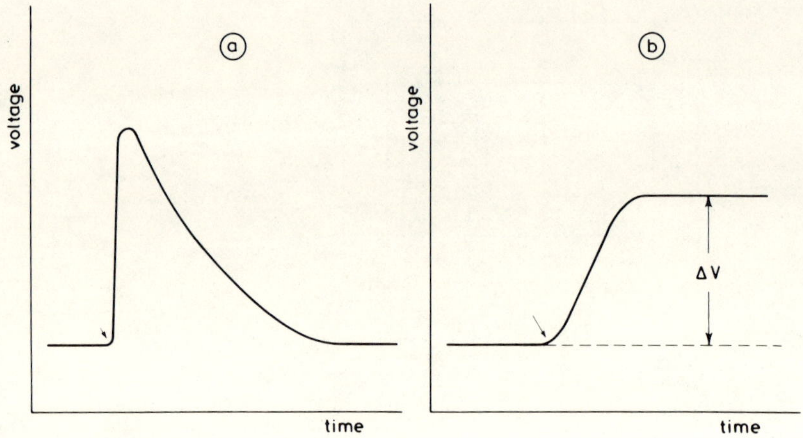

FIGURE 2. Recordings of a thermopile signal from (a) a batch and (b) a flow calorimeter. The arrows indicate the onset of mixing.

the reaction vessel is represented by the voltage shift

$$W(=Q/t) = \text{const.} \times \Delta V \tag{8}$$

The constants in Eqs. (7) and (8) are determined by calibration of the instruments.

Most of the adsorption enthalpies discussed in this chapter were obtained using a batch-type microcalorimeter.

The intention of this section is to give a general impression of the principles of (micro)calorimetry. For a more detailed treatment of this subject, with special attention to its application to the life sciences, the reader is referred to a review written by Spink and Wadsö.[14]

3. MICROCALORIMETRY OF PROTEIN ADSORPTION

As mentioned in Section 1, relatively little attention has been paid to the calorimetric determination of the enthalpy involved in the adsorption of proteins and protein analogues (e.g., polyamino acids). Yet, because of the apparent irreversibility of the adsorption of these substances, it is often the only route to a quantitative thermodynamic treatment of the adsorption process. The sorbent material is usually supplied as a fine dispersion so that sufficient surface area is available to obtain a measurable effect. In the following subsections the influence of the variation of various parameters of the system on the adsorption enthalpy will be discussed.

3.1. SURFACE COVERAGE

Let us first consider a system in which a flexible homopolymer (i.e., a polymer consisting of repeating units of only one type of monomer) adsorbs on a homogeneous surface. The enthalpy change arising from the adsorption process may be considered to consist of two contributions. One is the enthalpy associated with the disruption of sorbent-solvent and polymer-solvent contacts and the concomitant formation of polymer-sorbent and solvent-solvent bonds. This contribution is expected to be proportional to the number of contacts between segments of the polymer and the sorbent surface. The other contribution originates from a change in polymer-solvent interactions due to the difference in polymer-segment density between the adsorbed layer and the bulk solution. Often, this last term is considered to be negligible [e.g., Reference (15)]. This implies that the total adsorption enthalpy $\Delta_{ads}H$ reflects the process in the layer immediately adjacent to the sorbent surface. Consequently, $\Delta_{ads}H$, expressed per unit mass of adsorbed polymer, would be proportional to the fraction of the polymer segments that are attached at the sorbent surface. In order to employ $\Delta_{ads}H$ as an analytical measure for the attached fraction, the enthalpy per adsorbed segment has to be known. For this purpose, the calorimetric data have to be calibrated by determining the attached fraction via an independent method. Korn and Killmann[16] applied infrared spectroscopy to determine the attached fraction. Combining these data with calorimetric data they obtained enthalpy values in the range of −35 to −40 kJ per mole of attached

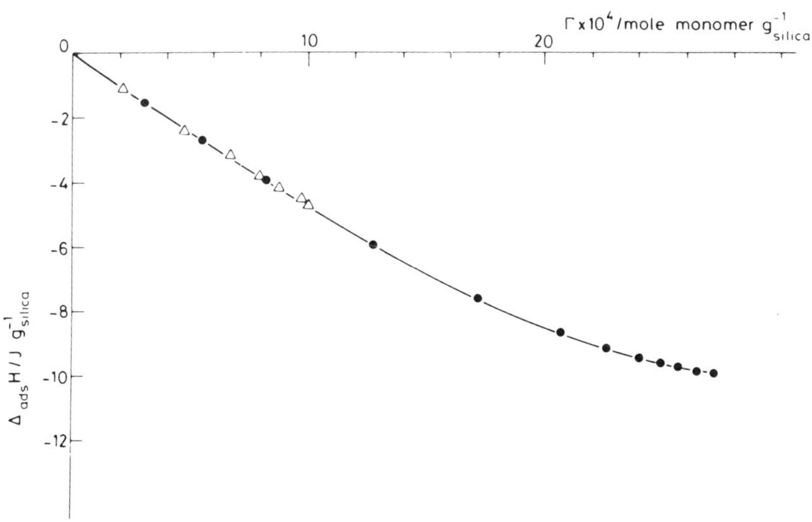

FIGURE 3. Enthalpy of the adsorption of uncharged polymers from carbon tetrachloride onto silica surfaces. $T = 25°C$. ● PVAc, △ PBMA.

monomer for the adsorption of various uncharged homopolymers (each of them containing an ester group in the monomer unit) from carbon tetrachloride onto silica surfaces. These exothermic effects were associated with the formation of a hydrogen bond between a monomer unit and an —OH group of the silica surface. Figure 3 presents $\Delta_{ads}H$ as a function of the amount of polymer adsorbed, Γ, as obtained by Korn and Killmann for poly(vinyl acetate) (PVAc) and poly(methacrylic acid *n*-butyl ester) (PBMA).

The curvature in the curve for PVAc is ascribed to a gradual conformational change of the adsorbed polymer on varying the surface coverage. More specifically, the fraction of monomers that are attached to the sorbent surface decreases with increasing amount of adsorbed polymer. For this particular system the attached fraction is found to decrease from 0.5 down to 0.2. This implies that the number of monomers in the "loops" and "tails" of the adsorbed polymer molecule increases with increasing adsorption. The results for PBMA show a constant value for the attached fraction, which was estimated to be *ca.* 0.4. As a result of this constancy the amount of PBMA that can be accommodated in the adsorbed layer is much smaller than that of PVAc.

Cohen Stuart[17] studied the adsorption of another uncharged polymer, viz. poly(vinyl pyrrolidone) (PVP), on silica surfaces, both from water and from dioxane. He also observed exothermic effects. Some of his calorimetric results are shown in Figure 4. Again, the curvature is associated with a reduced bound fraction with increasing amount of PVP adsorbed. It is interesting to note that the initial slopes are identical for dioxane and water. Based on this observation, Cohen Stuart concluded that the enthalpy effect per attached monomer is the same, *ca.* −10 kJ mol, in dioxane and in water. This similarity seems to be fortuitous.

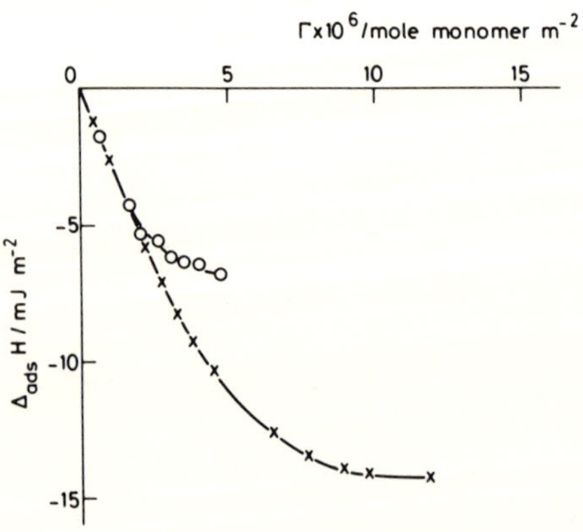

FIGURE 4. Enthalpy of the adsorption of the uncharged polymer PVP onto silica surfaces. $T = 25°C$. Adsorption from dioxane (×) and from water (○).

So far, experimental results for uncharged polymers have been discussed. They all show exothermic adsorption behavior. The adsorption enthalpies stem from hydrogen bonds and dipole–dipole, dipole-induced dipole, and dispersion interactions. On introducing electrical charges in the molecule coulombic interactions between the polymer and the charged sorbent surface as well as between adsorbed molecules also contribute to $\Delta_{ads}H$. Coulombic forces act over relatively large separations, so that charged groups in the loops and tails of the adsorbed polymer may also affect $\Delta_{ads}H$. Hence, this quantity cannot be regarded in this case as a reflection of the attachment of segments to the sorbent surface. Lamarche *et al.*[(18)] have compared adsorption enthalpies of poly(acrylic acid) (PAA) on surfaces of titanium dioxide (TiO_2) particles suspended in water under different conditions of charge on both the polymer and the sorbent surface. Measurements were carried out at pH 9.5 where both the TiO_2 surface and the PAA molecules are negatively charged, and at pH 3.5 where the TiO_2 surface is positive but the PAA molecules slightly negative. The curves in Figure 5 are constructed from their data. The maximum adsorption at pH 3.5 is about five times larger than at pH 9.5. This can to a large extent be ascribed to the swelling of the PAA molecules on charging them by changing the pH from 3.5 to 9.5. Moreover, at pH 9.5 the maximum surface coverage may be further limited by repulsion between the PAA molecules and the substrate and also by lateral repulsion between the adsorbed PAA molecules. The trend in the curve for $\Delta_{ads}H$ at pH 3.5 as a function of the amount adsorbed, Γ, for the slightly charged PAA molecules is similar to that found for uncharged polymers. The process is exothermic and the enthalpy change levels off as Γ increases. At pH 9.5, the adsorption of the negative polymer on the negative surface results in an endothermic effect, $\Delta_{ads}H > 0$. The value of $\Delta_{ads}H$ is more or less proportional to Γ, up to adsorption saturation. This suggests that the conformation and orientation of the adsorbed PAA

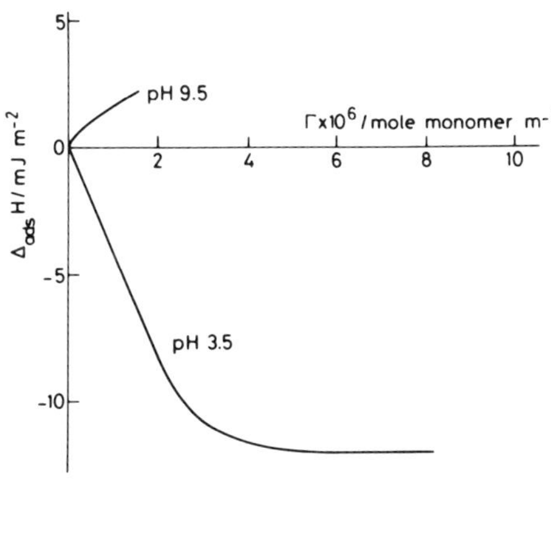

FIGURE 5. Enthalpy of the adsorption of the polyelectrolyte PAA from an aqueous medium onto titanium dioxide. Influence of the charge on the polymer and the sorbent surface. $T = 25°C$; ionic strength = 0.01 *M*.

molecules are independent of the surface coverage, or, alternatively, that leveling off due to conformational changes is compensated for by an endothermic contribution from electrostatic lateral repulsion. The mere fact that $\Delta_{ads}H > 0$, whereas the adsorption proceeds spontaneously ($\Delta_{ads}G < 0$), implies that the adsorption is entropically driven: $\Delta_{ads}S > 0$.

Since the structure of proteins is, as a rule, far more complicated than that of polyelectrolytes like PAA, interpretation of the trends in their $\Delta_{ads}H(\Gamma)$ will be even more ambiguous. If the protein is structured more or less like a flexible coil, explanation of $\Delta_{ads}H$ may still be relatively simple; the trend in $\Delta_{ads}H(\Gamma)$ would resemble that observed for the adsorption of a flexible charged homopolymer, although the chemical heterogeneity might cause some irregularities. In the literature no calorimetric data are available yet on the adsorption of such proteins. Another simple case would be the adsorption of compact globular protein molecules that adsorb in a constant orientation without any change in conformation. Then, $\Delta_{ads}H$ is expected to be proportional to the adsorbed amount up to a surface coverage where lateral interactions become operative. However, proteins usually undergo some conformational rearrangements that may depend on the degree of surface coverage. Also the orientations of the adsorbed molecules are likely to change on crowding at the sorbent surface.

In a series of papers[19,20,21] Nyilas *et al.* described microcalorimetric experiments on the adsorption of the blood proteins human γ-globulin (HGG) and human fibrinogen (HFb). Both proteins were adsorbed from an aqueous solution, pH 7.2, 0.05 *M* ionic strength, onto hydrophilic glass surfaces. In addition, to investigate the influence of the sorbent hydrophobicity, HFb was also adsorbed onto the more hydrophobic so-called low-temperature isotropic (LTI) carbon powder. At pH 7.2 the surfaces of both sorbents are negatively charged, the glass more strongly than the carbon. The proteins carry a net negative charge as well; the isoelectric point (i.e.p.) of HGG is in the pH range 5.8–7.0[22] and that of HFb at pH *ca.* 5.5.[22] In view of the fact that the charges on the protein and the sorbent have the same sign, it is, at first sight, surprising that the adsorption is exothermic. In Section 4 the quantity $\Delta_{ads}H$ will be further characterized.

Figure 6 shows the adsorption enthalpies as a function of the amount of HGG and HFb adsorbed on the glass surface. In the case of HGG, $\Delta_{ads}H$ becomes progressively more negative with increasing Γ up to completion of the first layer of protein molecules at the glass surface. This trend suggests that the lateral interaction between the HGG molecules in the first layer is attractive. It could be that these molecules, containing both positively and negatively charged groups, are arranged in such a way that intermolecular coulombic attraction prevails in this layer. At higher surface concentrations the variation in $\Delta_{ads}H(\Gamma)$ is much smaller, but it is still a decreasing function. This implies weak attraction between the first and the second (or subsequent)

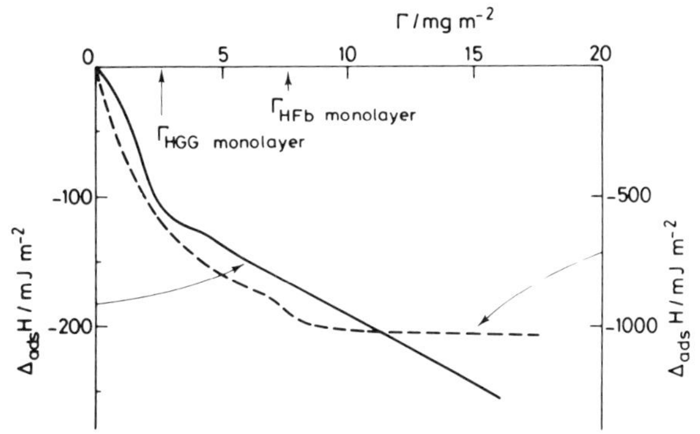

FIGURE 6. Adsorption enthalpy of HGG (——) and HFb (- - - -) on glass surfaces. $T = 25°C$; ionic strength = 0.05 *M*.

adsorbed layer, which may be caused by conformational changes in the adsorbing HGG molecules.[23,24] The enthalpy effect for HFb adsorption shows a different pattern. On approaching monolayer coverage $\Delta_{ads}H(\Gamma)$ levels off, just as is usually observed with synthetic (charged) polymers. Then, initial adsorption in the second layer is accompanied by an increase in the molar adsorption enthalpy, after which it levels off again.

If it is assumed that at low surface coverage lateral interactions between adsorbed protein molecules are absent, the initial slope of the $\Delta_{ads}H(\Gamma)$ curve reflects the enthalpy of interaction between the protein and the sorbent. Although this slope is difficult to establish, values of *ca.* 5×10^6 J/mol for HGG and *ca.* 135×10^6 J/mol for HFb seem to be realistic estimates. Morrissey and coworkers[23,25] studied the adsorption of γ-globulin and fibrinogen from bovine plasma onto hydrophilic silica at pH 7.4. This system, i.e., both the proteins and the sorbent surface, is similar to that studied by Nyilas *et al.* For the initially adsorbed protein molecules they reported a bound fraction of 0.2 for γ-globulin and 0.08 for fibrinogen. These values may be approximately valid for the systems studied by Nyilas. Then, for HGG the enthalpy change per mole of attached amino acid residues amounts to *ca.* 15.5 kJ. This is in the range of the enthalpy involved in the formation of a hydrogen bond.[26] For HFb a value of *ca.* 490 kJ per mole of attached amino acid residues is calculated, which suggests the formation of a chemical bond between the initially adsorbed HFb molecules and the glass surface.

The enthalpy effects for the adsorption of HFb on hydrophilic glass and on hydrophobic LTI carbon are compared in Figure 7. Since, under further identical conditions, the adsorbed amounts are different between the two sorbents, $\Delta_{ads}H$ is plotted against the relative surface coverage $(\Gamma/\Gamma_{monolayer})$ to allow a meaningful comparison.

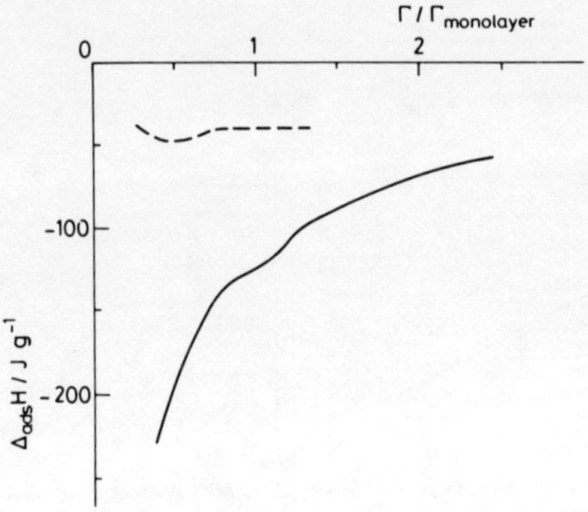

FIGURE 7. Comparison of the adsorption enthalpies of HFb on glass (——) and on LTI carbon surfaces (- - - -). $T = 25°C$; ionic strength = 0.05 *M*.

For LTI carbon $\Delta_{ads}H$ is much smaller than for glass. The authors concluded that the "intensity" of the interaction is reflected by the enthalpy effect. However, they ignored the contribution of the entropy change to the Gibbs energy of adsorption. Because of the difference in hydrophobicity, dehydration of the LTI carbon surface causes a much larger entropy increase than does dehydration of glass. Indeed, although for LTI carbon the enthalpy effect is less negative, the initial part of the adsorption isotherm for the sorbent is steeper, indicating a higher affinity of the protein for this surface than for glass. Moreover, in many systems, as will be shown in examples to be discussed, positive enthalpy changes are involved in the adsorption of proteins that show an—entropy-determined—high-affinity for the sorbent surface.

Nyilas *et al.* stated that the exothermic effect triggers changes in the structure of the adsorbed protein molecules so that the adsorption enthalpy may be considered as a proportionate measure of those adsorption-induced conformational rearrangements. This statement does not seem to be justified, however. If structural changes occur, they should be regarded as an integral part of the overall process that eventually favors the adsorption of the protein molecule. The enthalpy change resulting from such structured changes is an indistinguishable part of the net adsorption enthalpy $\Delta_{ads}H$ which is measured calorimetrically.

Koutsoukos *et al.*[27] measured enthalpies of the adsorption of human plasma albumin (HPA) from an aqueous solution onto hematite (α-Fe_2O_3) surfaces. In these systems the charge of the HPA and the hematite were varied by adjusting the pH. The i.e.p. of HPA is at pH 4.7 and that of the hematite at pH *ca.* 6.5. At a given temperature and electrolyte concentration $\Delta_{ads}H$ is determined by the pH and the adsorbed amount. In Figure 8 $\Delta_{ads}H$ is plotted against the relative adsorption θ of HPA, defined as Γ/Γ_P (Γ_P equals the plateau value of the adsorption isotherm).

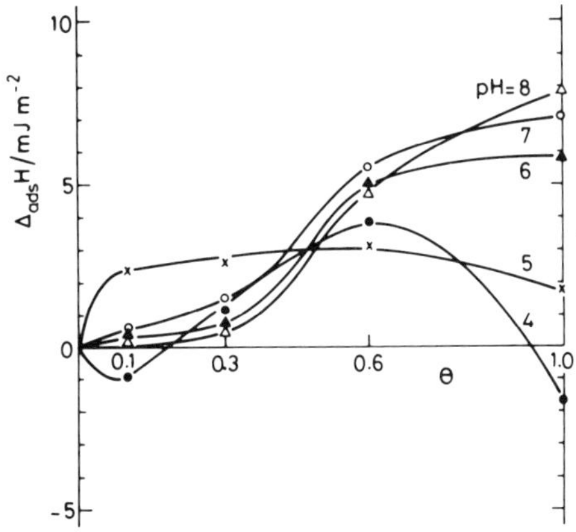

FIGURE 8. Adsorption enthalpy of HPA on hematite surfaces. $T = 25°C$; ionic strength = 0.01 *M*. The pH is indicated.

Under almost all conditions studied, the adsorption is endothermic, implying that it is driven by a gain in entropy, i.e., $\Delta_{ads}S > 0$. It can further be seen that the trend in $\Delta_{ads}H(\theta)$ strongly depends on the charge on the protein and the sorbent surface. At pH 5 ($\gtrsim$i.e.p. of HPA) $\Delta_{ads}H$ rises between $\theta = 0$ and $\theta = 0.1$, between $\theta = 0.1$ and 0.6 it remains more or less constant, and it shows a downward trend for higher surface coverages. From other experimental data[28,29] it was inferred that around the i.e.p. the conformational changes in the adsorbing HPA molecules are at a minimum. If all molecules adsorb in the same fashion $\Delta_{ads}H(\theta)$ would be a straight line through the origin. This is apparently not the case. The downward trend at high coverage may be due to lateral attraction between the adsorbed HPA molecules that, in the isoelectric state, contain equal numbers of positively and negatively charged groups. At pH $\neq$ i.e.p. of HPA the interaction between the protein molecules and the hematite surface is enthalpically more favorable than at the i.e.p. It could well be that at those pH values the protein molecule, which still contains both positive and negative groups, rearranges its structure, resulting in a more favorable interaction between the charged groups of the protein and the hematite surface. The rise in $\Delta_{ads}H$ between $\theta = 0.3$ and $\theta = 0.6$ is probably largely due to increased lateral repulsion between the adsorbed charged HPA molecules. At first sight, the deviation, at high θ, between the curves for pH < i.e.p. and pH > i.e.p., is difficult to explain. However, since at high θ, $\Delta_{ads}H$ is composed of contributions from protein-sorbent and protein-protein interactions, more insight into the relative importance of these contributions can be obtained by plotting $\Delta_{ads}H$ *per unit mass of protein* versus the pH. In Figure 9 this has been done for $\theta = 0.1$ and $\theta = 1.0$. Assuming that at $\theta = 0.1$ lateral interactions are negligible, the curve for $\theta = 0.1$ reflects only HPA–hematite interactions. An impression of the contribution from lateral interactions at full coverage is obtained by subtracting $\Delta_{ads}H$ for $\theta = 0.1$ from

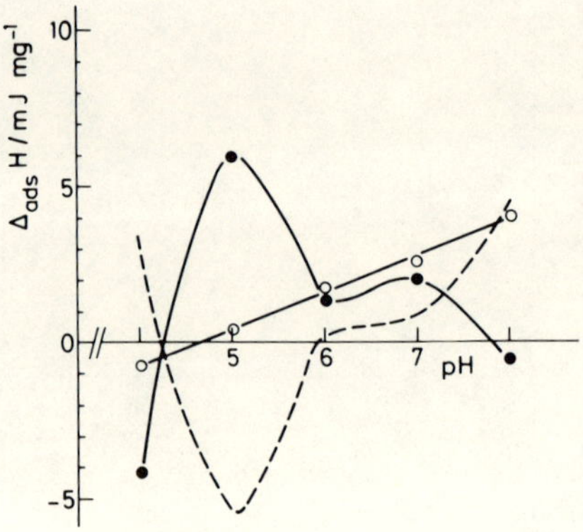

FIGURE 9. Adsorption enthalpy per unit mass of HPA on hematite surfaces. $T = 25°C$; ionic strength = 0.01 *M*. (●) $\theta = 0.10$; (○) $\theta = 1.00$. The dashed curve represents the difference between the curves for $\theta = 1.00$ and $\theta = 0.10$.

$\Delta_{ads}H$ for $\theta = 1.0$. The result is indicated by the dotted line in Figure 9. It clearly shows increasing lateral repulsion with increasing net charge on the adsorbed HPA molecules.

3.2. ELECTRICAL CHARGE

In the previous section the influence of charge on the protein molecules and the sorbent surface on the adsorption behavior and, consequently, on the enthalpy change has already been indicated. Here, this influence will

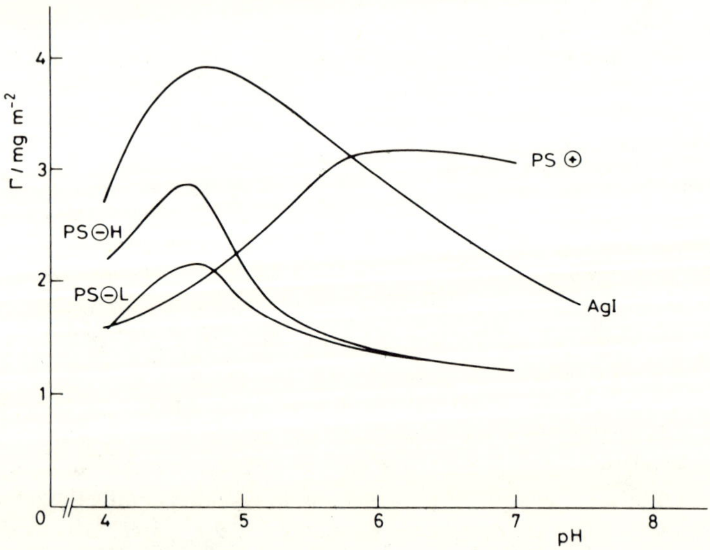

FIGURE 10. Plateau values of adsorption isotherms for HPA on various sorbents, as a function of the pH of adsorption. $T = 25°C$; ionic strength = 0.01 *M*.

be treated in more detail and the discussion will be restricted to conditions of surface saturation (plateau adsorption). A systematic study on the effect of charge has been undertaken by Norde and coworkers[(30,31)] for the adsorption of HPA on different substrates, i.e., particles of polystyrene (PS) and of silver iodide (AgI) that are finely dispersed in an aqueous medium. The charge on the HPA molecule varies with pH whereas that of the sorbent surfaces is essentially constant over the pH range studied. Two negatively charged PS latices were used, one (PS ⊖ L) having a low and the other PS ⊖ H having a high surface charge density. The experiments were performed in aqueous solutions containing 0.01 *M* KNO_3 or NaCl. In this medium the electrostatic interaction is much stronger than in most natural systems where it is screened by a higher ionic strength.

Figure 10 shows the plateau values Γ_P of the adsorption isotherms for HPA with the different sorbents. These adsorptions were attained in the systems that were examined calorimetrically. The resulting enthalpy effects are presented in Figure 11. The $\Delta_{\text{ads}}H(\text{pH})$ curves are qualitatively similar. At low pH, say pH 4, the adsorption is exothermic. On increasing the pH of adsorption it becomes less exothermic or, in the case of PS surfaces, it is even endothermic. On further increasing the pH the adsorption tends to be exothermic again. For the PS latices $\Delta_{\text{ads}}H > 0$, at least over a part of the pH range. Since the adsorption still occurs spontaneously and hence $\Delta_{\text{ads}}G < 0$, under and probably also beyond these conditions $\Delta_{\text{ads}}S > 0$. In line with this, there is no simple relation between $\Delta_{\text{ads}}H$ and the charge on either the HPA molecule or the sorbent surface. If interaction between the overall charges would control the adsorption process, the curves for $\Delta_{\text{ads}}H(\text{pH})$ would be gradually ascending for the negative and descending for the positive sorbent. In the case that electrostatic interaction between the adsorbed protein molecules were a major factor, a minimum in $\Delta_{\text{ads}}H$

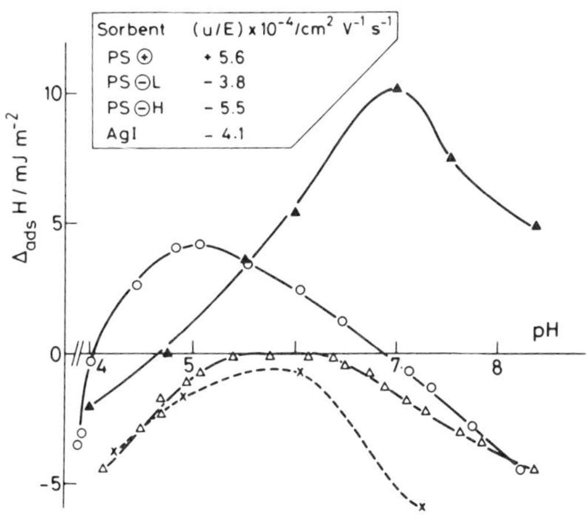

FIGURE 11. Adsorption enthalpy of HPA on various substrates. $T = 25°C$; ionic strength = 0.01 *M*. ▲ PS⊕; △ PS⊖L; ○ PS⊖H; × AgI. Insert: Electrophoretic mobilities (u/E) of the bare sorbents under the experimental conditions.

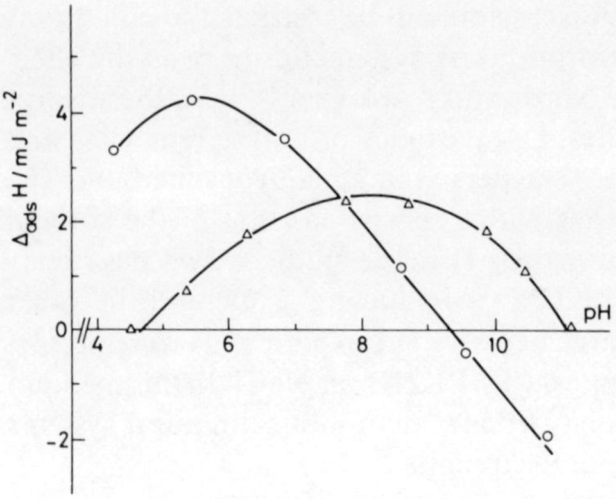

FIGURE 12. Adsorption enthalpy of RNase on negatively charged polystyrene surfaces. $T = 25°C$; ionic strength = 0.01 *M*. △ PS⊖L; ○ PS⊖H.

could be expected in the isoelectric region of the protein, i.e., pH *ca.* 4.7. Neither one of these trends is observed. Thus, although electrostatic interactions surely contribute to $\Delta_{ads}H$, they do not dominate this quantity. On the contrary, with all sorbents it is found that spontaneous adsorption occurs also when the charges on the protein and the sorbent have the same sign and that under these conditions the process is even exothermic.

The same kind of calorimetric measurements have been performed for the adsorption of bovine pancreas ribonuclease (RNase) on the two negatively charged PS surfaces.[30] The results are given in Figure 12. The isoelectric point of RNase is at pH *ca.* 9.3. For this protein too, the enthalpy effects do not directly reflect coulombic interaction and here again the adsorption proceeds by an increase in entropy.

3.3. TEMPERATURE

It has already been mentioned in Section 1 that variation in the enthalpy of a process on changing the temperature, which according to Eq. (6) equals the change in the apparent heat capacity ${}^{\phi}C_P$ of the system, may reflect structural alterations. The influence of the temperature on $\Delta_{ads}H$ for HPA and RNase on PS latices is shown in Figure 13. The curves clearly demonstrate that $\Delta_{ads}\,{}^{\phi}C_P$ does not follow a simple pattern. This again is indicative of the intricate nature of the adsorption process. Before discussing the trends any further, it is useful to recall the factors that contribute to $\Delta_{ads}\,{}^{\phi}C_P$: (a) Dehydration of the hydrophobic PS surface and of hydrophobic parts of the protein lowers ${}^{\phi}C_P$,[6] (b) transfer of ions from the aqueous solution into the adsorbed protein layer leads to an increase in ${}^{\phi}C_P$,[7] (c) changes in the protein structure also affect ${}^{\phi}C_P$. Since the perturbation of the protein structure is not known in detail its contribution to $\Delta_{ads}\,{}^{\phi}C_P$ cannot be predicted. An increase in motional freedom inside the protein molecule

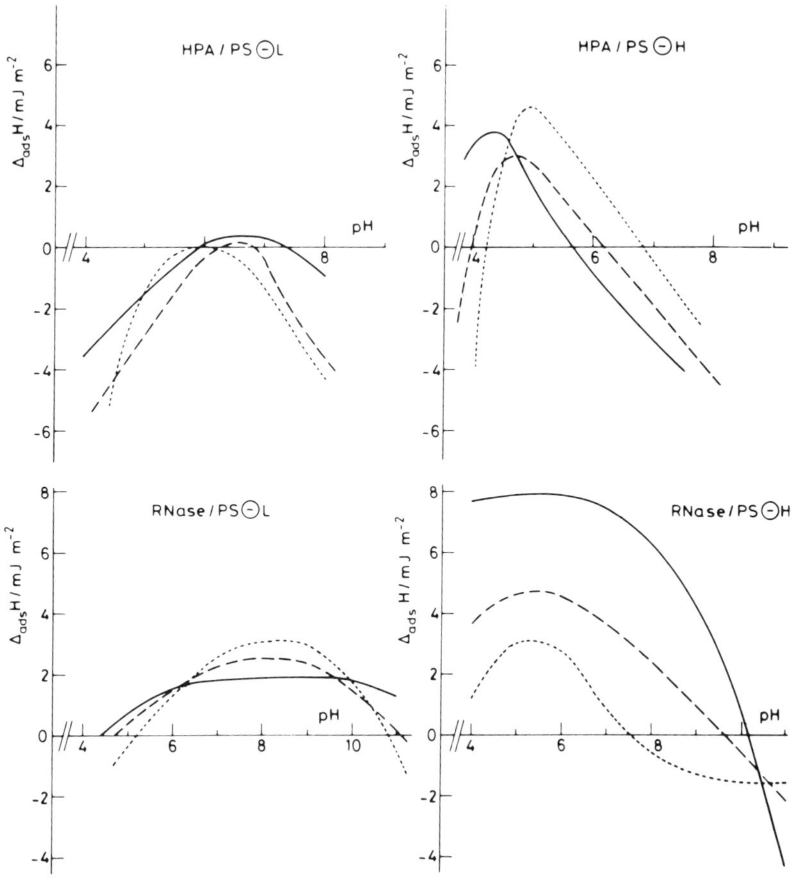

FIGURE 13. Influence of the temperature on the adsorption enthalpies of HPA and RNase on polystyrene surfaces. Ionic strength = 0.05 *M*. —— 9°C; - - - - 25°C; · · · · · 37°C.

(breakdown of secondary structure) would lead to higher ${}^{\phi}C_P$ values whereas the actual attachment of the protein molecule on the sorbent surface influences ${}^{\phi}C_P$ in the opposite direction. In addition to these intrinsic contributions (a), (b), and (c), there is the effect that, particularly for HPA, the adsorbed amount itself is sensitive to the temperature.[28] Therefore, $\Delta_{ads}H$ per unit area at various temperatures refers to different amounts of protein adsorbed (or $\Delta_{ads}H$ per unit mass of protein at various temperatures to different areas of sorbent surface covered).

For the adsorption of HPA on the low charge latex $\Delta_{ads}{}^{\phi}C_P$ derived from $\Delta_{ads}H$ at 9°C and 25°C is negative over the entire pH range studied. It is probable that dehydration of the hydrophobic PS surface dominates the change in ${}^{\phi}C_P$. At 37°C the curve for $\Delta_{ads}H$(pH) deviates between pH 4.5 and 5.8. This may be due to larger perturbations in the adsorbed protein structure in that pH region at 37°C. This suggestion is consistent with the different Γ_P(pH) profile at 37°C, as compared to those at 25°C and 9°C.[28]

The adsorption of HPA on the high charge latex shows a different pattern for $\Delta_{ads}\,{}^{\phi}C_P$. In the first place, because of the higher density of sulfate groups at the latex surface and, consequently, of the counterions (in these experiments K^+) in the slipping layer around the PS particles, a larger number of chaotropic ions are buried by the adsorbed HPA molecules. This gives a more positive contribution to $\Delta_{ads}\,{}^{\phi}C_P$ than in the case of adsorption on the low-charge PS latex. Secondly, as a result of the higher charge density, the hydrophobic fraction of the latex surface is smaller and, hence, the lowering effect on ${}^{\phi}C_P$ on dehydrating the latex surface is proportionally smaller. These considerations could very well explain the influence of the surface charge density of the sorbent on $\Delta_{ads}\,{}^{\phi}C_P$.

In the case of RNase adsorption on the low-charge latex, $\Delta_{ads}\,{}^{\phi}C_P < 0$ only at pH < 6.5 and at pH > 10. The less dominant contribution of hydrophobic dehydration, as compared to that for the adsorption of HPA, may be related to the smaller degree of structural changes in the adsorbing RNase molecules.[28,29] This results in a larger protein-void volume in the adsorbed layer and hence in less dehydration of the hydrophobic PS surface. Moreover, the RNase molecule itself is much less hydrophobic than the HPA molecule.[32] The number of ions transferred from solution into the adsorbed layer was found to be at a maximum in the intermediate pH range[33] and this could be responsible for the positive $\Delta_{ads}\,{}^{\phi}C_P$ value in this region. The increasing effect of the latex surface charge density on $\Delta_{ads}\,{}^{\phi}C_P$ can be explained along the same lines as for HPA adsorption. The sign reversal at pH > 10.5 may be due to conformational changes in the adsorbing RNase molecules in that pH region[28] that, in turn, would cause a further dehydration of the PS surface.

Thus, the influence of temperature on $\Delta_{ads}H$ for the adsorption of HPA and RNase on PS latices is in qualitative agreement with other experimental data, notably those concerning the adsorbed amount and the incorporation of ionic groups.

4. THERMODYNAMICS OF PROTEIN ADSORPTION

The adsorption of protein molecules, leading to the formation of a dense adsorbed layer, is schematically depicted in Figure 14.

It has already been mentioned in Section 1 that because of the irreversibility of protein adsorption thermodynamic quantities cannot be derived from isotherm constants.

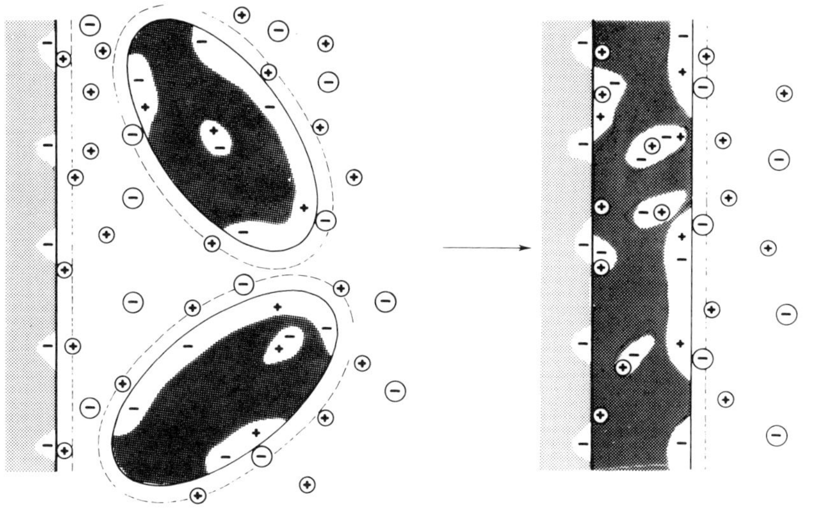

FIGURE 14. Schematic representation of protein adsorption on a negatively charged surface. +/− refer to the proton charge of the protein and to the sorbent surface charge; ⊕/⊖ refer to mobile ions. Shaded areas indicate hydrophobic regions; dashes indicate hydrodynamic slipping planes.

The following thermodynamic analysis is based on the consideration that the overall adsorption process may be divided into several subprocesses of which the contributions to the net change in the thermodynamic quantities can be estimated. The subprocesses that constitute the overall adsorption process are mentioned in Table 1. It is noted that these subprocesses occur more or less simultaneously but that, for the calculation of the changes in the functions of state, they may be considered as occurring consecutively. By subtracting the sum of the estimated constituents from the overall effect it is possible to find the most elusive contribution. This procedure has been followed for the enthalpy.[4] Specifically, attempts have been made to obtain the enthalpy change arising from structural alterations in the adsorbing protein molecules by subtracting the contributions from the subprocesses, listed under A and B in Table 1, from the calorimetrically measured overall adsorption enthalpy $\Delta_{\text{ads}}H$. In principle, the same procedure could be followed for the entropy and the Gibbs energy, provided that $\Delta_{\text{ads}}S$ and $\Delta_{\text{ads}}G$ are known.

In order to estimate the contributions from the subprocesses, information on the electrostatic interaction, ion transfer, and dehydration of the sorbent surface is required. Therefore, in addition to calorimetric measurements, other experiments, notably hydrogen ion titrations and electrophoresis, should be carried out. This has been done for the adsorption

TABLE 1
Subprocesses Involved in Protein Adsorption

A. Redistribution of charge	Chemical part: medium change of transferred ions	H^+
		other ions
	Electrical part: overlap of electric fields	
B. Changes in the state of hydration	Sorbent surface	
	Protein surface	
	Formation of protein–sorbent contacts	
C. Rearrangements in the protein structure	Intramolecular van der Waals interactions	
	Ion pairs, hydrogen bonds, hydrophobic interaction	
	Internal rotational mobility	
	Bond lengths and angles	

of HPA and RNase on PS latices.[29,33,34] Hence, the thermodynamic analysis given in Reference (4) is based on data for the systems mentioned above, under conditions of plateau adsorption.

HPA and RNase are very different proteins, having different isoelectric points, molecular weights, and susceptibilities to structural changes upon charging. The charge densities and, therefore, the hydrophobicities of the PS surfaces used vary over a wide range. As the analysis applies to these different systems it is likely to be applicable to the adsorption of other globular proteins at other surfaces as well. Below, a brief outline of the thermodynamic analysis is given.

4.1. FACTORS INVOLVED IN THE ADSORPTION PROCESS

4.1.1. Redistribution of Charged Groups

The distribution of charged groups undergoes drastic alterations during the adsorption process. When a protein molecule arrives in the electric field of a charged sorbent surface, protons are taken up or released, depending on the pH and the sign and strength of the electric field.[29] Like protons, other ions are also incorporated or released.[33,34,35] These ion exchange processes are reversible and hence the relations of Eqs. (4) and (5) may be applied to these parts of the adsorption process.

The distribution of ions between the solution and the adsorbed protein layer is determined by the molar Gibbs energy of transfer, i.e., by the difference between the molar Gibbs energy g_i of an ion of species i in the protein layer p and in the bulk solution s, respectively

$$g_i^p - g_i^s = (\mu_i^p - \mu_i^s) + z_i F(\phi^p - \phi^s) \tag{9}$$

where μ_i is the chemical potential of i and ϕ is the electrostatic potential in the indicated phase, F is Faraday's constant, and z_i is the valency of the ion, sign included. Obviously, the first term in the r.h.s. of Eq. (9) represents the chemical contribution and the second term the electrical contribution to the Gibbs energy of transfer.

The contribution $\Delta_{\text{ads}}G_{\text{H}}$ from the change in the degree of protonation of the protein can be calculated from the shift in the titration curve upon adsorption (see also Chapter 7). If titrations are performed at different temperatures $\Delta_{\text{ads}}H_{\text{H}}$ and $\Delta_{\text{ads}}S_{\text{H}}$ can also be obtained. In systems that do not contain additional buffer the protons taken up by the adsorbing protein molecules are delivered by the protein molecules that remain in solution. The resulting thermodynamic effects will therefore be small.

For the thermodynamic calculations of the transfer of ions other than protons, it is expedient to consider the chemical contribution, $\Delta_{\text{ads}}G_i$, and the electrical contribution, $\Delta_{\text{ads}}G_{\text{el}}$, separately.

The chemical contribution from the transfer of one mole of ion i from water into the adsorbed layer equals

$$\Delta_{\text{ads}}g_i(\text{chem}) = \mu_i^{0p} - \mu_i^{0s} + RT \ln \frac{c_i^p}{c_i^s} \tag{10}$$

where $(\mu_i^{0p} - \mu_i^{0s})$ is the standard molar Gibbs energy of transfer and c_i the concentration of i. Strictly speaking, Eq. (10) is only valid if μ_i^{0p} and c_i^p are independent of the distance from the sorbent surface. If not, a more general equation should be formulated. However, the experimental data usually do not provide sufficiently detailed information to justify such a refinement. In the case of the HPA/PS and RNase/PS systems a model for the adsorbed protein layer has been proposed in which three homogeneous regions are distinguished,[33] so that in each region Eq. (10) may be used without further modification. The concentrations of the ions in the adsorbed protein layer may be deduced from titration and electrophoresis data[33] or determined directly,[35] but $(\mu_i^{0p} - \mu_i^{0s})$ has to be estimated from model studies on the transfer of ions from water into various nonaqueous media.[36,37] With the more chaotropic ions, i.e., ions that cause disorder in their hydration shell, $(\mu_i^{0p} - \mu_i^{0s})$ attains more positive values by virtue of a greater loss of entropy. In Eq. (10) activity corrections have been omitted.

If the activity coefficients in the adsorbate and the solution are different it affects the result of the calculations. These activity coefficients are certainly different, but the main cause is the electric field in the adsorbed layer and this will be accounted for in the electrostatic contribution from the charge redistribution.

As long as the system may be treated as ideal, the standard molar enthalpy of transfer of ion i is independent of the concentration

$$\Delta_{\text{ads}} h_i(\text{chem}) = h_i^{0p} - h_i^{0s} \tag{11}$$

Again, the deviation from ideality due to the electric field is accounted for in the electrostatic contribution. Approximate values for $\Delta_{\text{ads}} h_i(\text{chem})$ can be derived from model systems.[37]

The electrostatic contribution $\Delta_{\text{ads}} G_{\text{el}}$ can be calculated by comparing the charge distributions before and after adsorption. It is composed of three terms, one stemming from the bare sorbent surface, one from the dissolved protein molecules, and one from the surface covered with a protein layer.

The electric part of the Gibbs energy of an electrical double layer equals the isothermal reversible work of charging it [References (38), (39), and Chapter 7].

$$G_{\text{el}} = \int_0^{\sigma_0} \phi_0' \, \mathrm{d}\sigma_0' \tag{12}$$

ϕ_0' is the potential and σ_0' the charge density at the sorbent surface during the charging process. To obtain ϕ_0' as a function of σ_0' assumptions concerning the charge distribution have to be made. For the three-region model[33] it can be derived that

$$\phi_0 = \frac{1}{\varepsilon_0} \sum_j \frac{A_j}{\varepsilon_j} \tag{13}$$

in which A_j is a function of the charge density in, and the thickness of, region j, ε_j the dielectric constant, and ε_0 the permittivity of free space. Substituting Eq. (13) in Eq. (12) one obtains

$$G_{\text{el}} = \frac{\sigma_0}{2\varepsilon_0} \sum_j \frac{A_j}{\varepsilon_j} \tag{14}$$

For a given charge distribution the temperature dependence of G_{el} is determined by the temperature dependence of ε_j. Application of the

Eqs. (4) and (5) yields

$$S_{\text{el}} = \frac{\sigma_0}{2\varepsilon_0}\left[\sum_j \frac{A_j}{\varepsilon_j^2}\frac{\delta\varepsilon_j}{\delta T}\right] \tag{15}$$

and

$$H_{\text{el}} = \frac{\sigma_0}{2\varepsilon_0}\left[\sum_j \frac{A_j}{\varepsilon_j}\left(1 + \frac{T}{\varepsilon_j}\frac{\delta\varepsilon_j}{\delta T}\right)\right] \tag{16}$$

It can be inferred from Eqs. (14), (15), and (16) that the thermodynamic effects of the redistribution of charged groups are very sensitive to the difference between the dielectric permittivities in the adsorbed protein layer on the one hand and in the hydration layer at the bare sorbent surface on the other.

4.1.2. Changes in the State of Hydration

The formation of an adsorbed protein layer (see Figure 14) may be thought of as occurring in two steps: (i) aligning and deformation of the protein molecules leading to a film and (ii) deposition of this film on the sorbent surface. The thermodynamic effects involved in the deformation, if any, will be accounted for in factor C of Table 1: "Rearrangements in the protein structure." Here, only the changes originating from the transfer of the protein from the solution to the sorbent surface will be considered.

Any changes involved in step (ii) result from the disruption of bonds between the sorbent (1) and the water (2) and between the protein (3) and the water (2) and the formation of bonds between sorbent (1) and protein (3) and between water (2) and water (2). Thus for the Gibbs energy

$$\Delta_{\text{ads}}G_h = G_{13} + G_{22} - G_{12} - G_{23} \tag{17}$$

and similar equations apply to $\Delta_{\text{ads}}H_h$ and $\Delta_{\text{ads}}S_h$. With hydrophilic surfaces $\Delta_{\text{ads}}G_h$ is expected to be positive, but it becomes more negative with increasing hydrophobicity of the sorbent. For hydrophobic surfaces a positive entropy change due to hydrophobic dehydration is the main contribution.

Because the various interactions between the components involved are not known in detail, it is impossible to assess the thermodynamic effects quantitatively. They may, however, to a first approximation be estimated from studies with model systems. For instance, the transfer of ethylbenzene from water to an organic medium could be considered a mimicry of the dehydration of PS and the concomitant covering of the PS surface with a protein layer. The only data available in the literature on such a transfer of ethylbenzene concern the dissolution of this substance from its own bulk liquid into water.[40] The ensuing change in Gibbs energy can be written as

$$\Delta_{\text{ads}} G_h = G_{22} - G_{12} \tag{18}$$

which deviates from Eq. (17). The difference between Eqs. (17) and (18) is $(G_{13} - G_{23})$. This difference reflects the formation of protein–sorbent contacts at the expense of protein hydration. This is essentially a protein "solvent effect" and it will also be counted in the contribution from the rearrangements in the protein structure.

4.1.3. Rearrangements in the Protein Structure

In view of the structure–function relationship, the question of whether or not the adsorbing protein molecules undergo structural alterations is of great biological importance. Especially regarding the application of biomedical materials, knowledge and control of the protein structure in the adsorbed state is desired. In the process of deposition of biological matter, adsorption of proteins is the fastest event. Subsequent adhesion of biological cells (platelets, erythrocytes, leucocytes, microorganisms, etc.) may be influenced by the structure of the adsorbed protein molecules.

The structure a protein molecule adopts is the result of the interplay between different types of interaction inside the molecule but also between the molecule and its environment. A change in the environment, e.g., by introducing an adsorbing interface, affects the interactions, which may lead to changes in the protein structure. For instance, if a globular protein molecule in solution is stabilized by intramolecular hydrophobic bonding, whereas the sum effect of the other structure-determining factors favors a more expanded conformation, a structural change upon adsorption may be expected. This is the more likely when the protein molecule has a larger charge density, i.e., at pH values away from the i.e.p. The reduction of intramolecular hydrophobic bonds would be compensated for by placing a fraction of the hydrophobic amino acid residues at the sorbent surface, thus shielded from water. Since hydrophobic interaction between side groups of the polypeptide chain supports the formation of α-helices and β-sheets, a conformational change as indicated above may be accompanied by a reduction of those secondary structures. This provides the polypeptide chain with a greater rotational mobility.

According to the literature, many proteins at various interfaces undergo structural alterations. It is, for the time being, impossible to give a detailed analysis of these rearrangements and hence the ensuing effects in the thermodynamic quantities cannot be evaluated from first principles.

However, the enthalpy change $\Delta_{ads}H_{str\ pr}$ may be approximated by subtracting from the overall adsorption enthalpy $\Delta_{ads}H$ (which can be well established by microcalorimetry) those contributions that can be estimated within certain limits,

$$\Delta_{ads}H_{str\ pr} = \Delta_{ads}H - (\Delta_{ads}H_H + \Delta_{ads}H_i + \Delta_{ads}H_{el} + \Delta_{ads}H_h) \quad (19)$$

It must be realized that $\Delta_{ads}H_{str\ pr}$ thus derived includes contributions from the change in hydration of the protein and from van der Waals interactions between the adsorbed molecules. Furthermore, the accuracy of the value obtained for $\Delta_{ads}H_{str\ pr}$ is determined by the uncertainties in the constituent terms. Nevertheless, the sign and the order of magnitude of $\Delta_{ads}H_{str\ pr}$ may give some information on the perturbations in the structure of the adsorbing protein molecules.

4.2. RESOLUTION OF THE ADSORPTION ENTHALPY

The enthalpy of the adsorption of HPA and RNase from an aqueous KNO_3 solution onto PS surfaces having different charge densities has been analyzed according to the procedure described in Section 4.1. Two adjustable parameters are chosen, i.e., the molar enthalpy of ion transfer h_i(chem) and the dielectric constant ε^{PS/H_2O} in the hydration layer at the bare PS surface.

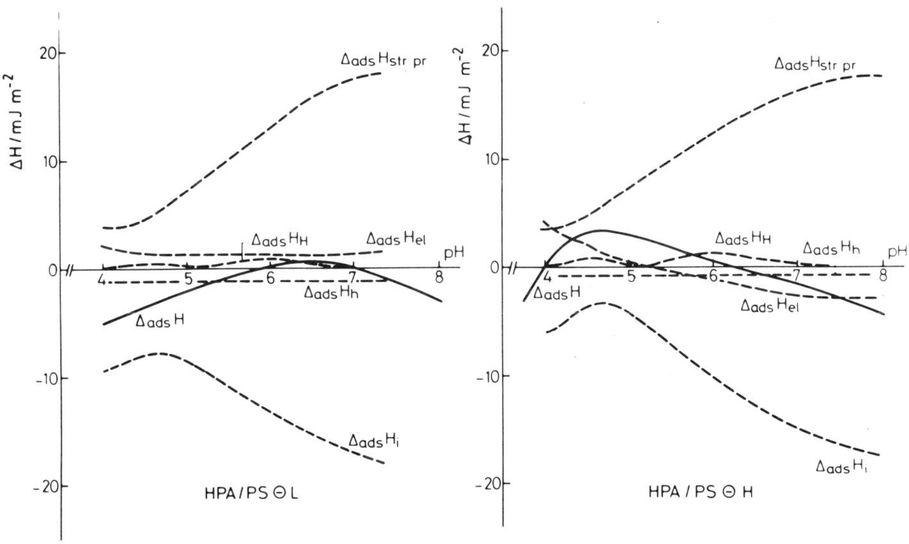

FIGURE 15. Adsorption of HPA on PS ⊖ latices. Resolution of the adsorption enthalpy (——) into the contributions (- - - -) from the various subprocesses. For further explanation see the text.

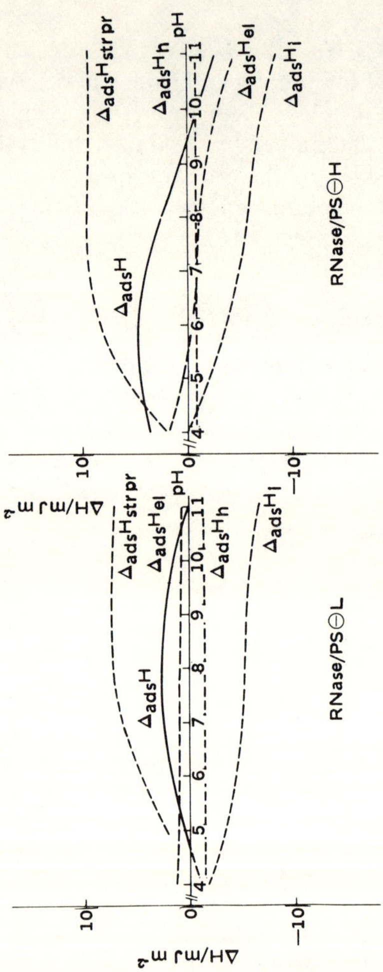

FIGURE 16. Adsorption of RNase on PS⊖ latices. Resolution of the adsorption enthalpy (——) into the contributions (- - - -) from the various subprocesses. For further explanation see the text.

The values for these parameters are adjusted such that the curves for $\Delta_{ads}H_{str\ pr}(pH)$ are similar for the two latices. The requirement of this similarity is based on the observation that the variation in $\Gamma_p(pH)$, which has been interpreted in terms of rearrangements in the protein structure, is also comparable for the two latices.[28] Since the sensitivity of Γ_p to the pH of adsorption is much larger for HPA than for RNase, h_i(chem) and ε^{PS/H_2O} are evaluated for the HPA/PS systems. The result is shown in Figure 15.

The curves for $\Delta_{ads}H_H$ have been derived from titration data[41] and the ones for $\Delta_{ads}H_h$ from dissolution studies of ethylbenzene in water.[40] The uncertainties in $\Delta_{ads}H_H$ and $\Delta_{ads}H_h$ may be large, but their contributions to $\Delta_{ads}H$ are small. The condition of comparable values for $\Delta_{ads}H_{str\ pr}$ for the two latices is met by positioning the curves for $\Delta_{ads}H_i(pH)$ and $\Delta_{ads}H_{el}(pH)$. The result is $\Delta_{tr}h_{K^+}(\text{chem}) \approx -7RT$ and $\varepsilon^{PS/H_2O} \approx 5$. These values fit well with those in the literature: *ca.* $-7RT$ has been reported for the molar enthalpy of transfer of K^+ from water into a series of organic media[37] and values between 5 and 10 have been mentioned for ε in the hydration layer at different surfaces.[42,43]

In the case of RNase adsorption there is no *a priori* reason to assume other values for $\Delta_{tr}h_{K^+}$(chem) and ε^{PS/H_2O} than those derived from HPA adsorption. The resolution of $\Delta_{ads}H$ for RNase/PS is shown in Figure 16. Since titration data at different temperatures are not available $\Delta_{ads}H_H$ cannot be calculated. For reasons mentioned in Section 4.1 neglecting this contribution would not severely affect the analysis.

It is noted that for both HPA and RNase $\Delta_{ads}H_{str\ pr}$ is found to be positive, but it is not clear whether this is a general feature of protein adsorption.

In Figure 17 $\Delta_{ads}H_{str\ pr}$ (expressed per unit mass of protein) is plotted versus Γ_p. As both these quantities are associated with structural rearrangements, the observed functional relationship supports the correctness of the analysis.

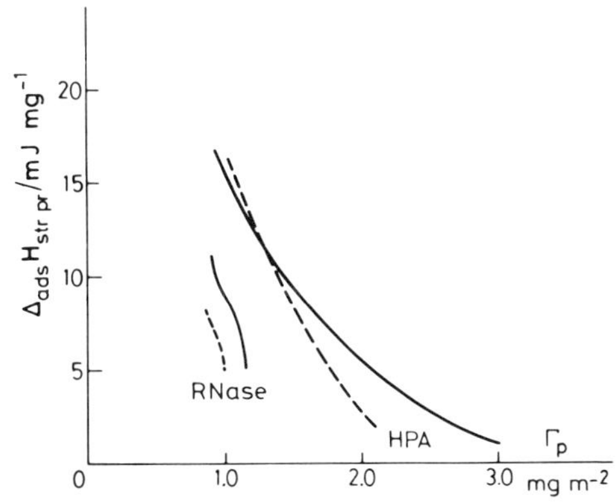

FIGURE 17. Adsorption of HPA and RNase on PS⊖H (——) and PS⊖L (- - - -) latices. Contribution of structural rearrangements in the protein molecule to the total adsorption enthalpy.

4.3. GIBBS ENERGY AND ENTROPY OF ADSORPTION

The observation that in many systems protein adsorption is irreversible implies a large negative value for $\Delta_{ads}G$. It is of prime interest to know how the various subprocesses contribute to $\Delta_{ads}G$. Such information indicates whether a particular subprocess favors ($\Delta G < 0$) or opposes ($\Delta G > 0$) the overall adsorption. The transfer of protons yields a very slight positive effect that will be neglected. The constituent terms stemming from the redistribution of charge and from changes in the state of hydration may be established as explained in Sections 4.1.1 and 4.1.2, considering the same model systems as for the analysis of $\Delta_{ads}H$. In view of the result arrived at in establishing $\Delta_{ads}H_i$, $\Delta_{tr}g_i$(chem) for the transfer of ions from water to organic media[36] has been taken to estimate $\Delta_{ads}G_i$. As mentioned in Section 4.1.3 direct assessment of $\Delta_{ads}G_{str\ pr}$ is impossible. Moreover, this quantity cannot be approximated in the same way as done with the enthalpy, because $\Delta_{ads}G$ is not known. Consequently, the Gibbs energy is unraveled in less detail than the enthalpy. The results are shown in Figure 18. For HPA $(\Delta_{ads}G - \Delta_{ads}G_{str\ pr}) > 0$ over almost the entire pH range studied. Since $\Delta_{ads}\ G < 0$ it follows that $\Delta_{ads}G_{str\ pr} < 0$. Hence, structural rearrangements in the HPA molecule (including its hydration layer) are in favor of the adsorption process. The results for RNase are not conclusive as to this point. It can further be concluded that dehydration of the PS surface largely supports the adsorption and that the electrical part of the charge redistribution may

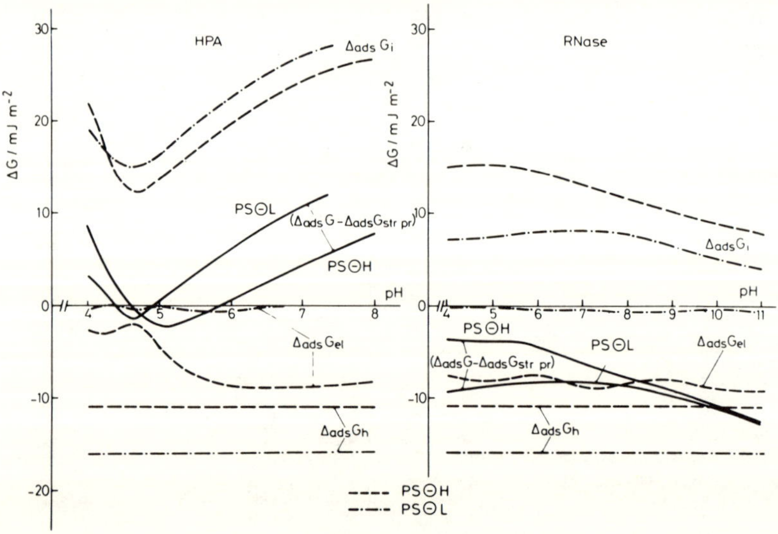

FIGURE 18. Adsorption of HPA and RNase on PS ⊖ latices. The figure shows contributions to the Gibbs energy of adsorption and the resulting value for $(\Delta_{ads}G - \Delta_{ads}G_{str\ pr})$.

do so to a minor extent. The chemical part of the redistribution of charged groups, i.e., the change of the medium of the ions transferred from water into the protein layer, strongly opposes protein adsorption. However, if no ions were incorporated in the adsorbed layer either $\Delta_{\text{ads}}G_{\text{el}}$ would attain large positive values or the adsorbed protein molecules would unfold into a very loose structure. In the latter case hydrophobic amino acid residues are exposed to the aqueous medium which gives rise to positive values for $\Delta_{\text{ads}}G_{\text{str pr}}$.

It is instructive to compare $\Delta_{\text{ads}}H$ and $\Delta_{\text{ads}}G$ of each constituent. The large negative value for $\Delta_{\text{ads}}G_h$ is almost entirely the result of a positive entropy change from the dehydration of the hydrophobic PS surface. It is obvious that dehydration of hydrophilic surfaces would not stimulate the adsorption process. Comparing $\Delta_{\text{ads}}G_{\text{el}}$ with $\Delta_{\text{ads}}H_{\text{el}}$ reveals that $\Delta_{\text{ads}}S_{\text{el}} > 0$. In other words, the degree of order in the distribution of charge lowers upon adsorption. Since $\Delta_{\text{ads}}H_{\text{str pr}}$ is positive throughout, the negative value for $\Delta_{\text{ads}}G_{\text{str pr}}$ (as is shown to be the case for HPA) is attained through an increase in entropy, $\Delta_{\text{ads}}S_{\text{str pr}} > 0$. This could originate from dehydration of hydrophobic patches at the protein surface but also from breakdown of secondary structures in the protein molecule leading to more rotational freedom along the polypeptide chain (cf. Section 4.1.3). Indeed, circular dichroism measurements[44] point to a reduction of the α-helical content of adsorbed proteins. Finally, the change in hydration of the transferred ions involves an entropy decrease that largely overcompensates the negative enthalpy effect.

The conclusion arrived at in Section 3.2 that under most conditions the adsorption of HPA and RNase on PS latices is entropically driven can now be specified with respect to the sources of entropy gain. From the thermodynamic analysis it can, furthermore, be inferred in which way some physicochemical properties of the system influence the affinity between the protein and the sorbent surface. The affinity is a measure of the strength of the adsorptive bond but it is not directly related to the amount of protein that can be accommodated in a saturated layer. It may thus be expected that the affinity between a (negatively charged) sorbent surface and a given protein increases if that surface is more hydrophobic and if it has a smaller electrokinetic potential. The influence of the surface charge density depends on the difference between the electrostatic potential at this surface before and after adsorption. The affinity of a protein for a given (negatively charged) surface would be higher the more hydrophobic the protein is, the stronger the tendency to rearrange its structure upon transfer from an aqueous to a nonaqueous medium, and the smaller the fraction of its negative groups that arrive in the inner regions of the adsorbed layer. Protein adsorption would further be facilitated by increasing the concentration and the valency and decreasing the chaotropicity of the small ions in the system.

5. CONCLUDING REMARKS

The enthalpy change measured calorimetrically reflects the overall adsorption process. Therefore, the interpretation of this quantity becomes progressively more difficult with increasing complexity of the adsorption mechanism. In the case of adsorption of flexible uncharged polymers on a homogeneous surface, the theory accounts fairly well for the trends observed experimentally, and the enthalpy effect may be interpreted in terms of segment–sorbent contacts. With proteins, however, interpretation of the calorimetric data is only possible if additional information is obtained from other, independent experiments about the various types of interactions operative in the adsorption process.

The basic assumptions made in the thermodynamic analysis of the adsorption of HPA and RNase on PS surfaces are not confined to any particular protein or sorbent surface. Therefore, a similar approach may, *mutatis mutandis*, apply as well to other systems. Although the analysis is not very rigorous it provides some insight into the roles played by various parameters, so that some general principles of protein adsorption can be traced.

REFERENCES

1. S. N. Levine, Thermodynamics of adsorbed protein films, *J. Biomed. Materials Res.* **3**, 83–94 (1969).
2. G. Deželić, N. Deželić, and Z. Telišman, Binding of human-serum albumin by monodisperse polystyrene latex particles. *Eur. J. Biochem.* **23**, 575–581 (1971).
3. E. C. Moreno, M. Kresak, and D. I. Hay, Adsorption of two human parotid salivary macromolecules on hydroxyapatites, fluorhydroxyapatites and fluorapatites, *Arch. Oral Biol.* **23**, 525–533 (1978).
4. W. Norde and J. Lyklema, Thermodynamics of protein adsorption. Theory with special reference to the adsorption of human-plasma albumin and bovine pancreas ribonuclease on polystyrene surfaces, *J. Colloid Interface Sci.* **71**, 350–366 (1979).
5. D. H. Everett, O. Y. Samoilov, D. D. Eley, J. S. Frank, T. A. Turney, S. Lengvel, R. Parsons, and J. F. B. Randles, Molecular mechanism of rate processes in solids: Ion-solvent interactions, *Disc. Faraday Soc.* **24**, 216–238 (1957).
6. C. Tanford, *The Hydrophobic Effect*, Wiley-Interscience, New York (1973).
7. H. S. Frank and W. Y. Wen, Structural aspects of ion–solvent interactions in aqueous solutions—A suggested picture of water structure, *Disc. Faraday Soc.* **24**, 133–140 (1957).
8. J. F. Brandts, in: *Biological Macromolecules*, Vol. II (G. Fasman and S. N. Timasheff, eds.), pp. 213–243, Dekker, New York (1969).
9. E. Calvet, in: *Experimental Thermochemistry*, Vol. 2 (F. D. Rossini, ed.), pp. 237–286, Interscience, New York (1956).
10. E. Calvet and H. Prat, in: *Rec. Prog. in Microcalorimetry* (H. A. Skinner, ed.), 35–50, Pergamon Press, London (1963).
11. T. Benzinger, in: *A Laboratory Manual of Analytical Methods of Protein Chemistry*, Vol. 5 (P. Alexander and H. P. Lundgren, eds.), pp. 95–149, Pergamon Press, New York (1969).

12. P. Monk and I. Wadsö, A flow micro reaction calorimeter, *Acta Chem. Scand.* **22**, 1842-1852 (1968).
13. I. Wadsö, in: *Calorimetry, Thermometry and Thermal Analysis*, p. 42, Society of Calorimetry and Thermal Analysis, Kagaku Gijitsu-sha, Tokyo (1973).
14. C. Spink and I. Wadsö, Calorimetry as an analytical tool in biochemistry and biology, *Methods Biochem. Analysis* **23**, 1-157 (1976).
15. E. Killmann, H. J. Strasser, and K. Winter, *VI Intern. Kongress Grenzflachenaktive Stoffe*, Zurich, Berichte, Band III, 221 (1972).
16. M. Korn and E. Killmann, Infrared and micro-calorimeter studies of the adsorption of polymers with ester groups in the main or side-chain, at the silica-carbon tetrachloride interface, *J. Colloid Interface Sci.* **76**, 119-131 (1980).
17. M. A. Cohen Stuart, Thesis, Agricultural University Wageningen, The Netherlands (1980).
18. J. M. Lamarche, A. Foissy, G. Robert, J. C. Reggiani, and J. Bernard, in: *Adsorption at the Gas-Solid and Liquid-Solid Interface* (J. Rouquerol and K. S. W. Sing, eds.), pp. 117-123, Elsevier, Amsterdam (1982).
19. E. Nyilas, T. H. Chiu, and G. A. Herzlinger, Thermodynamics of native protein/foreign surface interactions. I. Calorimetry of the human gamma-globulin/glass system, *Trans. Am. Soc. Art. Int. Org.* **20**, 480-490 (1974).
20. E. Nyilas, T. H. Chiu, and D. M. Lederman, Thermodynamics of native protein/foreign surface interactions. 2. Calorimetric and electrophoretic mobility studies on the human γ-globulin/glass system, in: *Colloid and Interface Science*, Vol. 5 (M. Kerker, ed.), pp. 77-89, Academic Press, New York (1976).
21. T. H. Chiu, E. Nyilas, and D. M. Lederman, Thermodynamics of native protein foreign surface interactions—Calorimetric and microelectrophoretic study of human fibrinogen sorption onto glass and LTI-carbon, *Trans. Am. Soc. Art. Int. Org.* **22**, 498-513 (1976).
22. H. E. Schultze and J. F. Hermans, *Molecular Biology of Human Proteins*, Elsevier, Amsterdam (1966).
23. B. W. Morrissey and C. A. Fenstermaker, Concentration of adsorbed gamma-globulin and beta-lactoglobulin—Effects of surface concentration, *Trans. Am. Soc. Art. Int. Org.* **22**, 278-284 (1976).
24. P. Bagchi and S. M. Birnbaum, Effect of pH on the adsorption of immunoglobulin-G on anionic poly(vinyltoluene) model latex particles, *J. Colloid Interface Sci.* **83**, 460-478 (1971).
25. B. W. Morrissey and R. R. Stromberg, Conformation of adsorbed blood proteins by infrared bound fraction measurements, *J. Colloid Interface Sci.* **46**, 152-164 (1974).
26. G. C. Kresheck and I. M. Klotz, Thermodynamics of transfer of amides from an apolar to an aqueous solution, *Biochemistry* **8**, 8-16 (1969).
27. P. G. Koutsoukos, W. Norde, and J. Lyklema, Protein adsorption on hematite, *J. Colloid Interface Sci.* **95**, 385-397 (1983).
28. W. Norde and J. Lyklema, Adsorption of human-plasma albumin and bovine pancreas ribonuclease at negatively charged polystyrene surfaces. 1. Adsorption isotherms, effects of charge, ionic strength, and temperature, *J. Colloid Interface Sci.* **66**, 257-265 (1978).
29. W. Norde and J. Lyklema, Adsorption of human-plasma albumin and bovine pancreas ribonuclease at negatively charged polystyrene surfaces. 2. Hydrogen-ion titrations, *J. Colloid Interface Sci.* **66**, 266-276 (1978).
30. W. Norde and J. Lyklema, Adsorption of human-plasma albumin and bovine pancreas ribonuclease at negatively charged polystyrene surfaces. 5. Microcalorimetry, *J. Colloid Interface Sci.* **66**, 295-302 (1978).
31. P. G. Koutsoukos, C. A. Mumme-Young, W. Norde, and J. Lyklema, Effect of the nature of the substrates on the adsorption of human-plasma albumin, *Colloids and Surfaces* **5**, 93-104 (1982).

32. C. C. Bigelow, On the average hydrophobicity of proteins and the relation between it and protein structure, *J. Theor. Biol.* **16**, 187–211 (1967).
33. W. Norde and J. Lyklema, Adsorption of human-plasma albumin and bovine pancreas ribonuclease at negatively charged polystyrene surfaces. 4. Charge distribution in adsorbed state, *J. Colloid Interface Sci.* **66**, 285–294 (1978).
34. W. Norde and J. Lyklema, Adsorption of human-plasma albumin and bovine pancreas ribonuclease at negatively charged polystyrene surfaces. 3. Electrophoresis, *J. Colloid Interface Sci.* **66**, 277–284 (1978).
35. P. van Dulm, W. Norde, and J. Lyklema, Ion participation in protein adsorption at solid surfaces, *J. Colloid Interface Sci.* **82**, 77–82 (1981).
36. M. H. Abraham, Ionic entropies of transfer from water to non-aqueous solvents, *J. Chem. Soc., Faraday Trans. 1* **69**, 1375–1388 (1973).
37. M. H. Abraham, E. Ah-Sing, A. F. Danil de Namor, T. Hill, A. Nasehzadeh, and R. A. Schulz, Heats of solution of electrolytes in ethanol and derived enthalpies of transfer from water, *J. Chem. Soc., Faraday Trans. 1* **74**, 359–365 (1978).
38. E. J. W. Verwey and J. Th. G. Overbeek, *The Theory of the Stability of Lyophobic Colloids*, Elsevier, Amsterdam (1948).
39. A. J. Babchin, T. Gur, and I. J. Lin, Repulsive interface forces in overlapping electric double-layers in electrolyte-solutions, *Adv. Colloid Interface Sci.* **9**, 105–141 (1978).
40. S. J. Gil, N. F. Nichols, and I. Wadsö, Calorimetric determination of enthalpies of solution of slightly soluble liquids. 2. Enthalpy of solution of some hydrocarbons in water and their use in establishing temperature dependence of their solubilities, *J. Chem. Thermodyn.* **8**, 445–452 (1976).
41. C. Tanford, S. A. Swanson, and W. S. Shore, Hydrogen ion equilibria of bovine serum albumin, *J. Am. Chem. Soc.* **77**, 6414–6421 (1955).
42. B. Vincent, B. H. Bijsterbosch, and J. Lyklema, Competitive adsorption of ions and neutral molecules in Stern layer on silver-iodide and its effect of colloid stability, *J. Colloid Interface Sci.* **37**, 171–178 (1971).
43. J. O. M. Bockris, M. A. V. Devanathan, and K. Muller, On the structure of charged interfaces, *Proc. Roy. Soc., Ser. A* **274**, 55–79 (1963).
44. M. E. Soderquist and A. G. Walton, Structural changes in proteins adsorbed on polymer surfaces, *J. Colloid Interface Sci.* **75**, 386–397 (1980).

9

Protein Adsorption Hysteresis

Herbert P. Jennissen

1. INTRODUCTION

Hysteresis was discovered over one hundred years ago by Ewing[1] who studied the influence of torsion and magnetization on the production of a transient current (i.e., polarization) in wire. He found that the change in polarization "lags behind" the change in torsion. To this phenomenon he applied the term "hysteresis" derived from the Greek (to lag behind). A few years later,[2,3] after having described the familiar magnetization-demagnetization loops in soft iron as a function of the applied magnetizing force, Ewing gave the following definition for hysteresis[3]:

> "Thus, when there are two qualities M and N such that cyclic variations of N cause cyclic variations of M, then if the changes of M lag behind those of N, we may say that there is hysteresis in the relation of M to N. The value of M at any given point of the operation depends not only on the actual value of N, but on all the preceding changes of N, and by properly manipulating those changes, any value of M within more or less wide limits may be found associated with a given value of N."

Soon afterwards in 1897 *adsorption hysteresis* was first reported by von Bemmelen[4] in a very extensive paper covering about nine years of work on the adsorption of water on to silica gel. In conclusion he wrote: "Es scheint mir nicht unpassend, die Erscheinungen bei der Kieselsäure, nämlich das *Zurückbleiben der Änderungen des (Wasser-) Gehaltes hinter den Änderungen des (Dampf-) Druckes* mit dem Namen *Hysteresis* zu belegen."*

* Translation: "It appears reasonable that the phenomena encountered on silica, namely *the lagging-behind of the changes in (water) content versus the changes in (vapor) pressure,* be called *hysteresis.*"

Herbert P. Jennissen • Institut für Physiologie, Physiologische Chemie und Ernährungsphysiologie, Ludwig Maximilians-Universität München, Veterinärstr. 13, D-8000 München 22, West Germany.

The work of von Bemmelen was followed by reports on the adsorption hysteresis of alcohol and benzene on silica gels[5] and ferric oxide gels.[6,7] Similarly the adsorption of water on proteins was reported to be of hysteretic nature [see References (8) and (9)].

This field, specifically adsorption hysteresis of gases on solids, was reviewed extensively in 1967 by Everett,[9] who also gave a more recent definition of hysteresis:

> "A process is said to exhibit *hysteresis* if, when the direction of change of an independent variable x is reversed, a dependent variable y fails to retrace the values through which it passed in the forward process; the dependent variable 'lags behind' in its attempt to follow the changes in the independent variable."

The above definitions are the basis for our understanding of hysteresis today when applied to the adsorption of macromolecules such as proteins onto interfaces from solution. In the case of *protein adsorption hysteresis* as employed in this paper the independent variable is defined as the apparent equilibrium concentration of the free solute (expressed either as c or c/c_0) and the dependent variable as the amount of adsorbed solute (expressed as the surface concentration Γ^s or fractional saturation Γ^s/Γ^s_0). c_0 and Γ^s_0 denote the respective concentrations at saturation.

Adsorption hysteresis was regarded for a long time as an artifact of the experimental technique (e.g., insufficient equilibration) and not as a fundamental phenomenon having physical bearing (see also ref. [9]). This may be one of the reasons why adsorption hysteresis of a macromolecule i.e., a protein, is a relatively recent discovery.[10,11] In addition, the advent of affinity chromatography[12] and hydrophobic chromatography[13] played an important role in the development of substituted hydrophilic gels which can be manipulated and well characterized for protein adsorption,[14,15,16] thus paving the way for such findings as protein adsorption hysteresis.

Besides differentiating *microscopic* (i.e., intramolecular) from *macroscopic* (intermolecular, e.g., adsorption) hysteresis,[17] the term adsorption hysteresis of proteins or enzymes should also be distinguished from other hysteretic phenomena. In enzymology the hysteresis concept of Frieden[18,19] supplemented by the mnemonical model of Meunier *et al.*[20] is best known. *Hysteretic enzymes* are defined according to their catalytic activity "as those enzymes which respond slowly to rapid changes in ligand concentration." The changes taking place in such enzyme systems are considered to be reversible[18] and not a result of nonequilibrium metastable states (see Section 3.2). Since in essence Frieden's term applies to a relaxation phenomenon (kinetics of regulatory enzyme systems) it will not be discussed here [for a review see Reference (19)]. The very interesting phenomenon termed "*functional hysteresis*" by Neumann,[17] which consists of salt-dependent inactivation–reactivation cycles of certain enzymes from halobacteria,[21,22] is also not within the general scope of this paper and can only be touched on briefly. On the other hand, a special form of *thermal hysteresis*[17] will be treated, since it appears directly connected to a protein showing adsorption hysteresis.

Since the only example of clear-cut protein adsorption hysteresis up to now is that shown for phosphorylase *b*,[10,11] the main emphasis of the article will be on this protein and to stimulate future research in this area. For methods see References (11), (14), (15), (23), and (24) and for definitions of terms References (11) and (25).

2. THE PHENOMENON OF PROTEIN ADSORPTION HYSTERESIS

2.1. TIME DEPENDENCE AND EQUILIBRIUM

Already Ewing[3] was aware of the importance of the time factor in hysteresis and differentiated a "viscous" from a "static" hysteresis, the former being characterized by a time lag in magnetization with the latter being practically time independent. Similarly von Bemmelen[4] had difficulties in obtaining true equilibria at low water contents of his gels and extended his desorption kinetics to times of up to 2–3 years before the changes in the water content apparently became zero. He speculated that the true equilibrium times might be infinitely long. Because slight permanent changes in gel structure also occurred during such long time spans, which probably were a cause of the difficulties, and for practical reasons von Bemmelen introduced "approximate" equilibria which at water contents of the gel of *ca.* 1 mol H_2O/mol SiO_2 he defined as states where the change in water content was less than 0.01 mol H_2O/mol SiO_2 per day. In this way he obtained approximate equilibria in several days.

Similar problems as described by von Bemmelen still exist today in the analysis of hysteresis, since it is now known that the equilibria obtained in such hysteretic systems are actually long-lived, nonequilibrium, metastable states showing time independence[17] (see also Section 3). In addition, the possibility of a conformational change or even the denaturation of a protein or enzyme in the bulk during the binding experiment may restrict the enzymologist in the length of the observation time of progress curves. In this connection Neumann[17] writes: "If within the limits of experimental accuracy this value remains constant for long observation times $t \gg \bar{t}$ ($\bar{t}$ = half-time of the initial reaction), we attribute longevity to the underlying metastability." Thus for practical purposes the nondetectability of a concentration change within experimental accuracy under the criteria above has been employed for the determination of "apparent equilibria"[11] in the protein adsorption hysteresis involving phosphorylase *b* (see Section 3.2).

Typical progress curves of the adsorption and desorption of [^{3}H]phosphorylase b_r on butyl-Sepharose at different bulk concentrations and dilutions are shown in Figure 1 as a change in surface concentration of adsorbed enzyme per unit time. Apparent equilibrium is obtained for adsorption after 60–90 min (Figure 1A–C) and for desorption after 60–120 min (Figure 1D–F). A more sensitive parameter for equilibration is the change in the free concentration of enzyme[11,24] (versus the control gel of unsubstituted agarose) which in the above case (Figure 1) also indicates apparent equilibrium [not shown, see Reference (11)]. Since the adsorption reaction is of

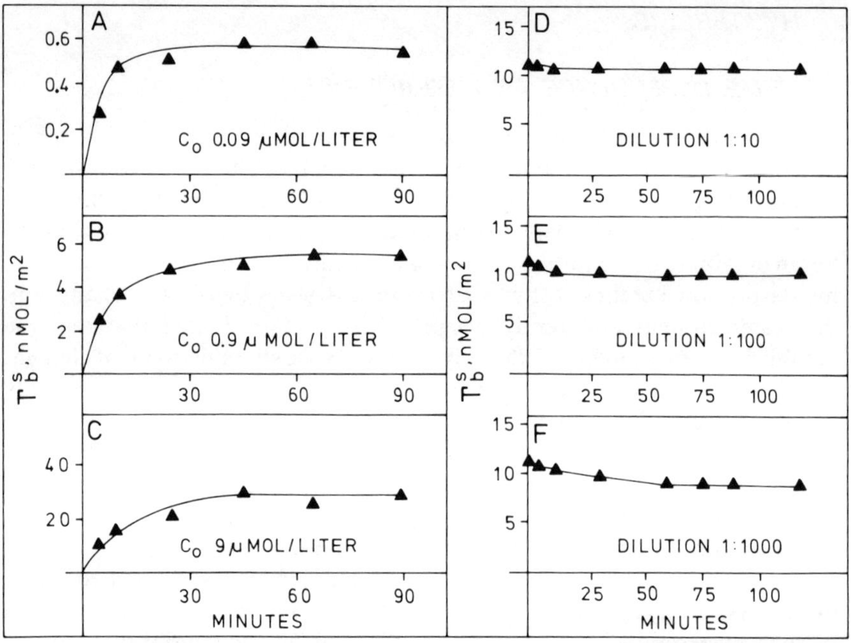

FIGURE 1. Progress curves of the adsorption (A–C) and desorption (D–F) of [^{3}H]phosphorylase b_r on butyl-Sepharose containing 269 nmol immobilized residues/m^2. The initial concentration of phosphorylase *b* (c_0) in the reaction mixture for adsorption varied between 0.09 μM and 9 μM (see figure). In order to obtain representative progress curves for the desorption isotherms, butyl-agarose gel loaded with 10.9 nmol enzyme/m^2 was diluted between 10- and 1000-fold. The stirring rate was controlled to yield adsorption and desorption rates unaffected by unstirred layers.[25] For the adsorption experiments the total volume of the incubation mixture was 60 ml (A–C) and in the dilution experiments the total volume varied between 17 and 500 ml. All experiments were corrected for nonspecific adsorption on unsubstituted Sepharose 4B. Γ_b^s denotes the surface concentration of [^{3}H]phosphorylase b_r. For further details see Reference (11) and the text. [Data from Reference (11) by permission of IPC Business Press Ltd.]

second order, the reaction rate depends on the concentration of each partner. Therefore, the time for attainment of apparent equilibrium has to be tested empirically in the concentration range for which the isotherms are to be determined (Figure 1). Under pseudo-first-order conditions the binding of [^{3}H]phosphorylase b_r to butyl-agarose is biphasic yielding a $t_{1/2} = 5$ min for the fast and a $t_{1/2} = 15$ min for the slow phase.[(24)] Similar equilibrium times (between 20–60 min) were reported for the adsorption of other proteins, e.g., bovine serum albumin and hemoglobin, on charged aminodecyl-Sepharose[(26)] and of the enzyme β-galactosidase on diaminodipropyl-Sepharose.[(27)]

The very complex kinetics of cooperative adsorption and desorption of phosphorylase *b* on butyl-agarose have been described in References (24) and (25).

2.2. ADSORPTION–DESORPTION LOOPS

Protein adsorption hysteresis is discussed according to the nomenclature[(9)] employed in Figure 2. The surface concentration (although for practical reasons other units are often employed) on the ordinate is plotted against a bulk concentration on the abscissa. The junction between adsorption and desorption isotherms at high bulk concentrations is called the *upper closure point*, the second junction, e.g., at the origin, the *lower closure point*

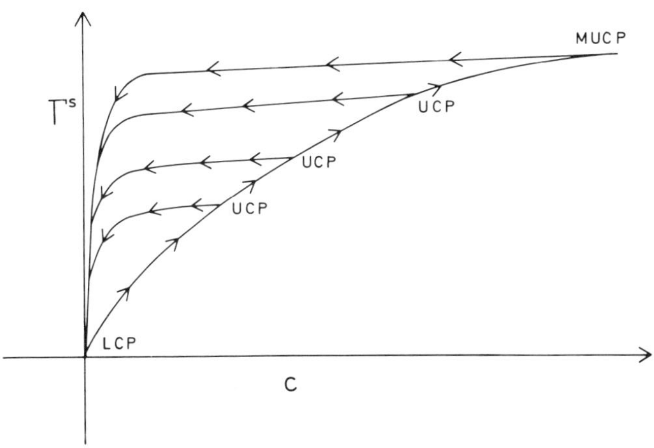

FIGURE 2. Theoretical adsorption hysteresis loops of a protein on a substituted agarose gel. Γ^s is the surface concentration of adsorbed protein and c the corresponding apparent free equilibrium concentration in the bulk. The surface concentration of immobilized residues is constant. The arrows indicate the direction of the closed adsorption–desorption loops. MUCP indicates the main upper closure point of the main hysteresis loop, UCP the upper closure points of the scanning curves, and LCP the lower closure point. For further details see the text.

of the loop. The upper closure point at gel saturation is called the *main upper closure point* and the corresponding loop the *main hysteresis loop.* All loops within the main hysteresis loop are called *scanning curves* [see also Reference (9)].

Figures 3 and 4 show adsorption–desorption loops of [^{3}H]phosphorylase b_r on butyl-Sepharose of two different degrees of substitution, i.e., different surface concentrations of immobilized butyl residue. The double logarithmic plots (Figures 3A and 4A) demonstrate that the isotherms are power functions, i.e., Freundlich isotherms.(11,15) The *adsorption* isotherms are not retraced by the *desorption* isotherms. Each concentration of free enzyme is associated with 2–3 different surface concentrations of enzyme and vice versa, depending on the number of scanning curves determined. Enhancing the surface concentration of immobilized butyl residue from 64 (Figure 3) to 269 nmol/m^2 results in: (a) a *ca.* 100-fold increase in the

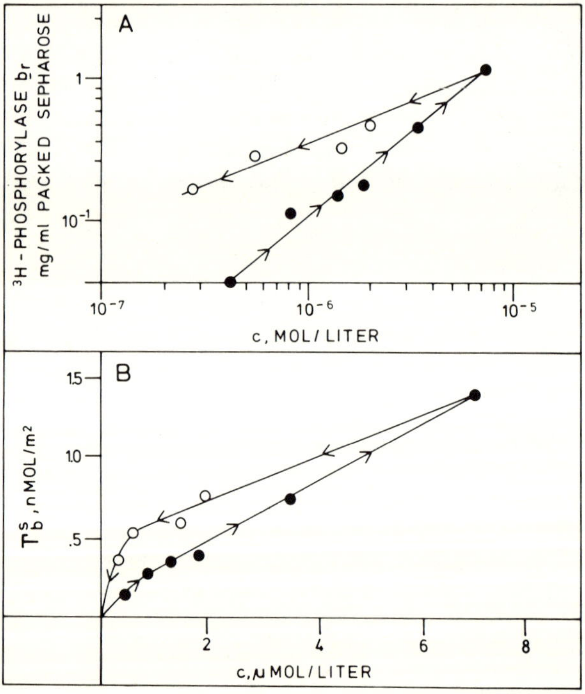

FIGURE 3. Adsorption and desorption isotherms (loop) of [^{3}H]phosphorylase b_r on butyl-Sepharose containing 64 nmol/m^2 of immobilized residues. A. Adsorption and desorption isotherms in double logarithmic coordinates. B. Adsorption–desorption loop in rectilinear coordinates. For further details see legend to Figure 1, Reference (11), and the text. (●) Adsorption isotherm, (○) desorption isotherm ($\Gamma^s_{b,0}$ = 1.25 nmol/m^2). [Data from Reference (11) by permission of IPC Business Press Ltd.]

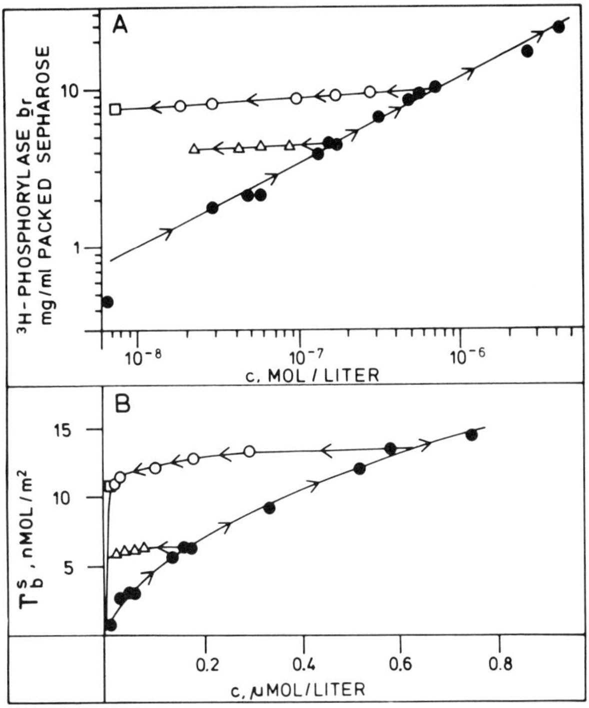

FIGURE 4. Adsorption and desorption isotherms (loops) of [^{3}H]phosphorylase b_r on butyl-Sepharose containing 269 nmol/m^2 of immobilized residues. A. Double logarithmic plot of the isotherms. B. Adsorption–desorption loops of the isotherms in rectilinear coordinates. The two uppermost data points of the adsorption isotherm (see A) are omitted for clarity. For further details see legend to Figure 1, Reference (11), and the text. (●) Adsorption isotherm, (○) desorption isotherm ($\Gamma^s_{b,0}$ = 10.9 nmol/m^2), (Δ) desorption isotherm ($\Gamma^s_{b,0}$ = 5.1 nmol/m^2). [Data from Reference (11) by permission of IPC Business Press Ltd.]

surface concentration of adsorbed enzyme at comparable apparent equilibrium concentrations of free enzyme, (b) an increase in loop area, (c) a decreased susceptibility of adsorbed enzyme to desorption: a 60-fold dilution of the enzyme–agarose complex leads to a release of *ca.* 75% of the adsorbed enzyme from the gel containing 64 nmol immobilized butyl residue/m^2 (Figure 3) but to the release of only ca. 5–10% from the gel containing 269 nmol/m^2 (Figure 4).

For a quantitative evaluation the data can be analyzed according to the Hill equation[15,16,28]:

$$\frac{\theta_2}{1-\theta_2} = K'_{H2}[\mathrm{P}]^{n_{H2}} \tag{1}$$

TABLE 1
Apparent Association Constants of Half-Maximal Saturation and State Functions of Irreversibility of the Adsorption–Desorption Loops of $[^3H]$Phosphorylase b_r on Butyl-Sepharose at 5°[a]

$\Gamma_s^{r(Aga-}$ (nmol m^{-2})	$K'_{a,0.5}$ (M^{-1})	$K'_{d,0.5}$ (M^{-1})	$\Delta_i S$ (J mol^{-1} K^{-1})	$\Delta_i G$ (kJ mol^{-1})
64	1.8×10^4	2.2×10^5	5.9	−1.6
269	1.6×10^5	4.2×10^9	33 to 42	−9 to −12

[a] $\Gamma_s^{r(Aga)}$ is the surface concentration of immobilized butyl residues, $K'_{a,0.5}$ and $K'_{d,0.5}$ the apparent association (half-saturation) constants of adsorption and desorption, respectively, $\Delta_i S$ the irreversible entropy, and $\Delta_i G$ the Gibbs free energy dissipated in the hysteresis loop of Figure 3B (●, ○) and Figure 4B (●, ○). The fractional saturation of the upper closure point is *ca.* 0.2 in both cases. For further details see Figures 3 and 4, the text, and Reference (11).

θ_2 is the fractional saturation of the gel surface at a specific surface concentration of immobilized butyl residues, K'_{H2} is a bulk Hill constant, n_{H2} a bulk Hill coefficient and [P] the free protein concentration at apparent equilibrium. For the adsorption of phosphorylase *b* the bulk Hill coefficient has generally been found to be smaller than unity.[11,15] From this behavior a negative cooperative mechanism of binding has been concluded[15] (see Section 4.2.1). The similarity between the Hill equation [Eq. (1)] and the SFE isotherm[29] has been noted before[15] and formally corresponding functions may be obtained at low surface coverages. Half-saturation constants ($K'_{0.5}$) which correspond to apparent association constants can be derived from the Hill constant[15] and are shown for the isotherms of Figures 3 and 4 in Table 1. The non-retraceability appears to be due to a drastic change in the half-saturation constant in the transition from adsorption ($K'_{a,0.5}$) to desorption ($K'_{d,0.5}$).

2.3. OTHER FORMS OF MACROSCOPIC HYSTERESIS INVOLVING PROTEINS

In this paragraph forms of protein hysteresis are described which strictly do not fall under adsorption hysteresis since the dependent and independent variables (see Section 1) are differently defined. On the other hand, these types of hysteresis may be of more or less direct importance to adsorption hysteresis, since they may all be the expression of a common principle associated with proteins in general.

2.3.1. Thermal Sol–Gel Hysteresis

Recently it has been demonstrated that, in addition to the hysteresis shown by phosphorylase after adsorption onto a solid phase, the enzyme also shows hysteresis in solution.[30] The hysteresis effect is found in connection with a temperature-dependent sol–gel transition[24] of the enzyme at high protein concentrations (*ca.* 10% w/v) in the same buffer as employed for the adsorption onto butyl-Sepharose. Gel formation of phosphorylase *b* can be viewed as a "self-adsorption" of protein molecules at bulk concentrations 100–1000-fold higher than in a usual adsorption experiment with butyl-Sepharose. As shown in Figure 5 the sol–gel transition of phosphorylase *b* can be followed by measuring the light scattering intensity of the protein solution as a function of the temperature. In the hysteresis loop the transition temperature decreases from *ca.* 25° on the increasing temperature branch to *ca.* 17° on the decreasing branch. The changes in light scattering intensity are paralleled by changes (not shown) of enzyme fluorescence intensity measured at 280/535 nm,[31] possibly indicating a conformational change of phosphorylase *b* concomitant with the phase change. Both the sol–gel transition and the hysteresis disappear at protein concentrations below *ca.* 50 mg/ml (not shown).

The occurrence of hysteresis during a sol–gel transition has been described for, among others, agarose gels[32] which liquefy at high temperatures (80°) and solidify at low temperatures (25°). The *exothermic*

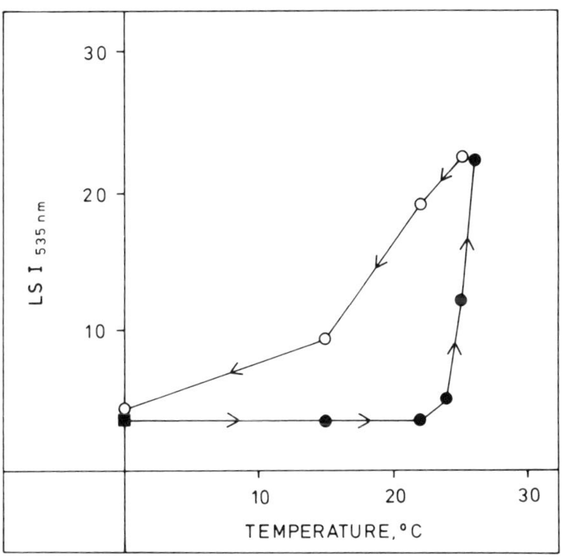

FIGURE 5. Temperature-dependent sol–gel transition of phosphorylase *b* at high ionic strength as measured by light scattering. The light scattering intensity at 535 nm was measured in a cuvette containing phosphorylase *b* (100 mg/ml). (●) Values measured during an increase in temperature from 0–26°, (○) values measured during a decrease in temperature from 26–0°, (■) separate experiment. For further details see References (24) and (30). [From Reference (30) by permission of Academic Press Inc., New York.]

gelation of agarose may be explained by the formation of hydrogen bonds at lower temperatures, leading to extensive stacked double helices which form the gel network. At high temperatures these hydrogen bonds are disrupted and the gel liquefies. In contrast, the *endothermic* formation of a phosphorylase *b* gel (solidification at higher temperature of *ca.* 25°, liquefaction at lower temperature of *ca.* 17°) may be explained by the formation of a polymerized network of phosphorylase molecules on the basis of entropically driven, hydrophobic interactions.

2.3.2. Ionic Strength-Dependent Hysteresis

Another hysteresis effect has been described for β-galactosidase adsorbed onto diaminodipropyl-Sepharose.[27] It appears that the ratio of bound to free enzyme (dependent variable) is a pathway-dependent function of the ionic strength of the buffer (independent variable). Hysteresis loops were not reported. Since the substituted gel contains charges, ion exchange effects are probably involved. It might be that the authors are indirectly observing an ionic strength dependence of a true adsorption hysteresis (see criteria in Section 1) of β-galactosidase in their system, analogous to the ionic strength dependence of the pH-induced hysteresis of the polyriboadenylate–polyribouridylate system.[33] However, this has to be verified by the measurement of isotherms. The effect may also be related to the functional hysteresis[17] of enzymes from extreme halophilic bacteria, where the independent variable is also the ionic strength.[21,22]

3. THERMODYNAMIC ASPECTS OF ADSORPTION HYSTERESIS

3.1. IRREVERSIBILITY

The second law of thermodynamics states that a spontaneous process in a closed system can only take place if the entropy change (ΔS) is positive:

$$\Delta S > 0 \tag{2}$$

This entropy term (ΔS) can be further resolved[34] into the two terms $\Delta_e S$ (entropy exchange with the *external* world, i.e., the reversible entropy change) and $\Delta_i S$ (*internally* created entropy of the system, i.e., the irreversible entropy change):

$$\Delta S = \Delta_e S + \Delta_i S \tag{3}$$

In a reversible process $\Delta_i S = 0$ and in an irreversible process $\Delta_i S > 0$. In their book Prigogine and Defay concluded[34]: "No change which takes place in nature is reversible." If irreversibility in spontaneous processes is therefore a matter of course, why is it stressed in hysteretic phenomena? Which property allows the distinction between two chemical binding reactions, one of which shows hysteresis and the other which does not?

An answer may be obtained from an example (Figure 6) derived from Prigogine and Defay[34] for the reversible dissociation of a gas (e.g., water) at high temperatures [see also Reference (30)]. The reversible change from state *A* to state *D* follows the trajectory *F–B–E–C*. For true reversibility the reverse transition from *D* to *A* is such that[34]: "(a) the variables characterizing the state of the system return through the same values but in the inverse order (i.e. *C–E–B–F*) and (b) exchanges of heat, matter and work with the surroundings are of the reverse sign and take place in the reverse order." The above trajectories characterize the *reversible transition line*. All changes which do not satisfy the two conditions and do not lie on this line are termed irreversible.[34] Such changes may follow the trajectories *A–B′–C′–D′–D* and *D–E′–F′–A′–A*.

If in a theoretical experiment the temperature is increased at point *A*, the path followed by the dissociating gas is from *A* to *B′* and not along the reversible transition line *A–B*. Only after allowing for equilibration at point *B′* does the system return to *B* on the reversible transition line. Alternatively if the temperature is increased continuously (temperature gradient) from point *B′* on, the trajectory followed is *B′–C′–D′*. Interrupting the gradient at *D′* and allowing for equilibration returns the system to *D*. In this case *D* is reached from *A* along a nonequilibrium trajectory. Similarly if the temperature is decreased from *D* and a temperature gradient is initiated until it is interrupted at *A′*, the system returns to *A* via the nonequilibrium trajectory *E′–F′–A′*. Since the reverse path *E′–F′–A′* does not return through the same states *D′–C′–B′* of the forward reaction, the process *B′–C′–D′* is

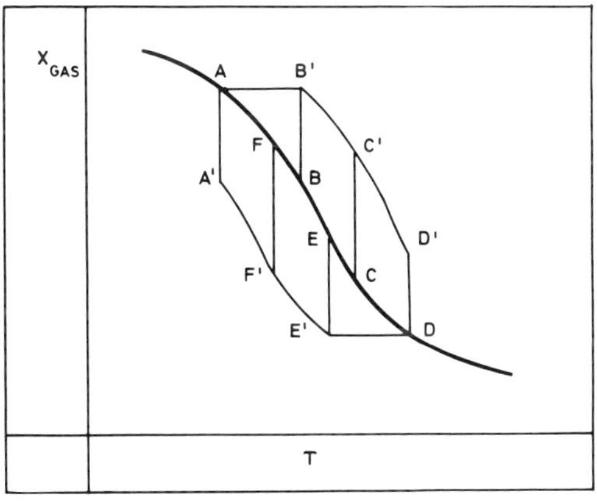

FIGURE 6. Schematic diagram of the reversible dissociation of a gas. *X* is the mole fraction of undissociated gas and *T* is the temperature. For further details see References (30, 34). [From Reference (30) by permission of Academic Press Inc., New York.]

not reversible. However, the reversible change ($A–F–B–E–C–D$ and vice versa) can be approximated very closely by carrying out the temperature changes in infinitesimally small steps close to equilibrium. Thus under certain conditions a real transition (irreversible change) can approximate a reversible change trajectory and we speak of an *equilibrium change* or transition. Ideally, equilibrium changes such as the dissociation of gases or other reactions can be *treated* thermodynamically as reversible phenomena if the conditions are so chosen that the nonequilibrium trajectories superimpose on the reversible change trajectory and $\Delta_i S$ approximates zero.

It has, therefore, been suggested that the persistence of a system on a trajectory distinct from the equilibrium change trajectory (in spite of conditions allowing for equilibration in finite times) leads to the phenomenon of hysteresis.[30] In such systems an experimental approximation of the reversible change trajectory is not possible. Two types of irreversible processes may therefore be differentiated: (a) processes based on *non-equilibrium changes*, not approximatable to a reversible change with finite values for $\Delta_i S$ and (b) processes based on *equilibrium changes*, approximatable to a reversible change with a concomitant disappearance of $\Delta_i S$ under ideal conditions. In the former type of processes (e.g. hysteresis) the dissipated irreversible entropy and a corresponding dissipated irreversible Gibbs free energy[17,35] can be calculated from the experimental data (loop area), as will be discussed below (Section 3.3). The negentropy inherent in the directionality of a hysteresis loop is equivalent to information and forms the basis for so-called memory effects.[17]

3.2. METASTABILITY

The persistence of a system on a trajectory distinct from the reversible change trajectory is a strong indication for the existence of states other than the equilibrium state. Such states have been termed "false equilibria"[34] ("states where no reaction proceeds even though the affinity* of the reaction is not zero") or *metastable states.*[9,17] The occurrence of a metastable state can be taken to indicate an *energy barrier*[36] which prevents the transition to the reversible change trajectory.

A major practical question is how these states can be treated and quantitatively evaluated. In an answer to this question Katchalsky and coworkers[35,36] proposed that the states exhibiting long-lived metastability can be treated thermodynamically in the same way as equilibrium states. A similar assumption was made by Everett and Whitton.[37] Therefore it has been reasoned[11] that the apparent equilibrium concentrations obtained in adsorption hnysteresis (see Section 2.1) may be treated as true equilibrium concentrations, allowing a quantitative evaluation of the binding data.

* Affinity here is a state function.

However, it should be emphasized that the parameters calculated under these conditions are valid only for the respective metastable states and not for the equilibrium state and should therefore be characterized, e.g., by the word "apparent."

3.3. TREATMENT OF EXPERIMENTAL DATA

In connection with protein adsorption hysteresis, evidence was presented that between 0° and 35° the adsorption of phosphorylase *b* on butyl-Sepharose at high ionic strength is an endothermic reaction.[11,15] The apparent, isosteric molar enthalpy of adsorption $\Delta_a H_i'$ lies between 16–56 kJ mol^{-1} and the apparent isosteric molar entropy of adsorption ($\Delta_a S_i'$) was estimated to lie between 160–327 J mol^{-1} K^{-1}.[11] These parameters, which depend on the fractional saturation of the gel, were derived from the Clausius–Clapeyron equation and the apparent association constants of half-maximal saturation under the conditions stated above (Section 3.2).

The irreversible entropy ($\Delta_i S$) which is produced under isobaric, isothermal conditions in such hysteresis loops (Figures 3 and 4) can be calculated from the area of the closed loop according to the cyclic integral[11,37]:

$$\Delta_i S = R \oint N \, \mathrm{d} \ln c \tag{4}$$

where N denotes the dimensionless quotient $\Gamma_b^s/\Gamma_{UCP}^s$ (Γ_{UCP}^s is the surface concentration of protein at the upper closure point), c is the molar concentration of free ligand at apparent equilibrium (prerequisite: low concentrations), and R is the gas constant. From the irreversible entropy the Gibbs free energy ($\Delta_i G$) dissipated irreversibly in the cycle can be calculated[17,35]:

$$\Delta_i G = -T \Delta_i S \tag{5}$$

From the area of the adsorption–desorption loops in Figures 3 and 4 the dissipated irreversible entropy and the dissipated Gibbs free energy have been calculated [for details see Reference (11)] and are shown in Table 1. An increase in the surface concentration of immobilized butyl residues from 64 to 269 nmol/m^2 leads to a 5–6-fold increase in irreversible entropy. Thus, in the case of protein adsorption hysteresis, the dissipated irreversible entropy, the dissipated Gibbs free energy, and concomitantly the position of the assumed energy barrier are a function of the surface concentration of lattice binding sites on the gel surface. The laws or rules governing the possible interdependence of $\Delta_a S_i'$ and $\Delta_i S$ in the case of hydrophobic protein adsorption hysteresis remain to be defined.[11]

3.4. MODEL OF A DOMAIN IN PROTEIN ADSORPTION HYSTERESIS

The domain concept,[38,39,40] when applied to adsorption, states that binding manifests itself in a series of microscopic steps which elude detection in the smooth isotherms. These microscopic steps occurring in units called domains are accompanied by the production of irreversible entropy. This is shown schematically in Figure 7 for the binding of phosphorylase *b* to a domain on the agarose surface. On the ordinate the extent of the ligand-matrix interaction is depicted. It is correlated to the fractional saturation and the number of interacting alkyl residues and is shown as a function of the free ligand concentration in the bulk (depicted on the abscissa as the independent variable). As the concentration of free ligand is increased (step **1–2**) the probability for the occurrence of nucleation is enhanced. After adsorption (nucleation) on a minimum of, e.g., four alkyl residues (E^4-*Aga*, see $1/K'_{a,0.5}$) the cooperative reaction is propagated in a second reaction until *n* residues have reacted with the enzyme (path **2–3**). Subsequent to

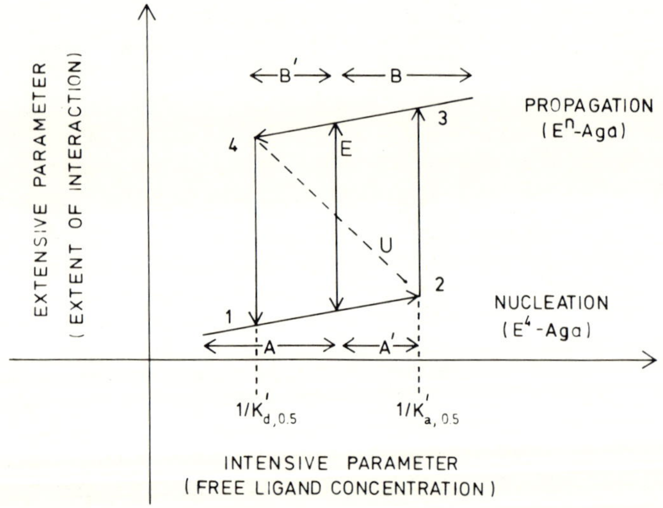

FIGURE 7. Schematic representation of a domain for a hysteretic adsorption–desorption cycle of phosphorylase *b* on butyl-Sepharose. *EQ* symbolizes the equilibrium transition, *U* the unstable transition, *A* the first stable state (e.g., an unnucleated form of phosphorylase *b*), *A'* the metastable state of *A* (e.g., nucleated form of the enzyme, E^4-*Aga*), *B* second stable state (e.g., enzyme after propagation, E^n-*Aga*), *B'* the metastable state of *B*. *E* symbolizes the enzyme and *Aga* the substituted agarose matrix; the superscript on *E* denotes the assumed valence involved in binding per binding unit. The reciprocal of the half-saturation constants of desorption ($K'_{d,0.5}$) and adsorption ($K'_{a,0.5}$) are indicated. For further details see References (9), (11), and (40) and the text.

propagation (E^n-*Aga*, see $1/K'_{d,0.5}$) desorption occurs via the path **3–4–1**. Two metastable states A' (for adsorption) and B' (for desorption) are assumed. The dotted line U (path **4–2** or **2–4**) denotes unstable transition states which are not accessible to experiments. The arrow E denotes the equilibrium transition approximating a reversible change (see Section 3.2.1). Thus in adsorption hysteresis of phosphorylase *b* on butyl-Sepharose the trajectory **1–2–3–4** is postulated for a closed hysteresis loop. The increase in the half-saturation constant ($K'_{0.5}$) in the transition from adsorption (path **2–3**) to desorption (path 4–1) is indicated on the abscissa by $1/K'_{a,0.5}$ for adsorption and by $1/K'_{d,0.5}$ for desorption. However the combination of an irreversible transition (e.g., **1–2–3**) in one direction (e.g., adsorption) with a reversible transition via E in the other (e.g., desorption) is also possible and cannot be excluded.

4. MOLECULAR BASIS OF ADSORPTION HYSTERESIS

4.1. CAPILLARY CONDENSATION THEORY OF GASES

One of the earliest theories of adsorption hysteresis was put forward for vapors by Zsigmondy,[(41)] who applied the laws of capillarity to the isotherms of von Bemmelen,[(4)] thus introducing the *capillary condensation theory of hysteresis*. According to this theory, the water adsorbed on silica gel condenses in the pores (assumed to be cylindrical capillaries) of the gel. Adsorption hysteresis is explained by assuming that the mean radius of the curvature of the concave meniscus of the water column is smaller during dehydration (desorption) than during hydration (adsorption). Since the vapor pressure over the meniscus of smaller mean radius is lower than over the meniscus of larger radius (Kelvin equation[(9,42)]), a difference in the curvature of the menisci would suffice to explain hysteresis. Zsigmondy assumed that the difference in curvature could be due to an incomplete wetting of the pores during adsorption with a concomitant larger radius of curvature than during desorption, when the pores have been completely wetted.

Since the time of Zsigmondy very elaborate capillary condensation theories incorporating entropy production and metastability have been put forward,[(9,39)] some of which predict the form of the hysteresis loop from the geometry of the capillary, i.e., pore.[(43)] Thus the capillary condensation theory is still one of the major theories for the mechanism of adsorption hysteresis on porous media today. For a detailed discussion of this and other theories on the adsorption hysteresis of gases the reader is referred to Reference (9).

4.2. MECHANISM OF PROTEIN ADSORPTION HYSTERESIS

The noncovalent binding forces generally involved in protein adsorption can be of various types: (a) ionic interactions (ion exchange[44,45]), (b) hydrogen bonds,[45] (c) hydrophobic interactions,[13,46,47,48] (d) charge transfer interactions,[49] and (e) van der Waals forces.[50] From the endothermic adsorption of phosphorylase *b* on butyl-Sepharose at high ionic strength it has been concluded that this enzyme is adsorbed mainly by hydrophobic interactions.[15] Hydrophobic interactions are based on the entropy increase resulting from the extrusion of ordered interfacial water into a less ordered bulk form. In contrast to van der Waals energy, which decays as a function of r^{-6} (r = distance between interacting groups), the energy of hydrophobic interactions appears to follow an exponential decay function.[48]

Under appropriate conditions other solutes besides water may also be extruded during polymer adsorption. One of the first observations of an endothermic adsorption reaction for a polymer in which solute extrusion played a major role was reported by Koral *et al.*[51]

The following discussion applies to the adsorption of proteins from dilute solutions, e.g., maximally 1–2 mg/ml, corresponding to $1\text{–}2 \times 10^{-5}$ *M* monomer units in the case of phosphorylase *b*.

4.2.1. Multivalent Protein Binding

The classic isotherm of adsorption is the Langmuir isotherm. According to Langmuir[52] the following postulates must be fulfilled for hyperbolic adsorption: (a) the adsorption sites (i.e., surface cells) are all identical; (b) each surface cell can hold only one molecule; and (c) the potential energy of a surface cell is independent of the presence of molecules in other cells.

Applying these postulates to the adsorption of proteins on surfaces substituted homogenously with identical residues, two basic mechanisms can be differentiated: (a) monovalent independent binding; (b) multivalent cooperative binding.

In the first mechanism the shortest distance between the immobilized residues is greater (>2 fold) than the largest diameter of the protein. In addition, the free energy change of binding between a single immobilized residue (e.g., coenzyme) on the interface and a single binding site (e.g., active site) on the protein suffices for adsorption. The protein is bound independently to the surface by a single interaction site. This type of binding may be found in certain types of affinity chromatography [see References (12), (15), and (16)] and should (e.g., in the absence of induced conformational changes) correspond to a Langmuir isotherm.

In the second mechanism the largest distance between the immobilized residues is smaller than the diameter of the protein molecule. The protein therefore covers an area of the gel surface (i.e., *binding unit*[25]) containing $n > 1$ residues. If hydrophobic alkane derivatives are immobilized on the surface a single interaction with the protein will be so weak that only the cooperative interaction of several residues (multivalence) will lead to adsorption. This number (valence) of residue–protein interactions can be controlled during gel synthesis via the surface concentration of immobilized residues. In addition to multivalence, binding is influenced by another important parameter. There is an unavoidable overlap of binding units on the gel surface so that the binding of one protein molecule is no longer independent of the binding of other protein molecules in neighboring binding units. As a result the binding of the protein molecules occurs in a "non-independent," i.e., cooperative, manner. Thus two forms of cooperativity must be differentiated: (a) the cooperative interaction of multiple immobilized residues with the protein (multivalence); (b) cooperativity due to non-independent binding of the protein to overlapping binding units (see also negative cooperativity in Section 2.2).

Experimental evidence has been provided for both mechanisms.[15,16,25,53] The effect of multivalence is illustrated in a typical experiment in Figure 8. The surface concentration of adsorbed enzyme is measured as a function of the surface concentration of immobilized residues under the condition of an identical apparent equilibrium concentration of free enzyme in the bulk for each concentration of immobilized residue tested. At low surface concentrations of immobilized butyl residue there is practically no binding of protein since only single residues may interact independently with the protein. As the surface concentration of residues is enhanced the amount of adsorbed enzyme increases in a sigmoidal fashion since multiple residues can now interact simultaneously, i.e., in a positive cooperative manner, with the protein and the affinity of binding increases by

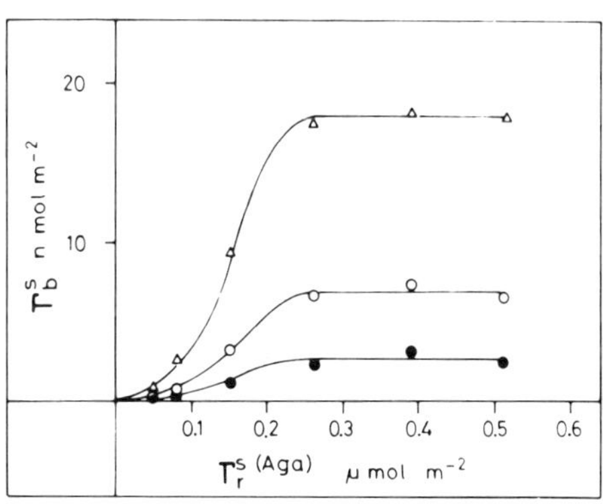

FIGURE 8. Surface concentration of phosphorylase $b(\Gamma_b^s)$ as a function of the surface concentration of immobilized butyl residues ($\Gamma_r^{s(Aga)}$) at 5°. The data points were calculated for the denoted constant apparent free bulk equilibrium concentration (see below) from corresponding adsorption isotherms measured for each butyl-Sepharose at a given surface concentration (abscissa). Constant apparent concentration of enzyme: (●) 0.1 μM, (○) 0.7 μM, (△) 5 μM. For further details see Reference (25). [From Reference (25) by permission of Pergamon Press Ltd., Oxford.]

magnitudes. Thus there is a "threshold value" ("critical concentration"[14]) of the surface concentration of butyl residues at which multivalent protein adsorption occurs. At high surface concentrations of immobilized residues, saturation of the protein with the residues occurs. The sigmoidal curves can again be described according to the Hill equation[16,28,53]:

$$\frac{\theta_1}{1-\theta_1} = K'_{H1}(\Gamma_r^{s(Aga)})^{n_{H1}} \tag{6}$$

where θ_1, the fractional saturation ($\Gamma_b^s/\Gamma_{b,0}^s$), reflects the saturation of protein with immobilized residues, K'_{H1} is the surface Hill constant, n_{H1} the surface Hill coefficient, and $\Gamma_r^{s(Aga)}$ the surface concentration of immobilized residues. From the Hill coefficients obtained ($n_{H1} > 3$) it has been concluded that at least 3–4 immobilized butyl residues are involved in the initial binding reaction of phosphorylase *b* on butyl-agaroses [see Reference (25)]. As stated above, saturation here corresponds to a saturation of the protein with immobilized residues. The level of saturation under these conditions (see Figure 8) is, however, also a function of the constant free protein equilibrium concentration (0.1–5 μM) in the bulk[16]:

$$\Gamma_{b,0}^s = A_p[P]_c^\alpha \tag{7}$$

where $\Gamma_{b,0}^s$ is the surface concentration of protein at saturation and $[P]_c$ is the bulk protein held at a constant concentration. A_p and the exponent α are constants. The sigmoidal curves of Figure 8 reflect a nonlinear increase in the apparent association constants of the enzyme as a function of the surface concentration of immobilized butyl residues. Thus the apparent binding constant for the adsorption step is seen in Table 1 to increase from $1.8 \times 10^4\ M^{-1}$ to $1.6 \times 10^5\ M^{-1}$ as the surface concentration of butyl residues is increased from 64 to 269 nmol/m^2, respectively.

From the very strong binding of immobilized (i.e., two-dimensional concentration) butyl residues to phosphorylase *b* one could conclude that soluble (i.e., three-dimensional concentration) alkyl residues might also bind strongly in solution. However, this is not the case as was shown in flow dialysis experiments[24] with *n*-[1-^{14}C]-butylamine and *n*-[1-^{14}C]-hexylamine at high ionic strength. A binding of *n*-[1-^{14}C]-hexylamine to phosphorylase *b* in solution could not be detected (see Figure 9). Thus the binding constant of the protein–hexylamine interaction in solution is so low that an equimolar (in this case mM) mixture of the binding partners does not lead to a significant saturation of putative hydrophobic binding sites on the enzyme. It was therefore concluded that there are no high-affinity alkyl binding sites on phosphorylase *b*.[24] This result is in agreement with the model of a multivalent binding of immobilized alkyl residues to a finite

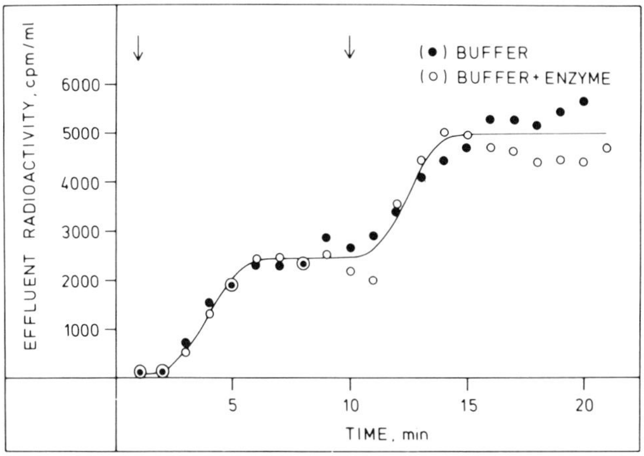

FIGURE 9. Analysis of binding of *n*-[1-^{14}C]hexylamine to phosphorylase *b* at high ionic strength in a flow dialysis cell. The arrows indicate the addition of 10 μl of 0.050 *M* *n*-[^{14}C]hexylamine (5×10^7 cpm) to the upper chamber containing 0.5 ml solution as indicated. The concentration of enzyme was *ca.* 1 m*M*. Samples were taken at the indicated times from effluent fractions (0.3 ml) of the lower chamber (1 fraction per min). The radioactivity of the fractions is given on the ordinate. For further details see Reference (24). [From Reference (24) by permission of Elsevier Scientific Co., Amsterdam.]

number of low-affinity ($K'_{r,0.5} \sim 10\text{–}100\ M^{-1}$) binding sites on the enzyme.(15,16,25) Accordingly, the simultaneous binding of, e.g., three residues on the surface to the protein (without correcting for the nucleation parameter) leads to a binding constant of the magnitude $(K'_{r,0.5})^3$ and can explain the high binding constants observed with immobilized residues. Thus the results of both experiments—the sigmoidal binding curve (Figure 8) and the absence of binding in solution (Figure 9)—lead to the conclusion that phosphorylase *b* is multivalently adsorbed on the agarose surface.

Since the experiments also demonstrate (see Figure 8 and Table 1) that the affinity of the phosphorylase *b*-butyl-agarose reaction is a function of the surface concentration of butyl residues (i.e., the number of residues interacting with a protein molecule), it has been concluded that the increase in the apparent binding constant during the transition from adsorption to desorption in a hysteresis loop also in some way involves an increase in the number of residues interacting with phosphorylase *b*.(11)

4.2.2. Models of Protein Adsorption Hysteresis

Adsorption hysteresis of proteins is a young field and no general theories for this type of hysteresis exist. However two models have been suggested(30): (a) the capillary gelation hypothesis and (b) the surface dynamic hypothesis.

4.2.2.1. Capillary gelation hypothesis. Sepharose 4B spheres are agarose beads with a mean pore radius of *ca.* 80 nm.[11] Since it can be shown in solution that phosphorylase *b* can exist either in a sol or a gel state (see Section 2.3.1), it was reasoned that the mechanism involved in protein adsorption hysteresis may be similar to capillary condensation as described for gases (see Section 4.1). Instead of a gas (vapor) condensing to a liquid in a pore of the adsorbent, the adsorbed protein molecules could solidify to a gel in a pore of the agarose bead. Such a mechanism would be possible if either the sol–gel transition temperature were reduced for the adsorbed molecule (e.g., from 25° to 5°, see Figure 5) by the adsorption process itself or as a result of high local concentration of the enzyme on the gel surface. Concerning the latter possibility, it has, for example, been shown for agarose that the *exothermic* sol–gel transition temperature increases as a function of the agarose concentration.[54] For the *endothermic* gelation of phosphorylase *b* one could, therefore, assume that the gelation temperature decreases as a function of the protein concentration in solution and on a surface. One or more metastable states could be involved, in a manner similar to that in solution, leading to adsorption hysteresis.

This hypothesis would, however, also imply that nearest neighbor interactions between enzyme molecules (i.e., attraction), which would lead to an affinity increase on adsorption, should play an important role in binding. This has, however, been shown to be very improbable under the conditions of adsorption from dilute solutions.[15] In addition, in this model the restrictions of adsorption occurring in two dimensions would be eliminated, since gelation in pores having a *ca.* 13-fold larger pore diameter than the molecular diameter of the enzyme[11] would involve three dimensions (e.g., multilayers).

4.2.2.2. Surface-dynamic hypothesis. In a simplified way two aspects of the dynamics of a protein adsorbed to a surface can be differentiated: (1) movements of the whole molecule, e.g., surface diffusion, surface rotation (reorientation reaction) and (2) intramolecular movements, e.g., conformational changes (conformational reaction). Under these conditions hysteresis originates in a two-dimensional protein monolayer on the interface.

In both cases the increase in the half-saturation constant observed in adsorption hysteresis (Table 1) would be due to an increase in the number of butyl residues interacting with the phosphorylase molecule. This could result either from an optimal reorientation yielding a maximum number of interaction sites or a conformational change exposing new hydrophobic contact sites. It has been suggested[30] that the dissipated irreversible entropy could originate from the conformational change of the protein (decrease in order) or from the additional new hydrophobic interactions between protein and alkyl residues, both processes leading to an entropy increase, in part by water extrusion. In the case of conformational reactions the state

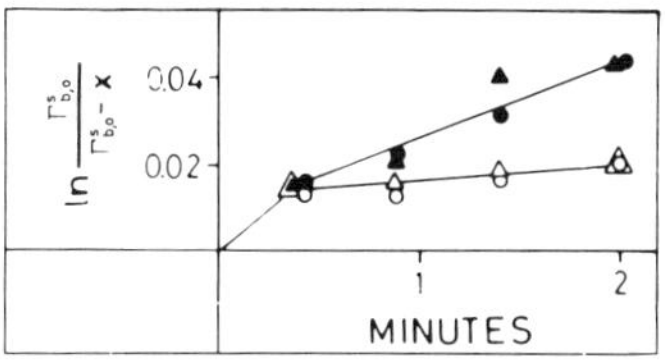

FIGURE 10. Effect of dilution on the initial dissociation rate of the [^{3}H]phosphorylase *b*-butyl-agarose complex in the absence and presence of unlabeled reduced enzyme in the bulk. In the semilogarithmic plot $\Gamma^s_{b,0}$ is the surface concentration of enzyme at the time $t = 0$. $\Gamma^s_{b,0} = 4.6\text{–}5.7\ \text{nmol/m}^2$. X corresponds to the amount of labeled enzyme released in free form per m^2 of gel surface (immobilized butyl residue concentration: 380 nmol/m^2). Desorption was initiated by addition of 0.2–1.0 ml of packed enzyme-loaded gel to the incubation mixture varying between 60 and 20 ml. The stirring velocity was 700 rpm. The initial concentration of unlabeled enzyme in the bulk was 1×10^{-5} *M*. For further details see Reference (25). Dilution 1:20 (○), 1:300 (△) in the *absence* and (●) 1:20, (▲) 1:300 in the *presence* of unlabeled enzyme in the bulk. [Data from Reference (25), by permission of Pergamon Press Ltd., Oxford.]

functions of irreversibility (Table 1) would be linked to the partition of the different conformational states of the protein on the surface.

4.2.3. Experimental Evidence in Favor of the Surface-Dynamic Model

Desorption experiments show that the capillary gelation model of adsorption hysteresis of [^{3}H]phosphorylase b_r on butyl-Sepharose is very improbable. Such an experiment is shown in Figure 10. The initial desorption rate constant ($k'_{\text{off}} = 0.0036 \pm 0.0003\ \text{min}^{-1}$) of the enzyme obtained after 300-fold dilution in buffer can be increased several fold (to $k'_{\text{off}} = 0.016 \pm 0.0027\ \text{min}^{-1}$) if the [^{3}H]phosphorylase b_r-agarose complex is diluted in buffer to which unlabeled ("cold") phosphorylase b_r has been added. Under these conditions, the enhancement of the off-rate constant is not due to an inhibition of readsorption by unlabeled enzyme or to a simple exchange reaction.[25] In fact, the off-rate constant enhancement effect depends on the surface concentration of butyl residues[25] and the fractional saturation of the gel (Figure 11). This strongly indicates that there is a direct interaction (negative cooperativity,[15,25] see Section 2.2) between the secondarily adsorbed, unlabeled and the primarily adsorbed, labeled molecules. This effect could be mediated by the surface diffusion of freshly adsorbed molecules.

If the capillary gelation model were correct, a displacement of labeled molecules by unlabeled molecules should be very unlikely because of pore obstruction. In addition, one would expect intermolecular attraction (see Section 4.2.2.1) to enhance the gelation process by a "capture" of new molecules, thus inhibiting the displacement. Finally, the observation of adsorption hysteresis at very low fractional saturation[11] makes the occurrence of gelation very improbable.

It has, therefore, been suggested that conformational reactions, possibly in combination with reorientation reactions, yield a plausible model for the mechanism of adsorption hysteresis of phosphorylase *b* on butyl-agarose.[30]

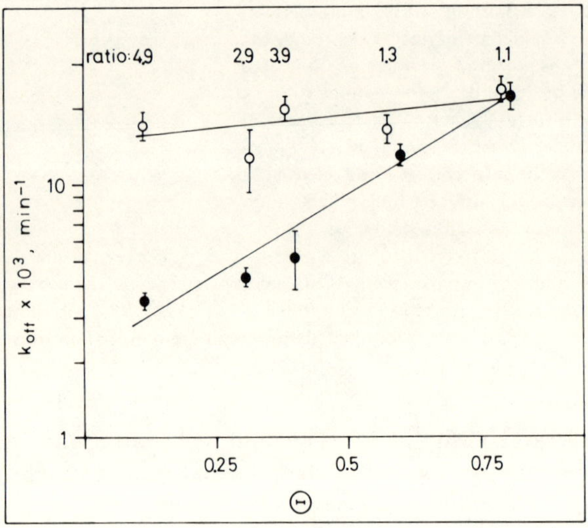

FIGURE 11. Semilogarithmic plot of the apparent off-rate constants in the presence (○) and the absence (●) of unlabeled enzyme as a function of the fractional saturation (θ) of the gel with enzyme. Desorption was initiated by the addition of *ca.* 1 ml packed butyl-agarose, fractionally saturated with [^{3}H]phosphorylase b_r as indicated, to the incubation mixture of 20 ml. The initial concentration of unlabeled enzyme in the bulk was 1×10^{-5} *M*. The off-rates with their standard errors are given. For further details see Reference (25). [From Reference (25) by permission of Pergamon Press Ltd. Oxford.]

Evidence for the existence of long-lived metastable conformational states of proteins has been obtained from experiments with enzymes from extreme halophilic bacteria,(21,22,55) where it could be shown that the ionic strength-dependent hysteretic inactivation–reactivation cycle is accompanied by conformational changes. In the case of malate dehydrogenase(21) it appears that inactivation is accompanied by minor conformational changes which lead to a dissociation of the dimer into monomers. For alanine dehydrogenase(55) it could be demonstrated that major conformational changes accompany salt-dependent inactivation of the enzyme with a practically total loss of α-helix and a *ca.* 50% loss of β-structure. This loss of secondary structure is completely reversed on reactivation at high ionic strength. Metastable states have also been reported for other protein interactions.(56–59)

Concerning phosphorylase *b* there is as yet no direct information on the extent of the conformational changes occurring on the gel. However the enzyme appears not to be denatured, since it was shown that phosphorylase *b*(60) and other enzymes(61,62) express activity in the adsorbed state and can be eluted from alkyl-Sepharoses without a major activity loss.(13,14,60) In addition, if phosphorylase *b* is extensively unfolded and denatured (e.g., by freezing[(24,63)]), an insoluble, inactive precipitate is formed irreversibly. This has not been found in phosphorylase *b* solutions eluted from hydrophobic gels. On the other hand the specific activity of leucine aminopeptidase[(64)] (which does not bind to alkyl-agaroses) decreases by 15–40% after elution from tyramine-agarose, indicating a conformational change sufficient to inactivate the enzyme irreversibly. Thus reversible and irreversible conformational changes of proteins adsorbed to interfaces are very likely, depending on the protein structure and the affinity, valence, and time dependence of the interaction.

5. FUTURE ASPECTS OF PROTEIN ADSORPTION HYSTERESIS

One main aim of further research should be to put the phenomenon of adsorption hysteresis of proteins on a much wider basis than is now the case by making adsorption-desorption measurements with a large number of proteins. One example (phosphorylase *b*) does not suffice for a final statement on the various aspects of this phenomenon.

Another major aim should be to gain further insight into the mechanisms which produce protein adsorption hysteresis (metastability) and to understand the laws governing it. One possible way would be an attempt to apply the theory of critical phenomena in phase transitions[(65–67)] to the data. The order parameter of the postulated phase transition on the gel during adsorption and the dimensionality should be determined. Similarly, the conditions for measuring the critical point must be defined.

In the case of the sol–gel transition of phosphorylase *b* a few statements can be made: The order parameter of the sol phase is zero ($M = 0$) and of the gel phase one ($M = 1$). The sol–gel phase change corresponds to a first-order transition of a three-dimensional system. Here also the critical point (absence of discontinuity of the phases) has to be determined.

Only if some general law(s) capable of describing a large number of adsorption–desorption loops of many proteins can be found, possibly by reducing the adsorption phenomena to a phase change which can be described by the theory of critical phenomena, can we hope to understand and predict protein adsorption hysteresis.

ACKNOWLEDGMENTS

This work was supported by the Deutsche Forschungsgemeinschaft (Je 84/6-6) and the Fonds der Chemie.

REFERENCES

1. A. Ewing, On the production of transient electric currents in iron and steel conductors by twisting them when magnetised or by magnetising them when twisted, *Proc. Roy. Soc. (London), Ser. A* **33**, 21–23 (1881).
2. A. Ewing, On the production of transient electric currents in iron and steel conductors by twisting them when magnetised or by magnetising them when twisted, *Proc. Roy. Soc. (London), Ser. A* **36**, 117–135 (1883).
3. A. Ewing, Experimental researches in magnetism, *Phil. Trans. Roy. Soc. (London)* **176**, 523–624 (1885).
4. J. M. von Bemmelen, Die Adsorption. Das Wasser in den Kolloiden, besonders in dem Gel der Kieselsaure, *Z. Anorg. Chem.* **13**, 233–356 (1897).
5. J. S. Anderson, Die Struktur des Gels der Kieselsaure, *Z. Physik. Chem. (Liepzig)* **88**, 191–228 (1914).
6. A. G. Foster, The sorption of vapors by ferric oxide gel. 1. Aliphatic alcohols, *Proc. Roy. Soc. (London), Ser. A* **147**, 128–140 (1934).
7. B. Lambert and A. M. Clark, Studies of gas-solid equilibria. Part II. Pressure-concentration equilibria between benzene and (a) ferric oxide gel, (b) silica gel, directly determined under isothermal conditions, *Proc. Roy. Soc. (London), Ser. A* **122**, 497–512 (1929).
8. W. P. Bryan, Thermodynamics of water-protein interactions, *J. Theor. Biol.* **87**, 639–661 (1980).
9. D. H. Everett, in: *The Solid–Gas Interface*, Vol. 2 (E. A. Flood, ed.), pp. 1055–1113, Dekker, New York (1967).
10. H. P. Jennissen, Hysteresis and thermodynamic irreversibility in protein adsorption on hydrophobic agaroses, *Hoppe-Seyler's Z. Physiol. Chem.* **359**, 1101–1102 (1978).
11. H. P. Jennissen and G. Botzet, Protein binding to 2-dimensional hydrophobic binding-site lattices—Adsorption hysteresis on immobilized butyl-residues, *Int. J. Biol. Macromol.* **1**, 171–179 (1979).
12. P. Cuatrecases and C. B. Anfinsen, Affinity chromatography, *Methods Enzymol.* **22B**, 345–385 (1971).
13. S. Schaltiel, Hydrophobic chromatography, *Methods Enzymol.* **34B**, 126–140 (1974).
14. H. P. Jennissen and L. M. G. Heilmeyer, Jr., General aspects of hydrophobic chromatography—Adsorption and elution characteristics of some skeletal-muscle enzymes, *Biochemistry* **14**, 754–760 (1975).
15. H. P. Jennissen, Evidence for negative cooperativity in the adsorption of phosphorylase-*b* on hydrophobic agaroses, *Biochemistry* **15**, 5683–5692 (1976).
16. H. P. Jennissen, Immobilization of residues on agarose gels—Effects on protein adsorption-isotherms and chromatographic parameters, *J. Chromatogr.* **215**, 73–85 (1981).
17. E. Neumann, Molecular Hysteresis and Its Cybernetic Significance, *Angew. Chem., Int. Ed. Engl.* **12**, 356–369 (1973).
18. C. Frieden, Kinetic aspects of regulation of metabolic processes—Hysteretic enzyme concept, *J. Biol. Chem.* **245**, 5788–5799 (1970).
19. C. Frieden, Slow transitions and hysteretic behavior in enzymes, *Ann. Rev. Biochem.* **48**, 471–489 (1970).
20. J. C. Meunier, J. Buc, and J. Ricare, Enzyme memory. 3. Effect of the glucose-6-phosphate and temperature on the molecular transition of wheat germ hexokinase, *Eur. J. Biochem.* **97**, 573–583 (1979).
21. M. Mevarech and E. Neumann, Malate dehydrogenase isolated from extremely halophilic bacteria of dead sea. 2. Effect of salt on the catalytic activity and structure, *Biochemistry* **16**, 3786–3792 (1977).

22. D. Keradjopoulos and A. W. Holldorf, Purification and properties of alanine dehydrogenase from halobacterium-*Salinarium*, *Biochim. Biophys. Acta* **570**, 1-10 (1979).
23. H. P. Jennissen, Multivalent adsorption mechanisms in hydrophobic chromatography, *J. Solid Phase Biochem.* **4**, 151-165 (1979).
24. H. P. Jennissen, A. Demiroglou, and E. Logemann, in: *Affinity Chromatography and Related Techniques* (T. C. J. Gribnau, J. Visser, and R. J. F. Nivard, eds.), pp. 39-49, Elsevier, Amsterdam (1982).
25. H. P. Jennissen, The binding and regulation of biologically active proteins on cellular interfaces: Model studies of enzyme adsorption on hydrophobic binding-site latices and biomembranes, *Adv. Enzyme Regul.* **19**, 377-406 (1981).
26. R. J. Yon and R. J. Simmonds, Adsorption of proteins and protein-dodecyl sulfate complexes on aminodecyl-Sepharose, *Biochem. J.* **177**, 417-424 (1979).
27. C. T. Chang, B. J. McCoy, and R. G. Carbonell, Hydrophobic chromatography of beta-galactosidase, *Biotechnol. Bioeng.* **27**, 377-399 (1980).
28. A. V. Hill, Combinations of haemoglobin with oxygen and with carbon monoxide, *Biochem. J.* **7**, 471-480 (1913).
29. R. Simha, H. L. Firsch, and F. R. Eirich, Adsorption of flexible macromolecules, *J. Phys. Chem.* **57**, 584-589 (1953).
30. H. P. Jennissen, in: *Affinity Chromatography and Biological Recognition* (I. M. Chaiken, M. Wilchek, and I. Parikh, eds.), pp. 281-290, Academic Press, New York (1983).
31. M. Cortijo, I. Z. Steinberg, and S. Shaltiel, Fluorescence of glycogen phosphorylase-b—Structural transition and energy-transfer, *J. Biol. Chem.* **246**, 933-938 (1971).
32. S. Arnott, A. Fulmer, W. E. Scott, I. C. M. Dea, R. Moorhouse, and D. A. Rees, Agarose double helix and its function in agarose gel structure, *J. Mol. Biol.* **90**, 269-284 (1974).
33. M. Spodheim and E. Neumann, Ionic-strength dependence of hysteresis in polyriboadenylate-polyribouridylate system, *Biophys. Chem.* **3**, 109-124 (1975).
34. I. Prigogine and R. Defay, *Chemical Thermodynamics*, pp. 32-47, Longmans, Green and Co. Ltd., London (1954).
35. A. Katchalsky and E. Neumann, Hysteresis and molecular memory record, *Int. J. Neurosci.* **3**, 175-182 (1972).
36. A. Katchalsky and R. Spangler, Dynamics of membrane processes, *Quart. Rev. Biophys.* **1**, 127-175 (1968).
37. D. H. Everett and W. I. Whitton, A thermodynamic study of the adsorption of benzene vapour by active charcoals, *Proc. Roy. Soc. (London), Ser. A* **230**, 91-110 (1955).
38. P. Weiss, L'Hypothèse du Champ Moléculaire et la Propriété Ferromagnétique, *J. Physique*, **6**, 661-690 (1907).
39. D. H. Everett, General approach to hysteresis. III. Formal treatment of the independent-domain model of hysteresis, *Trans. Faraday Soc.* **50**, 1077-1096 (1954).
40. A. Katchalsky, in: *Biology and the Physical Sciences* (S. Devons, ed.), pp. 267-298, Columbia University Press, New York (1969).
41. R. Zsigmondy, Über die Struktur des Gels der Kieselsäure. Theorie der Entwässerung. *Z. Anorg. Allgem. Chem.* **71**, 356-377 (1911).
42. W. Thomson (Lord Kelvin), On the equilibrium of vapour at a curved surface of liquid, *Proc. Roy. Soc. (Edinburgh)* **7**, 63-68 (1870).
43. J. H. de Boer, in: *The Structure and Properties of Porous Materials*, Vol. 10 (D. H. Everett and F. S. Stone, eds.), pp. 68-94, Butterworth, London (1958).
44. S. Moore and W. H. Stein, Column chromatography of peptides and proteins, *Adv. Prot. Chem.* **11**, 191-236 (1956).
45. A. D. McLaren, Adsorption and reactions of enzymes and proteins on kaolinite, *J. Phys. Chem.* **58**, 129-137 (1954).

46. W. Kauzmann, Some factors in the interpretation of protein denaturation, *Adv. Prot. Chem.* **14**, 1-63 (1959).
47. C. Tanford, *The Hydrophobic Effect*, Wiley-Interscience, New York (1980).
48. J. Israelachvili and R. Pashley, Hydrophobic interaction is long-range, decaying exponentially with distance, *Nature* **300**, 341-342 (1982).
49. J. Porath, Explorations into the field of charge-transfer adsorption, *J. Chromatogr.* **159**, 13-24 (1978).
50. P. J. W. Debye, *Molecular Forces*, Wiley, New York (1967).
51. J. Koral, R. Ullman, and F. R. Eirich, Adsorption of polyvinyl acetate, *J. Phys. Chem.* **62**, 541-550 (1958).
52. I. Langmuir, Monolayers on solids, *J. Chem. Soc.* **1940**, 511-543.
53. H. P. Jennissen, Multivalent adsorption of proteins on hydrophobic agaroses, *Z. Physiol. Chem.* **357**, 1201-1203 (1976).
54. N. M. K. Ng Ying Kin and W. Yaphe, Properties of agar—Parameters affecting gel formation and agarose-iodine reaction, *Carbohyd. Res.* **25**, 379-385 (1972).
55. D. Keradjopoulos and A. W. Holldorf, Salt dependent conformational changes of alanine dehydrogenase from halobacterium-*Salinarium*, *FEBS Lett.* **112**, 183-185 (1980).
56. A. Lambeir and Y. Engelborghs, A quantitative analysis of tubulin-colchicine binding to microtubules, *Eur. J. Biochem.* **109**, 619-624 (1980).
57. J. Herzfeld, T. Ichiye, and D. Jung, Molecular symmetry and metastable states of enzymes exhibiting half-of-the-sites reactivity, *Biochemistry* **20**, 4936-4941 (1981).
58. D. J. States, C. M. Dobson, M. Karplus, and T. E. Creighton, A conformational isomer of bovine pancreatic trypsin-inhibitor protein produced by refolding, *Nature* **286**, 630-632 (1980).
59. H. W. Chang, E. Bock, and E. Neumann, Long-lived metastable states and hysteresis in the binding of acetylcholine to *Torpedo californica* acetylcholine receptor, *Biochemistry*, **23**, 4546-4556 (1984).
60. S. Shaltiel, in: *Metabolic Interconversion of Enzymes* (E. H. Fischer, E. G. Krebs, H. Neurath, and E. R. Stadtman, eds.), pp. 379-392, Springer Verlag, Berlin (1974).
61. B. H. J. Hofstee, Immobilization of enzymes through noncovalent binding to substituted agaroses, *Biochem. Biophys. Res. Commun.* **53**, 1137-1144 (1973).
62. K. D. Caldwell, R. Axen, M. Bergwall, and J. Porath, Immobilization of enzymes based on hydrophobic interaction. 1. Preparation and properties of a beta-amylase adsorbate, *Biotechnol. Bioeng.* **18**, 1573-1588 (1976).
63. B. Schobert and H. P. Jennissen, Stabilization of phosphorylase-*b* against temperature-induced inactivation and denaturation by ionic and non-ionic solutes, *Z. Physiol. Chem.* **361**, 329-330 (1980).
64. R. Jost, T. Miron, and M. Wilchek, Mode of adsorption of proteins to aliphatic and aromatic amines coupled to cyanogen-bromine activated agarose, *Biochim. Biophys. Acta* **362**, 75-82 (1974).
65. K. G. Wilson and J. Kogut, Renormalization group and the ε expansion, *Physics Reports* **12C**, 75-200 (1974).
66. M. E. Fisher, Renormalization group in the theory of critical behavior, *Rev. Mod. Phys.* **46**, 597-616 (1974).
67. F. J. Wegner, Critical phenomena and the renormalization group, *Lecture Notes in Phys.* **37**, 171-196 (1975).

10

Modeling of Protein Adsorption

A. Silberberg

1. INTRODUCTION

We shall employ two kinds of models: theoretical models which will allow us to see the principles of what to expect when macromolecules are adsorbed and experimental models which will illustrate certain cases. Proteins, of course, are much too varied and much too special for any particular case to serve as a truly broad model. When, therefore, we shall discuss certain experimental cases they will be models not so much for classes of proteins but for the techniques that have been employed and for the vast variety of possibilities that can occur.

2. THEORETICAL MODELS

Macromolecules tend to adsorb extremely well. Even from very dilute solutions macromolecules will tend to move into a surface zone if, as is usual, there is a slightly favorable interaction between the groups on the molecule's surface and the groups which characterize the interface to which it becomes attached. Even a small effect, a small fraction of kT, but taken over the large number of the groups involved, becomes a large energy effect per macromolecule.[1-3] On the other hand, the free energy increase due to the removal of a single solute molecule, the macromolecule, from a solution, even of extreme dilution, is in general kT times a much smaller number. Hence the adsorption of macromolecules is favored under conditions where adsorption of low molecular weight material would be unimaginable. Once adsorbed, the macromolecule will tend to stay. This is because the chances of gathering enough activation energy to remove all contact between

A. Silberberg • Weizmann Institute of Science, Polymer Department, Rehovot 76100, Israel.

molecule and surface, and to do so for a time sufficient to allow the molecule to diffuse away, are extremely small. Hence many systems appear to be irreversibly adsorbed because the desorption reaction is so slow. Macromolecular surface layers will stay in place even when the solution is replaced by pure solvent.[4]

It does not matter too much for this general effect whether the macromolecule is an open, freely coiling random chain or whether it is a specifically folded single conformation of, say, a globular protein, i.e., of a specifically collapsed (on itself) chain. Even in such a case the number of contacts with the surface will be high enough to make the system adsorb well.[5] The adsorption of macromolecules has, however, been mainly studied from the point of view of linear, flexible macromolecular chains and the configuration changes they undergo upon adsorption onto a plane, impenetrable interface.[1-3,6-10]

2.1. THE ADSORPTION OF LINEAR FLEXIBLE MACROMOLECULES

Seen as a problem involving conformational rearrangement, the point of main interest will be the character of the segment distribution which results and the nature of the parameters which determine the extent of adsorption. The polymer segment is seen as a material point interacting energetically only with its closest neighbours, whether these are solvent molecules, other polymer segments, or groups on the surface.[1,3] There is assumed to be only one kind of polymer segment present and it has no preferred direction of interaction. There is thus no directional constraint imparted to the adsorbing chain by the segments which touch the surface. The surface too is treated as a uniform atomically smooth layer. Only one kind of binding site occurs and it is uniformly present over all the surface. In fact all sites on the surface are binding sites. Both the solution and the surface phase are of finite concentration.

Given these assumptions, great strides have been made in understanding how chain macromolecules adsorb and how indeed they behave at surfaces.[3,6,7] It can be shown that solvent power is one of the basic influences which determines adsorption. Adsorption is prevented if too good a solvent is chosen. It is, in general, possible to clean surfaces from adsorbed polymers by choosing a good enough solvent. Once adsorption is favored to occur at all, saturated surface phases will arise even when the solution from which the macromolecules are adsorbed is very dilute indeed. A limited degree of adsorption will occur only when the solution is so dilute that the total amount of polymer present is inadequate to provide extensive coverage. From the very dilute to relatively concentrated solutions, the amount adsorbed on the surface is constant (or relatively so). The highest molecular weight polymer is generally the one which adsorbs best. Adsorption is temperature-dependent, but only moderately so, and this small temperature coefficient is mainly due to the effect of temperature on solvent power. The small temperature coefficient also expresses the fact that the energy of adsorption per segment is rather small. What is large is the number of segments held to the surface, not the effect per segment. While it is necessary to break all contact with the surface before the adherent macromolecule can diffuse away, it is not necessary to supply all this energy in one package, i.e., at the same instant. It is, in principle, possible to "peel" the polymer off the surface segment by segment. Yet, though the energy for each such step may not be large, the fact that in a succession of events the chain will each time have to remove itself from the surface is very improbable. Hence polymer desorption, even into pure solvent, is a highly unlikely event. The time required for desorption is indeed generally so long that on a practical time scale the phenomenon of macromolecular adsorption appears to be irreversible.

The loss in free energy on collapsing a three-dimensional into a two-dimensional random walk is of the order of $\frac{1}{2}\,kT$ per segment.[2] Hence the energy effect required, the reduction in internal energy per segment, has just to be a little larger than that to drive the polymer totally into the surface. Indeed insofar as adsorption of an isolated macromolecule is concerned, this is exactly what is expected to happen. Only if there is "competition" amongst many macromolecules, all wanting to be in the surface, is a situation of compromise attained. Hence at finite concentration only half of the segments of each chain will be in contact with the surface, in the form of segment trains of different length; the other half will protrude into the solution in the form of loops or dangling tails.[1,3] The resulting distribution of segments is well understood theoretically and so is the influence on this distribution of the presence of segments derived from the macromolecules near the surface in the solution.[7] Experimentally it can be shown that the surface layer is indeed thick. The overall dimensions in the only cases tested so far (those of dilute solutions) when divided by the radius of gyration of the unperturbed macromolecules in bulk space produce superimposable results.[7,11]

The thick surface layer which results is of great importance in many technological applications. It modifies the surface and intersperses a "gel" phase of the adsorbed polymer between the original surface and the solvent.[12,13] The presence of the surface gel can totally alter the particle/particle interaction. Where this interaction may originally have led to precipitation, it now provides for stabilization; where the system was originally stable, it can now be brought to flocculation.

The large conformational changes that are required to produce the surface phase, the problem of entanglement of the chains, the necessary crowding of the polymer segments, particularly those in the layers very close to the surface, all tend to slow down the rates at which the polymers at the surface reach internal and mutual equilibrium. Days or weeks may be required. The factor which paces the time of rearrangement is the time required to detach cooperatively all of a train of segments between two loops. This slow intermediate step makes total desorption in particular so very tardy.

These insights, gradually gathered over the years, but mainly associated with flexible open coil polymer chains, may have resulted in the impression that flexibility was a property critically involved. This is not at all the case[5] and the reason for this has indeed already been partially discussed. It was worthwhile, however, to emphasize the essentials of adsorption of flexible chains so that the case of the adsorption of macromolecules in general and of proteins and other biopolymers in particular can be better understood.

2.2. THE ADSORPTION OF MACROMOLECULES IN GENERAL[5]

Probably the antithesis to the open, linear chain, flexible polymer is a compact structure where all the segments are either covalently or otherwise bonded to their neighbors so that a particle of regular or irregular, but mainly compact, shape arises. We can visualize many intermediate situations as this compact structure is gradually loosened and allowed to fill up with solvent. One such structure, the statistically coiling linear chain, is the case that we have discussed above. If, however, we stiffen this chain and convert the coil into a rigid rod we will again have attained a particle of invariant configuration whose adsorption might be of interest. We will now investigate what all of these shapes and structures, rigid or flexible, have in common with respect to the adsorption problem.

Let us measure macromolecular size by the number P, where P is the ratio between the volume directly occupied by all the segments of the macromolecule and the volume of a solvent molecule. Let us, moreover, think of each of these P segments as homogeneous and representative of the unit which will interact with the surface. In the case of a flexible chain or a rigid rod, it is possible for the chain to make P such contacts with a flat surface. In the case of a compact sphere, however, only some $P^{2/3}$ segments will be on the surface. Hence the number of contacts will be proportional to $P^{2/3}$, but if $P = 10^3$, such a sphere would still have on the order of 10^2 interactions with the surface. This is smaller than 10^3 but still a very large number. The compact particle, moreover, need not undergo an unfavorable conformation change on the surface so that even a small supportive energy effect per segment constitutes a large net gain. The free energy required to remove a particle from a very dilute solution of volume fraction ϕ in which it is ideally dissolved is $kT \ln X$, with X the mole fraction of the solution.

$$X = (\phi/P)/[(1-\phi)+(\phi/P)] \cong \frac{\phi}{P} \qquad \text{for } \phi \ll 1$$

Hence the free energy change when a compact particle of size P is adsorbed and $P^{2/3}$ solvent molecules are released from the surface is given by

$$-kT\ln(\phi/P) - \chi_s kT P^{2/3} - kT P^{2/3}\ln[1+(\phi/(1-\phi)P)]$$
$$\cong kT[\ln P - \ln\phi - P^{2/3}\chi_s]$$

where $\chi_s kT$ is the energy change when a segment of the polymer displaces a solvent molecule on the surface. Since $\ln P$ is going to be small as compared to $P^{2/3}$, we see that the free energy change will be favorable (i.e., negative as long as

$$\phi > e^{-P^{2/3}\chi_s} = 10^{-P^{2/3}0.43\chi_s}$$

If, therefore, χ_s, which is non-dimensional, is about 1 or larger, essentially irreversible adsorption will occur, even for macromolecules of compact shape. The surface will become saturated and we can expect essentially total coverage of any available opening giving rise to a layer of thickness $P^{1/3}$. Of course, the energy change is even larger when instead of $P^{2/3}$ we can write P. Nevertheless, provided P is large enough, the effect will strongly favor adsorption irrespective of macromolecular shape and irrespective of whether the macromolecule undergoes a conformational change at the surface or not.

Even a surface layer of compact particles will be very thick when P is large enough. When the macromolecule is an open coil its radius of gyration varies as $P^{1/2}$ to $P^{3/5}$ depending upon solvent power. When it is compact the thickness is on the order of $P^{1/3}$. The interfacial zone is thus thick, many times thicker than the monolayer of segments which are in direct energetic contact with the surface. Seen from this point of view and counting layers in terms of such unit layers, the macromolecular surface film is certainly a "multilayer." It is not, however, from the point of view of the more usual concept of multilayer adsorption where subsequent layers of adsorbed molecules are layers of molecules adsorbing onto molecules of their own kind already adsorbed on the surface. Such molecules in a multilayer do not, therefore, make contact on their own with the original surface but with the first, second, etc. layers of adsorbed molecules which result. Can this happen in the macromolecular case?

2.3. THE ADSORPTION OF MACROMOLECULES IN MULTILAYERS[14]

In the case of long-chain flexible polymers, it is generally true that multilayer adsorption in the usual sense does not occur. What would be multilayer adsorption in the "usual sense" in this case? Since the statistical ensemble of all macromolecules adsorbed on the surface in a first layer has in common contact with the surface, and since the number of these attachment points will fluctuate, we can define as a characteristic of the first layer of these macromolecules that they have at least one segment in contact with the surface. A second layer of a multilayer would then be composed of macromolecules which have a tendency to stay close to the surface at a concentration above that of the adjoining bulk solution and in conformations which differ on the average from the average bulk conformation. A characteristic of these macromolecules would be that they do not have any of their segments in contact with the surface. The second layer develops at the surface only because close to the surface there is a zone of high polymer segment concentration.

If one may look upon the adsorption of macromolecules onto the surface as a facilitated phase separation,[12] multilayer formation in the sense defined above can be seen to be possible. Most polymer/solvent systems beyond a certain temperature, and beyond a certain solute concentration, become unstable. Two phases, a highly concentrated phase and a very dilute phase, are generally formed. Hence, given the energetic inducement of the surface, the concentrated polymer phase which is generated near the surface can be seen to anticipate the tendency of the system to give rise to such a phase spontaneously in bulk under slightly worse solvent conditions. Under these circumstances the usual monolayer is produced. If, however, the temperature is such that the polymer/solvent system is inherently unstable, but the system from which adsorption occurs is so dilute as still to be stable, the concentrated surface phase produced on the surface can act to attract more macromolecules. These macromolecules will then have no contact with the surface, but will produce a "second" layer. It can be shown, moreover, that for each segment held to the surface in the first layer, it is possible to add at most one segment in the second layer.[14]

This result has been derived for a homopolymer, i.e., for the case of a polymer and a surface where all segments and all sites are totally equivalent and where no directional effects occur. The result also depends upon the chain being flexible so that in principle each segment can be in contact with the surface, either singly or in trains of two, three, etc., segments, or with the segments in the loops. This is not possible for compact structures.

2.4. THE ADSORPTION OF MACROMOLECULES OF SPECIFIC STRUCTURE IN DIRECTED WAYS. CONFORMATIONAL CHANGES AND SURFACE ACTIVATION

If a compact homopolymer adsorbs and completely covers the surface, the surface so created will now be totally like the polymer in solution.

Again, we may assume that the solution is stable and that a "precipitation" of the polymer upon the monolayer of trains will only occur from systems where the solvent environment at higher concentrations would have favored a phase separation.

With compact structures, however, it is almost always true that we are dealing with copolymers. There are thus questions of packing and questions of maximizing the contact area. Polymers, compact or not, are built up out of one or more chains. Even where several separate chains are involved, their number remains small. Branching and cross-linking may occur, but the sections between branch points or cross-links are still linear chains. Hence if these structures, at one and the same time, "close up" in order to create molecular compactness and remain stable in solution, this must be because some segments which are strongly attractive compensate each other while others more soluble stay on the outside. At the same time, the surface of the compact particle will not be uniform. Some of its faces will be stickier than others. It will be the least "soluble" face which by preference interacts with the surface.[13] The surface is an energy well into which more segments might like to sink than the compact conformation adopted in solution actually permits. Hence, if the energy to be gained is large enough, a change in the conformation of the macromolecule would be induced by the surface, resulting in segments in the molecule becoming accessible and exposed which are not open to interaction in the solution.

In such a case, in particular, the adsorption of a first layer of compact structures will have created a surface which bears no resemblance to the surfaces the macromolecule experiences in its encounters in bulk. In such a case a further layer could become adsorbed and could in turn become activated. Multilayer adsorption of another kind thus becomes a possibility. It is reasonable to assume, moreover, that the conformational change might be less perfect in the second layer than the first and that, therefore, the effect would gradually decline, limiting the number of layers that are added, or making it depend upon the concentration of protein in the bulk.

In particular with proteins, therefore, we can see not only the possibility of effective adsorption, but also the possibility of adsorption in multilayers. It is further clear that compact globular proteins may be expected to adsorb in some preferred orientation and that they may lose specificity of function in this state because some essential group has become unavailable. If a conformational change is induced, energetically, function may be lost, but may also be acquired. We must, moreover, consider the details of the surface onto which adsorption is studied. This is assumed to be uniform but is most probably highly structured. Not all sites will be equivalent and a variety of cooperative effects can be envisaged in consequence. Proteins will not only bind well to surfaces, they may do so also in altered form. Proteins already bound may specifically be influential in activating proteins bound at adjoining sites.

We shall now consider some specific cases to illustrate these effects and ideas. Adsorption onto model surfaces from solutions of single proteins will be involved. It may be assumed that the effects of adsorption onto specifically structured surfaces in the presence of other proteins and biopolymers would be correspondingly more complex. It will be seen, however, that what is needed now is not new ideas about what might occur, but the development of good techniques for the accumulation of data in order to measure effects and to characterize them. Both novel approaches and sound systematic work are needed.

3. EXPERIMENTAL MODELS AND APPROACHES

3.1. THE ADSORPTION OF PROTEINS ON GLASS AND MICA

Neither the substrate glass nor the substrate mica have any special interest from the point of view of biology. The choice of these model surfaces is dictated solely by instrumental considerations. In both instances, as we shall see, there is as yet no alternative to this choice of material for the construction of the critical parts of the apparatus. On the other hand, no nonbiological surface is *a priori* more suitable than another.

When proteins are manipulated *ex vivo*, they must, particularly when they are in solution, be contained in some vessel. Almost inevitably, adsorption will occur for the reasons already given. In many or even most such instances the containing vessels are made of glass. *A priori*, certainly, a glass surface is in no way more or less exceptional than any other surface and it may be assumed that a protein handled in the laboratory without any special care will have had some contact with glass. This can be avoided when it is deemed necessary but in the type of measurements described here it would be impossible to circumvent.

3.2. MEASUREMENT OF THE THICKNESS OF THE LAYER OF ADSORBED MACROMOLECULES[15,16] BY TIME OF FLOW

An Ostwald U-tube viscometer is a very convenient tool for measuring adsorption layers which are not too thin.[16] The principle of the Ostwald viscometer depends upon the flow of a fixed volume of fluid from one side of a U-tube to the other through a cylindrical tubular section under a reproducible pressure head. This can be achieved with great precision if the viscometer is a one-piece glass instrument. In such a case, if a fixed volume of fluid is put into the U-tube, the viscometer is set up vertically, and the fluid is sucked up on one side and allowed to descend under gravity into the other, a very well-defined pattern of flow will result. The volume of fluid passing through the capillary tube is measured between two narrow constrictions with circular marks on them, one above and one below a bulb in that arm of the U-tube from which the fluid is dispensed. The time of passage of the fluid meniscus from the upper to the lower mark is measured with a stopwatch. Since the volume of fluid inside the U-tube is fixed and the whole viscometer is a rigid piece of instrumentation, the driving force (head) on the fluid will follow precisely the same history. If the measurements are conducted in a well-thermostated environment, the time of flow will then depend only upon the viscosity of the fluid. If the viscosity is the same, then the results should also be the same, unless the radius of the cylindrical glass capillary section of the U-tube has been changed by adsorption. If R is the unaltered radius and a layer of hydrodynamic thickness, a, exists, the time of flow of fluid over the surface will have been changed by a factor of $1 - (4a/R)$, where the factor 4 derives from the fact that flow through a cylindrical tube varies as the fourth power of R and $a \ll R$.

There are various errors which could creep in but can be eliminated. Besides good thermostating, it is important that the U-tube always be placed in the same vertical position. The arrangement should be such that the bulb with the two marks at either end is situated directly above the meniscus of the receiving bulb on the other side of the U-tube. For this purpose the lower part of the U-tube, notably the capillary section, is displaced sideways at an angle. The glass surface must be clean and the time of flow must be such that the passage of the meniscus can be observed with 0.1-s precision. As the fluid flows down the upper bulb, it leaves behind a film of liquid wetting the glass. The pressure drop across the meniscus due to capillarity is important as is the interaction with the glass which determines the contact angle between the meniscus and the wall. The contact angle affects the driving head. It is important, therefore, that for a comparison in terms of time of flow, the same fluid should pass through the U-tube at the same temperature and under the same contact conditions between meniscus and glass which existed before as well as after polymer adsorption.

If adsorption takes place onto the glass, the protein will coat not only the capillary walls but all the walls with which it comes in contact. Hence the surface properties of the glass both in the measuring bulb and in the narrow sections at either of its ends will have been changed. The time of flow will be affected not only by the change of radius from R to $R - a$ in the capillary, but also by the fact that the fluid now has a contact angle characteristic of the adsorbed protein layer rather than that of the glass. This problem will arise even more when (taking advantage of the effective irreversibility of adsorption) the solution from which adsorption to equilibrium had taken place is replaced by pure solvent.

To overcome this difficulty two modifications of the technique must be introduced[16]: The glass surface over which the upper meniscus passes between the marks is cleaned of protein. One follows the same procedure (exposure to sulfochromic acid, etc.) as when the entire viscometer surface is cleaned. To do so, the U-tube with its capillary section kept filled with buffer, but otherwise empty, is inverted and the upper part of the U-tube, now at the bottom, is cleaned by pumping cleaning fluid up and down following the standard cleaning procedure exactly. The glass surface is then washed with buffer and a certain amount of buffer is introduced after the viscometer is again put the normal way up.

Since it is clearly impossible to start with a dry viscometer in this case, it is impossible to have an exact volume in the viscometer. To overcome this problem we start instead with an arbitrary volume of fluid in the instrument, the whole viscometer and its contents are weighed, and the time of flow is measured. The contents of the viscometer are increased, it is again weighed, and the time of flow is measured and plotted against fluid volume by way of calibration. A certain mass of fluid is then chosen, say 7 g, and everything standardized to that.[16] Since the same protein-free fluid is flowing at all times, its viscosity and density are the same, the instrument is the same and so is the state of its internal surface, the driving pressure at each position of the meniscus will be the same.

Two quite different proteins have been studied in particular by this approach and two very different results have been obtained.

3.3. ADSORPTION OF SOLUBLE COLLAGEN

Collagen is based on three peptide chains of rather special composition and sequence.[17] These three chains wind around each other and create a triple helical structure which only the end sections of the chains are incapable of joining. The triple helix configuration is made possible because every third amino acid in the helix-forming central part of the three chains is glycine. The central regions of the three chains also contain proline and hydroxyproline in very large amounts. Since the glycine residues are on the inside of the triple helix, the molecule seen from the outside has over half of the exposed amino acids on the triple helical cylinder, either proline or hydroxyproline. The ends of the three chains do not follow this special sequence and triple helix formation is structurally impossible. The three chains there are randomly coiled and the appearance of the so-called collagen molecule with its central triple helix is that of a rod 3000 Å long, 15 Å in diameter, with rather frayed and bushy ends. When exteriorized in this form in embryonic tissue, collagen molecules will assemble first into microfibrils, then fibrils, and finally into fibers. Collagen constitutes the major component by weight of connective tissue and is its load-bearing element. The formation of microfibrils and indeed the entire process *a priori* depends upon secondary bond interactions which are, however, subsequently strengthened by covalent cross-link formation under the influence of special enzymes. It is thus impossible to obtain the free collagen molecule from mature tissue, except in uninterestingly minute amounts. In embryonic tissue, however, the covalent cross-linking process has in general not gone far enough. Hence a reasonable amount of the material can be solubilized from out of the fiber. Alternatively the soluble collagen molecule can be obtained from tissue where, for genetic or other reasons, the enzymatic system which introduces the cross-links is either not functioning at all or not well.

In the cold in acid buffer of high salt concentration the collagen is stable in solution as a particle of the shape and size described above. When such a solution is heated (to 37°C) rapid fibril formation follows a rather long initiation period. The assembly involves a process whose details have not yet been finally elucidated. Kept at temperatures below 10°C, the solutions are, however, rather stable. The results to be described here were obtained with soluble calf skin collagen (Worthington CLCS, 7 mg/ml in 0.75 *M* Na citrate buffer, pH 3.7). To start with, pH 4.0 was chosen.(16)

A solution was prepared in the cold and then spun for one hour at $40{,}000 \times g$ in a refrigerated centrifuge in order to free the solution of any collagen molecule oligomers that might have been formed during preparation or storage.

Collagen molecule adsorption studies using ^{125}I-labeled soluble collagen and glass slides of well-defined area had shown that, under the above conditions, a 10 μ/ml solution of collagen would produce saturation within some 8 h in the cold.(4) A solution of approximately this concentration was thus filled into the U-tube viscometer and kept there overnight in the cold with the fluid sucked up into the capillary and filling the measuring bulb to well above the upper mark.

It had also been shown that soluble collagen did not desorb, even against pure buffer.(4) We thus replaced the collagen solutes by buffer, by repeated dilution taking care that at no time an air/water interface is allowed to pass through the capillary of the viscometer. The viscometer was then weighed and a graph of time of flow against mass of fluid loaded into the viscometer was obtained. The result, extrapolated to 7 g, was noted for purposes of comparison with the time of flow of 7 g of pure buffer in the viscometer before it was exposed to soluble collagen (Figure 1). The cleansing procedure as described earlier was then instituted, care being taken, as before, that no air entered the capillary during this time. Then after washing

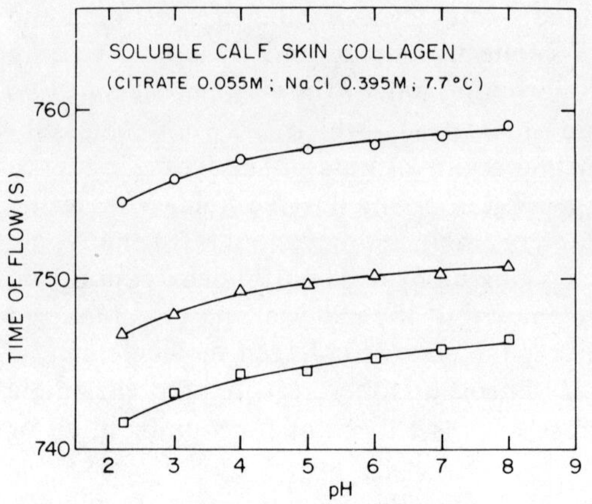

FIGURE 1. Time of flow of citrate buffer after adsorption of soluble calf skin collagen at various pH's. ○: Protein coat on bulb; △: Protein coat removed from measuring bulb; □: clean surface.

with buffer another set of readings was undertaken and the result for 7 g buffer found (see Figure 1).

It is apparent that the presence of the collagen film on the surface has an enormous influence on the time of flow.

The radius of the capillary of the viscometer was measured and the thickness of film, a, was calculated from

$$a = [(t - t_0)/t_0]R/4$$

It was found (Figure 2) that a = 2800 Å at pH 4 and this value was unaffected

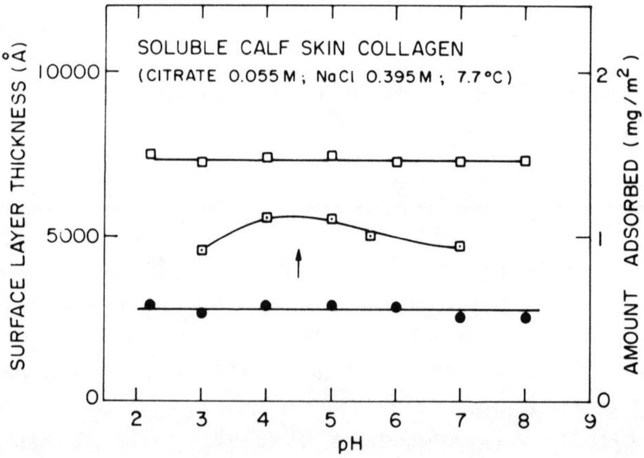

FIGURE 2. Layer characteristics of soluble calf skin collagen at various pH's. □: Apparent thickness (effect of contact angle); ●: true thickness; ⊡: amount adsorbed; ↑: isoelectric point.

by pH over the range of pH from 2 to 8. Had the surface of the bulb not been cleaned, a result of 7300 Å would have been found.

If the viscometer is heated with collagenase no surface film remains.[18] The collagen has been digested away. If the viscometer with the surface film is heated (60°C for 30 min) denaturation occurs, the triple helix unwinds, and the separate chains either collapse onto the surface or partially dissolve. The film that remains is too thin (~700 Å) to be measured precisely by the present technique.[18] This film thickness corresponds to a difference in time of flow of about 1 s.

A layer thickness of about 3000 Å clearly suggests that the collagen molecules are adsorbed end-on. If we combine this with the fact that the surface saturates when 1 mg/ml^2 is adsorbed[4] (Figure 2) we can conclude that the upright triple helical rods are rather far apart (some 200 Å) which may account for the ease with which unraveling takes place and for the fact that the denatured collagen collapses into a very thin layer. There are indirect indications that the glass surface remains totally covered by the separated, now randomly coiled, collagen chains after denaturation.[19]

The result for the thickness of the surface layer in the case of adsorbed collagen molecules could be confirmed using adsorption onto mica and a method of direct force–distance detection between approaching surfaces.[16,20] Mica has the property to be split and produce centimeter-sized atomically smooth sheets. These sheets bent into cylinders and put at right angles give a point-to-point contact. The distance between the surfaces can be measured by multiple beam interferometry. The outer side of each mica sheet is half-silvered. White light is then multiply reflected between these mirrors. Most of the light is eliminated by destructive interference. Only wavelengths for which the distance between the mirrors, optically, is an exact multiple of half-waves will pass through. These wavelengths are very sharply defined and give a very precise definition of distance once the order of the fringes is known. The force is measured by mounting one of the cylinders of mica on a soft calibrated spring and determining the displacement. It is found that when two surfaces with soluble collagen adsorbed upon them are brought close to each other, the force between them stays at zero until a distance of ~6000 Å (2 × 3000 Å) is reached. At that point the force starts going up very rapidly.[16] The thickness of the adsorbed zone of soluble collagen can thus be looked upon as very well established.

3.4. ADSORPTION OF BOVINE SERUM ALBUMIN

The thickness of absorbed serum albumin was studied using the U-tube viscometer and the same technique as described above. At room temperature and pH 4.0 and 7.0, the results of Figure 3 were obtained. As with collagen, it is apparent that the adsorption is irreversible in all essentials.

Two important differences will be noted, however:

1. There is hardly any change in the time of flow when the glass surface is cleaned with sulfochromic acid. The contact angle of buffer with bare glass and with serum albumin on the glass must thus be closely similar.
2. There seems to be no saturation. The amount of serum albumin, as judged by adsorbed layer thickness, increases proportionately with the overall concentration (Figure 3) over the entire range so far studied.

The resulting film thicknesses are enormous and adsorption in multilayers must be assumed. A serum albumin molecule is of the order of 50 Å in size. The micron-thick surface layers measured here are very many times this size indicating that a major effect is involved. The surface deposit is neither in reversible equilibrium nor is it uncontrolled. As in the case of soluble collagen (and many other adsorbed macromolecular deposits), replacing the original solution (of finite concentration) by pure buffer does not produce a diminution of the layer thickness or at any rate produces this only very gradually. Hence the reverse reaction, desorption, must be tremendously inhibited in this case as well. On the other hand, the forward

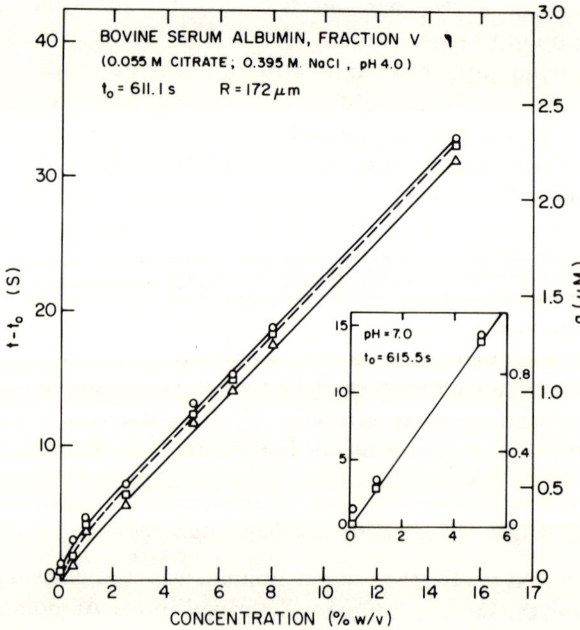

FIGURE 3. Thickness of bovine serum albumin layers as a function of concentration. ○: With protein film on bulb; □: cleaned bulb; △: cleaned bulb, after repeated washes.

reaction, adsorption, is still limited, and very reproducibly so, by concentration. Despite the enormous excess of albumin in solution, say at larger than 1% concentration, proportionately less albumin is deposited at 1% than at 5% or at 15%. A precise, concentration-dependent factor is at work. We may assume that the probability of adding each successive layer to the surface is steadily reduced as the distance from the surface is increased. Hence the effective chemical potential of the albumin on the surface rises with distance from the glass surface and deposition will stop when there is no further free energy to be gained from adsorption. The free energy effect actually involves a net reaction: a deposition of albumin and a corresponding release of solvent. It is for this exchange that the net free energy effect should be zero. (Note: This does not mean that the chemical potential of each individual species by itself is not balanced. The individual chemical potential, however, also involves a contribution from the increase in surface area multiplied by the interfacial tension. The interfacial tension is not easily assessed in this case. By considering an exchange the interfacial tension effect cancels out.)[1,3]

4. DISCUSSION

The examples brought from our own experience have been selected for the purpose of demonstrating the complexities that accompany the adsorption of compact copolymers such as proteins, particularly in cases when the molecules are asymmetric and have different chemical reactivities associated with different faces or sites. One can thus obtain such varied responses as those observed with collagen and albumin, even when using an essentially (on the scale of an amino acid) homogeneous glass surface.

There is a strong tendency to saturate the surface once adsorption is favored at all. Even in the case of collagen, it is easily calculated that the randomly coiling non-triple helical sections at the ends fully occupy the area between the coated rods. Near the isoelectric points, where these regions are more collapsed, slightly higher amounts are indeed adsorbed (Figure 2). Why the collagen molecule does not adsorb side-on is probably due to the high proline content of the cylinder surface. Much larger amounts of collagen per unit area can indeed be adsorbed on more hydrophobic surfaces than glass but in these cases we know nothing about thickness and molecular arrangements. If we are right that the coated end case, in the event of the surface being glass, is a fully saturated situation, the much higher levels found, say with Teflon, must be due to a mixed end-on/side-on situation.

The soluble collagen on glass surface has many interesting properties with respect to activation of blood platelets.[21] If the collagen is adsorbed on a plane glass surface, the platelet sees it as though it were subendothelial basement membrane collagen. The platelets stick but do not form thrombi. The interaction with collagen on adsorbed thin glass fibers (2 μ diameter or less) is as with the collagen fibers in the connective tissue space surrounding an injured blood vessel. Profound thrombus formation occurs. The response is different again when the collagen is heat-denatured on the surface. In this case the binding of platelets to the surface is considerably reduced; the platelets in fact interact less effectively with such a surface than with a bare glass interface.

The case of albumin illustrates the possible existence under various circumstances of very thick protective coats. The eventual compatibility of some prosthetic materials used for contact with blood may depend upon the presence of such extensive surface covers. This might indeed be an inherent effect of albumin and may be the way the body guards against the adverse effects which the absence of a glycoprotein coat may have at lipid plaques or other aberrations on the endothelial surface resulting from disease or age.

While concepts which we have developed from basic principles apply to the adsorption of macromolecules in general, we have here illustrated the very complex situations that the specific structural features of proteins can produce even in only two cases. Indeed, each protein/substrate pair must be examined on its own. Our general understanding of why macromolecules are so surface-active will help us only over the most preliminary stages of any particular case. Above all, good methods are required so that all details of the surface phases formed can be acquired.

REFERENCES

1. A. Silberberg, The adsorption of flexible macromolecules II. The shape of the adsorbed macromolecule; the adsorption isotherm; surface tension and pressure, *J. Phys. Chem.* **66**, 1884–1906 (1962).
2. A. Silberberg, Adsorption of flexible macromolecules III. Generalized treatment of the isolated macromolecule; effect of self-exclusion, *J. Chem. Phys.* **46**, 1105–1114 (1967).
3. A. Silberberg, Adsorption of flexible macromolecules IV. Effect of solvent-solute interactions, solute concentration and molecular weight, *J. Chem. Phys.* **48**, 2835–2851 (1968).
4. G. Penners, Z. Priel, and A. Silberberg, Irreversible adsorption of triple-helical soluble collagen monomers from solution to glass and other surfaces, *J. Colloid Interface Sci.* **80**, 437–444 (1981).
5. A. Silberberg, The effect of conformation on the interaction of macromolecules with surfaces. The adsorption of compact macromolecules. *Isr. J. Chem.* **11**, 153–160 (1973).
6. J. M. H. M. Scheutjens and G. J. Fleer, Statistical theory of the adsorption of interacting chain molecules, *J. Phys. Chem.* **83**, 1619–1635 (1979); **84**, 178–190 (1980).
7. P. G. de Gennes, *Scaling Concepts in Polymer Physics*, Cornell University Press, Ithaca (1979).

8. P. G. de Gennes, Polymer solutions near an interface 1. Adsorption and depletion layers, *Macromolecules* **14**, 1637–1644 (1981).
9. R.-J. Roe, Multilayer theory of adsorption from a polymer solution, *J. Chem. Phys.* **60**, 4192–4207 (1974).
10. C. A. J. Hoeve, Adsorption isotherms for polymer chains adsorbed from θ solvents, *J. Chem. Phys.* **44**, 1505–1509 (1966).
11. J. Klein, Forces between mica surfaces bearing adsorbed macromolecules in liquid media, *J. Chem. Soc., Faraday Trans I* **79**, 99–118 (1983).
12. A. Silberberg, Macromolecular surface phases and the stability of colloidal dispersions, in: Colloid stability, *Faraday Disc. Chem. Soc.* No. 65, pp. 194–201 (1978).
13. A. Silberberg, The role of membrane bound macromolecules and macromolecules in solution in cell/cell encounters in flowing blood, *Ann. N.Y. Acad. Sci.* **416**, 83–91 (1983).
14. A. Silberberg, Multilayer adsorption of macromolecules, *J. Colloid Interface Sci.* **38**, 217–226 (1972).
15. Z. Priel and A. Silberberg, The thickness of adsorbed polymer layers at a liquid-solid interface as a function of bulk concentration, *J. Polym. Sci., Polym. Phys.* **16**, 1917–1925 (1978).
16. A. Silberberg and J. Klein, Structure and properties of surface adsorbed biopolymer layers—soluble collagen adsorbed to glass, *Biorheology* **18**, 589–599 (1981).
17. W. Traub and K. A. Piez, The chemistry and structure of collagen, *Adv. Prot. Chem.* **25**, 243–352 (1971).
18. A. Silberberg, in: *Adhesion and Adsorption of Polymers* (L. H. Lee, ed.), pp. 837–845, Plenum Press, New York (1980).
19. F. A. Meyer and Z. Weisman, Adhesion of platelets to collagen. The nature of the binding site from competitive inhibition studies, *Thrombosis Res.* **12**, 431–446 (1978).
20. J. Klein, Forces between mica surfaces bearing layers of adsorbed polystyrene in cyclohexane, *Nature* **288**, 248–250 (1980).
21. F. A. Meyer and Z. Weisman, Collagen-induced platelet aggregation. Dependence on triple helical structure and fiber diameter, *Thrombosis Res.* **22**, 1–14 (1981).

Index